SPACE AND TIME IN THE MICROWORLD

D. I. BLOKHINTSEV

SPACE AND TIME IN THE MICROWORLD

D. REIDEL PUBLISHING COMPANY
DORDRECHT-HOLLAND/BOSTON-U.S.A.

PROSTRANSTVO I VREMYA V MIKROMIRE

First published by Nauka Publishers, Moscow, 1970

Translated from the Russian by Zdenka Smith

Library of Congress Catalog Card Number 72–77871

ISBN-13: 978-94-010-2510-2 e-ISBN-13: 978-94-010-2508-9
DOI: 10.1007/978-94-010-2508-9

Published by D. Reidel Publishing Company
P.O. Box 17, Dordrecht, Holland

Sold and distributed in the U.S.A., Canada and Mexico
by D. Reidel Publishing Company, Inc.
306 Dartmouth Street, Boston,
Mass. 02116, U.S.A.

to Shifro Drabkina

PREFACE

A way of understanding the laws which govern the world of elementary particles has not been found yet. Present-day theoretical physicists have to be satisfied with compromises which, at the best, promise some success at the expense of generality and unity.

Under these circumstances a critical analysis of the basic concepts of modern quantum theory may be timely and useful. It is hoped that the value of such an analysis may be preserved even if, in the near future, new ways of understanding the basis of elementary particle physics are discovered. In this monograph one specific aspect of this analysis is treated, namely the problems of geometry in the microworld. An outline of geometrical measurements in the macroworld was given previously. These measurements seem to be clear enough for at least a certain set of problems to be considered as a starting point for discussing the situation in the microworld. The concepts and methods which are useful in the macroworld may only indirectly be carried over into the microworld and they require a high degree of abstraction.

In comprehending the physical content of dynamic variables which have geometric meaning, for example, the space-time particle coordinates x, y, z, t it is often necessary to have recourse to gedanken experiments which, although not feasible in practice, can nevertheless be compatible with the basic principles of geometry and quantum mechanics. In a desert sea of abstract constructions there is a still larger distance between macroscopic concepts of space-time and the way of employing the coordinates x, y, z and t in relativistic quantum field theory.

It is shown in this monograph that if elementary particles have a structure it is doubtful whether the coordinates of elementary particles, x, y, z, t, can even be defined exactly, let alone the coordinates of the elements which make up these particles (if they do not exist only in our imagination). This important fact is revealed in even the most favourable gedanken experiments.

From this fact, doubt arises about the logical validity of using the symbols x, y, z, t as the space-time coordinates to describe phenomena inside elementary particles. This allows theoreticians a certain freedom of choice of space-time and causal relationships within elementary particles; in other words, an arbitrariness of choice of the geometry in the small.

The last chapters of this book describe some models used to illustrate the situation described above. In concluding, experimental data and experimental possibilities relating to geometric and causal problems in the microworld are discussed.

Since the appearance of the Russian edition in 1970 the situation in the experimental physics of elementary particles has not changed noticeably as regards the problems this book deals with. The derivation of asymptotic theorems (§30, §47) has been found to be less limiting with respect to the problem of nonlocality than was thought at that time. The assertion that no discrepancy with local theory predictions up to scales of the order 10^{-15} is observed still seems to be valid.

The present monograph is based to a great extent on the author's publications at different times. A large number of papers have been prepared jointly with his young colleagues at the Laboratory of Theoretical Physics at Dubna, B.M. Barbashov, N.A. Chernikov, G.I. Kolerov, G.V. Efimov and V.G. Kadyshevsky. The author was very glad to take the opportunity of reflecting in this monograph some of their original investigations. The author is very grateful to his colleagues for useful discussions which have greatly contributed to clarifying many controversial questions. The author is especially grateful to M.A. Markov and B.M. Barbashov for their invaluable advice and comments and for fulfilling the task of critically reading the manuscript, and to G.I. Kolerov for his assistance in preparing it.

The author is greatly indebted to Reidel Publishing Company, thanks to whom the ideas of this work were, not for the first time, made available to a large number of physicists.

The assistance of Mrs I.S. Zarubina, collaborator of the Joint Institute for Nuclear Research, in editing the English version is also appreciated.

I hope that this book may prove helpful to all those who are concerned with fundamental problems of modern atomic physics.

February 1972 Dubna – Moscow D. BLOKHINTSEV

TABLE OF CONTENTS

INTRODUCTION

> The interpretation of Geometry presented here must not be applied directly to submolecular spaces... such an extrapolation might prove to be as incorrect as applying the concept of temperature to bodies of molecular dimensions.[†]
>
> A. EINSTEIN

In modern philosophy, space and time are thought of as forms of existence of some material. Thus it is natural to consider that the laws of geometry are the most general and that their region of applicability covers all *events* and phenomena in the world which we know. We shall not define the term 'event' exactly here, but take it to mean some physical phenomenon which is expressed in some change in material state.

Since the time of Einstein and Minkowski [2, 3], geometry has united investigations of space and time and thus the concept of motion is firmly embedded in modern geometry. Actually this concept, together with the concept of time, was always implicitly included in geometry. One of the basic operations in Euclidean geometry, the matching of geometric figures, introduces the concept of displacement, or motion, indirectly into geometry. In the well-known "Erlangen program" [4], Klein introduced the group approach into geometry, which can also be considered as an introduction of the concept of motion into geometry.

Putting aside this aspect of the question, we can make the most general assertion that geometry entails the *ordering of events*. Each event will be issued a "passport" giving its "place of residence". The united geometry of space and time allows us to consider the neighborhoods of events in both space and time. Thus it becomes possible to base the order of events on causality, i.e., on the *genetic relationship* between events.

Present-day experience shows that such an approach is possible and is based on laws that are inherent in our world. But we do not, strictly

[†] See [1].

speaking, know the limits of this approach. The world of very large scales, *cosmology*, and that of very small scales, *the world of elementary particles*, have not been sufficiently well investigated for us to know if causality has any meaning there.

We are used to considering x as the spatial coordinate and t as the time coordinate and using them as such even inside of elementary particles. However, we do not know what they represent in reality. Should not the space-time relationships change radically in such small scale regions as those of the elementary particle world?

The problems of space and time in the microworld are the topic of this book.

CHAPTER I

GEOMETRICAL MEASUREMENTS IN THE MACROWORLD

1. The arithmetization of space-time

For the purpose of ordering events, we assume that it is possible to assign to each elementary event P certain fixed numbers $x_1, x_2, \ldots, x_n$– coordinates of the event. The number of necessary coordinates n is determined by the number of dimensions of the space-time manifold being considered. (For example, physical space is four dimensional; there are three spatial coordinates $x_1 = x$, $x_2 = y$, $x_3 = z$, and the fourth coordinate, $x_4 = t$, is time.) The set of these n numbers $(x) \equiv x_1, x_2, \ldots, x_n$ constitutes a description or "passport" of the elementary event.

By an elementary event P, we mean an event to which it is possible to assign only one set of numbers (x). Therefore, the term "elementary event P", and "point P" are equivalent in our space-time. We shall call this process of assigning coordinates to each event *arithmetization* of points in space-time, in order not to use the word passportization of points in space-time, which is not popular in science. Some mathematical (logical) requirements have to be given for the process of arithmetization in space-time, for example, the requirement of a one-to-one correspondence of points in space-time P to the coordinates (x). In addition, the physical methods have to be shown, at least in principle, by which this suggested arithmetization can be carried out. In other words, we have to give the physical processes on which the methods of arithmetization can be based.

Points in empty space do not, by the very definition of the term "empty", differ from each other in any way.† Therefore, a real, physical method of differentiating between points in space-time may be based only on real events at these points. In other words, it is possible to arithmetize real events, but not points in space-time themselves. By accepting the

† We are not discussing here the question of whether it is possible to consider real, physical space as empty.

equivalence of elementary events and points in space-time, we exclude from the very beginning undefined operations in a vacuum.

In this manner arithmetization of points in space-time can be based on methods of differentiating between the points which are independent of arithmetization itself. The point, equivalent to an elementary event, has to have individual characteristics, which delineate the situation in which the event will occur. In this formulation of the problem, arithmetization is still rather arbitrary. The nature of the arbitrariness may be seen from the following example. Let the space under consideration be filled with material whose color changes continuously (see the drawing on the cover). Each point contains a certain quantity of blue and yellow color. We can compare every combination of colors, to the extent to which the color of each point is different, with any possible pair of numbers x_1, x_2 only if the set of numbers x_1 and x_2 is as dense as the set of color points. The great arbitrariness of such a comparison may be seen from the fact that we should, in going from one color region to another, assign any value to the size of the steps between numbers x_1, x_2 without affecting the one-to-one correspondence between x_1, x_2 and the color points or "elementary events".

Let us now assume that we have a method available for measuring the concentration of the blue and yellow color at any given point, letting that of blue be a, and that of yellow be b. Then the concentration could be taken to be a pair of numbers (a, b) for the arithmetization of our events. Such arithmetization would directly reflect the structure of our manifold and the continuous character of the measurements of the color points. The choice of any other system of arithmetization such that $x_1 = X_1(a, b)$ and $x_2 = X_2(a, b)$, [we assume a one-to-one relationship between the sets of points (x_1, x_2) and (a, b)], would have a firm physical basis in a "prefered" or "natural" system of arithmetization (a, b).

The "preferedness" or "naturalness" is based on the existance of a physical method of defining the color composition of the "event", which is not intrinsically related to arithmetization.

When selecting another system of arithmetization (x_1, x_2), particularly a discontinuous one, we have to remember in the analysis of events in our color space that jumps in the system of passportization of points are artificial. In the same way it would be artificial, for example, to introduce on the same street several numbering systems for the houses so that the number of each house does not follow that of its neighbor, but is different and arbitrary.

This simple example shows that the method of arithmetization of points in space-time has to be based on real processes, and has to represent objectively the structure of space-time in the most direct manner possible. Arithmetization is not possible without using material processes, and therefore it is not possible to separate physics from geometry [5, 6].

2. The physical methods of arithmetization of space-time

Before Einstein, theoretical physics was based on Euclidean geometry and on the concept of some absolute time. This formed the basis on which any occurrance could be described. The idea that material phenomena could be associated with the geometry of space and time was foreign to physicists of that period.

Newton defined space and time in his *Mathematical Principles of Natural Philosophy* thus:

> Absolute, true, and mathematical time, of itself, and from its own nature, flows equably without relation to anything external, and by another name is called duration .... Absolute space, in its own nature, without relation to anything external, remains always similar and immovable. [7]

The complete independence of geometry of physics is stressed in these definitions. However, as was first shown by Helmholtz [8], Euclidean geometry is based on the concept of the existance of rigid bodies. The basic assumptions are:

(i) N-dimensional space is an n-fold extended manifold $\mathfrak{R}_n(x)$. In other words each element of the manifold (point) is defined by n continuously varying quantities – the coordinates of the element: $x_1, x_2, \ldots, x_n$.

(ii) The existence of moving, but unalterable (rigid) bodies or systems of points is assumed. Between the $2n$ coordinates of each pair of points that belong to a rigid body there exists an equation that is independent of the motion of the bodies and is the same for all mutually coincident pairs of points.

(iii) Complete independence of motion of rigid bodies is assumed. It is assumed that each point can continuously change from one to another in so far as this is not prevented by the bond imposed by assumption (ii).

(iv) Two superimposed bodies remain superimposed even after one of them undergoes a rotation about some axis (uniqueness).

Considering these statements for the case where $n=3$ and, assuming infinite space, Helmholtz showed that the quantity

$$ds^2 = \sum_{i=1}^{3} dx_i^2 \tag{2.1}$$

remains constant under all rotations of a rigid system about the point $dx_i=0$. Therefore, this quantity may be taken as a measure of distance in space – of points x_1, x_2, x_3 and $x_1+dx_1, x_2+dx_2, x_3+dx_3$.†

In the conclusion of his research, Helmholtz stressed

> that the possibility of our making spatial measurements depends, as the preceeding has shown, on the existence in nature of bodies which approximate sufficiently closely our concept of rigid bodies. The independence of compatibility from location and direction of coincident forms, and on the path by which they are brought into coincidence, is the fact on which the possibility of making measurements in space is based.

A modern summary of Helmholtz's research may be given by a statement that *Euclidean geometry is based on group motion of rigid bodies.* This group contains six continuous parameters, including both operations of rotation $\hat{O}_3$, and of translation $\hat{T}_3$.

Instead of motion of rigid bodies, we may consider the transformation of points in the Euclidean manifold itself,

$$x' = Ax, \tag{2.2}$$

where A is some linear operation on the coordinates $x(x_1, x_2, x_3)$ which keeps the geometry of a rigid body unchanged and transforms the point $P(x)$ of the manifold $\mathfrak{R}(x)$ to the point $P(x')$ of the manifold $\mathfrak{R}(x')$. This wider point of view allows us to bring into consideration, not only continuous transformation corresponding to motion, but also discrete changes in coordinates. In particular, the inverse transformation changes the sign of the coordinates

$$x' = -x. \tag{2.3}$$

† Helmholtz's result leads to the Riemann distance metric.

All such operations retain the invariant fundamental form

$$L(x, x') = (x_1 - x'_1)^2 + (x_2 - x'_2)^2 + (x_3 - x'_3)^2, \tag{2.4}$$

where (x_1, x_2, x_3) and (x'_1, x'_2, x'_3) are the coordinates of two points, $P(x)$ and $P(x')$. This form is the same as the square of the distance between the points $P(x)$ and $P(x')$.

From this discussion of the ideas concerning the nature of Euclidean geometry, we see that in the basic definition of measure, an ideally rigid indexed scale may be used. This ancient method of measuring distances still has great practical value in our time.

Arithmetization of points in space may now be carried out with the help of operations in which the rigid scale is plotted. The occurrence here or there of some other index on the scale is in itself a material event, and in this case the process of arithmetization of points in empty space automatically follows the filling of the space with material. A point in space P, with coordinates x [we will often denote such points with the symbol $P(x)$], will indicate those points which mark the position of the end of the single rigid scale l, if it has been plotted x times. The rigidity of the scale also assumes that it does not vary with time. Therefore, if the rigid rules is plotted again at some later time t, and if this is repeated a further x times, then by definition we obtain the identical point in space $P(x)$.

We note that there exists an elegant possibility of verifying the Euclidean character of n-dimensional space by measuring with the aid of the rigid scale an n-dimensional tetrahedron stretched out to $n+1$ points which are its vertices.

Let us take one of these points for the origin and assign coordinates x^i_α to the remaining points, where $i = 1, 2, \ldots, n$ (the number of points) and $\alpha = 1, 2, \ldots, n$ (the number of coordinates). The volume of such a polyhedron V_n is expressed in the form

$$V_n = \frac{1}{n!} \begin{vmatrix} x^1_1 & x^1_2 \ldots x^1_n \\ x^2_1 & x^2_2 \ldots x^2_n \\ \cdot & \cdot \quad \cdot \quad \cdot \quad \cdot \\ x^n_1 & x^n_2 \ldots x^n_n \end{vmatrix} . \tag{2.5}$$

Interchanging rows and columns and multiplying (2.5) the resultant we obtain

$$V_n^2 = \frac{1}{(n!)^2} \begin{vmatrix} (\mathbf{x}^1\mathbf{x}^1) & (\mathbf{x}^1\mathbf{x}^2) \ldots (\mathbf{x}^1\mathbf{x}^n) \\ (\mathbf{x}^2\mathbf{x}^1) & (\mathbf{x}^2\mathbf{x}^2) \ldots (\mathbf{x}^2\mathbf{x}^n) \\ \ldots & \ldots \\ (\mathbf{x}^n\mathbf{x}^1) & (\mathbf{x}^n\mathbf{x}^2) \ldots (\mathbf{x}^n\mathbf{x}^n) \end{vmatrix} . \tag{2.6}$$

The general term of the determinant has the form

$$a_{ik} = \sum_{a=1}^{n} x_a^i x_a^k \equiv \mathbf{x}^i\mathbf{x}^k = \tfrac{1}{2}[(\mathbf{x}^i - \mathbf{x}^k)^2 - (\mathbf{x}^i)^2 - (\mathbf{x}^k)^2] . \tag{2.7}$$

In other words, the volume V_n is a function only of the distances between the vertices of the polyhedron taken in pairs. For n-dimensional Euclidean space, $V_{n+1}=0$. If $V_{n+1}\neq 0$ for this space, this would mean that the geometry of the space is non-Euclidean.

In order to carry out an exhaustive arithmetization of events, it is necessary to have a standard time as well as a standard scale length. Such a time must also be "rigid", i.e., it must have a constant period τ. Thus the interval of time t will denote the number of measured periods τ, which we consider as unity. This assumed "rigidity" of time is mathematically equivalent to uniformity of time, to which the previously discussed group of transformations in Euclidean space has to be supplemented by the operation of translation in Time $\hat{T}_4$, which retains invariant the length of the time interval $t=t_2-t_1$. Thus the general group of transformations, "characteristic" of the geometry of Euclid and Newton, may be written as the product of the operations

$$\hat{g}=\hat{O}_3\hat{T}_3\hat{T}_4\hat{P} . \tag{2.8}$$

The treatment of higher transformations is related to the specific frame of reference, by which we mean the ensemble of rigid bodies and rigid times, which are motionless with respect to each other, and which serve for the arithmetization of points in space and time.

It is possible to imagine another system moving with a uniform velocity v with respect to the other. Let us set the direction of this velocity along the axis Ox. Then the transformation of the coordinates and time may be written as

$$x_1' = x_1 + vt, \qquad x_2' = x_2, \qquad x_3' = x_3, \qquad t' = t , \tag{2.9}$$

which is the *Galilean transformation*. Strictly speaking, this already goes beyond the bounds of geometry because it operates separately in space $\mathfrak{R}_3(x)$ and time $\mathfrak{T}(t)$. The meaning of this transformation is revealed if we go over to mechanics.

The Galilean transformations retain the invariance of acceleration,

$$\frac{\mathrm{d}^2x'}{\mathrm{d}t'^2} = \frac{\mathrm{d}^2x'}{\mathrm{d}t^2} = \frac{\mathrm{d}^2x}{\mathrm{d}t^2}, \tag{2.10}$$

and, therefore, the basic law of mechanics – Newton's second law – remains unchanged. This invariance is expressed in the Galilean principle of relativity; *all mechanical phenomena occur in the same manner in all inertial frames of reference.*

Let us now formally combine the space $\mathfrak{R}_3(x)$ and time $\mathfrak{T}(t)$ in one general manifold $\mathfrak{R}_4(x, t)$. By denoting the Galilean transformation operation (2.9) by $\hat{g}_3$, we may write a more general operation in the manifold space-time $\mathfrak{R}(x, t)$ as the product of the operations

$$\hat{g} = \hat{g}_3\hat{O}_3\hat{T}_3\hat{T}_4. \tag{2.11}$$

This is, strictly, the Galilean transformation. It may be extended to an *improper* transformation, if the operation $\hat{g}$ is extended by additional operations of the inverse $\hat{P}$ of the coordinates and operations of the reversed time $\hat{T}$. This last operation denotes the changes

$$t' = -t. \tag{2.12}$$

Thus the general operation of improper Galilean transformation is written in the form

$$\hat{g}_{\mathrm{H}} = \hat{g}\hat{P}\hat{T}. \tag{2.13}$$

In conclusion, we shall make a few remarks concerning the selection of a coordinate system. In considering Euclidean geometry, we used the Cartesian coordinate system x_1, x_2, x_3, in which the fundamental form $L(x, x')$ has the form (2.4). In few coordinate systems are the inner characteristics expressed in a natural manner. A number of paradoxes in the general theory of relativity are due to the fact that natural phenomena do not differ sufficiently from "pseudophenomena" which are caused by the sometimes unfortunate choice of the coordinate system, or in general, by the method of arithmetization of events.

To explain this, let us consider a method of arithmetization of points in space with measurements made using a stretched elastic thread so that the displacement x will be expressed in units of force f of the stretched thread. Insofar as Hooke's Law applies to small extensions,

$$x = \alpha f \tag{2.14}$$

where α is a coefficient. Since changes in geometrical lengths are proportional to the applied forces, all Euclidean geometrical relationships are preserved. However, it was found experimentally that Hooke's Law is applicable only up to a certain limiting value of f, beyond which the expression for the displacement takes on the form

$$x = \varphi(f) = \alpha f + \beta f^2 + \cdots, \tag{2.15}$$

or

$$f = \varphi^{-1}(x), \tag{2.15'}$$

where φ^{-1} is the inverse of the function φ.

The relationship between the circumference and the diameter of a circle would now be expressed as

$$\pi_f = \frac{f_{\text{circ}}}{f_{\text{diam}}} = \frac{\varphi^{-1}(2\pi r)}{\varphi^{-1}(D)} \neq \pi. \tag{2.16}$$

The number π_f would depend on the dimensions of the ring. Similarly Pythagoras' theorem would become more complicated and have the form

$$f_{AB}^2 + f_{BC}^2 \neq f_{AC}^2. \tag{2.17}$$

The well-known inequality or the sides of a triangle ABC,

$$AB + BC \geqslant CA, \tag{2.18}$$

would also not be satisfied for triangles whose longest side is CA, because, as the extension $x = CA$ increases, the force of tensile strength f_{CA} becomes relatively smaller. (The length of the thread tends to infinity for terminal values of f.)

Let us consider how uniform motion would appear if we assume that we have a good, rigid clock. We obtain

$$\frac{d^2 f}{dt^2} = \left(\frac{dx}{dt}\right)^2 \frac{d^2\varphi^{-1}(x)}{dx^2} < 0. \tag{2.19}$$

In other words, a uniformly moving body would appear to be slowing down.

This method of arithmetization of space, using a stretched elastic thread, is not in itself meaningless and should not be rejected. In fact, because of the one-to-one correspondence between the displacement x and the force f, it is one of the allowed methods of ordering points in space. Moreover, on small scales, in regions where Hooke's law applies, this method is equivalent to the usual one, of measurements made with a rigid scale. Why, then, must we nevertheless prefer the use of a rigid measuring rod?

The answer to this question may be given by comparing (a) the degree of universality of the method and (b) the degree of generality of the regularities which are revealed by some or other method of arithmetization of space.

Let us first consider point (a). All real rods depart in various ways from the ideal "absolute rigidity". However, there exists a vast domain of scales and temperatures in which these deviations are not large and may be accounted for by suitable corrections.

Real elastic thread also deviates in various ways from ideal elastic thread, which obeys Hooke's law strictly, but these deviations can not be considered in the form of corrections because they significantly affect the result of the measurements. Thus the generality of a "dynamometer" as an instrument for defining lengths is rather small. Each dynamometer would introduce its individual characteristics into the space-time relationship of events.

The second point (b) also does not speak in favor of measuring distances with a stretched elastic thread. If we accepted this method, we would determine that small circles have constant relationships and large circles have various relationships [see (2.16)]. The additivity of segments, generally speaking, would not hold; $AB+BC \neq AC$. For some triangles, the inequality of their sides would be fulfilled, and for others it would have no meaning. A body, not subjected to a real force, would slow down, etc. This is what a world would look like to us in which geometrical measurements are made with a stretched thread and a dynamometer.

If we were to cross over at an opportune time to arithmetization with a rigid scale, we would find the well-known laws of Euclidean geometry and the law of inertia. It would appear that, for all circumferences $2\pi r/D=\pi$, for all triangles $AB+BC \geqslant CA$, for all bodies that are not subjected to a force, $d^2x/dt^2=0$, etc. The laws that applied only to small scales before, $f \ll f_0$ [f_0 being the value of the extension for which the function $\varphi(f)$ begins to differ noticably from αf where α=constant], in the new system of arithmetization would extend by themselves over the whole region of importance of x and become *universal.*

Thus not all systems of arithmetization of space and time exhibit internal consistency in the space-time relationship. Amongst the possible systems and methods of arithmetization there exist methods possessing *the maximum universality* and the *maximum capability of exhibiting the most general regularities in the interrelationship of events.*

3. On Dividing the Manifold of Events into Space and Time

If some event P is able to exert influence on a second event P', it is possible to use such an event as a signal propagating from the point P to the point P' in the manifold of events $\mathfrak{R}_n(P)$.[†]

The set of events P', on which the events originating in P are able to exert influence, forms the manifold $\mathfrak{T}_+(P, P')$. We shall call this manifold the *absolute future*, with respect to the point P. The set of events P' that are able to influence the events in P form the manifold $\mathfrak{T}_-(P, P')$. We shall call this manifold the *absolute past* with respect to the point P. If an ordering of causes and events occurs, then the domains $\mathfrak{T}_+$ and $\mathfrak{T}_-$ do not have common points except for the one point P:

$$\mathfrak{T}_+ \cap \mathfrak{T}_- = 0. \tag{3.1}$$

The manifold formed by the union of the domains $\mathfrak{T}_+$ and $\mathfrak{T}_-$,

$$\mathfrak{T}_+ \cup \mathfrak{T}_- = \mathfrak{T}(P, P'), \tag{3.2}$$

will be called the *time domain*, with respect to the point P. We shall call the remaining part of the manifold

$$\mathfrak{S}(P, P') = \mathfrak{R}(P) - \mathfrak{T}(P, P'), \tag{3.3}$$

[†] $\mathfrak{R}_n(P)$ denotes an n-dimensional manifold of events; $\mathfrak{R}_n(x)$ denotes a manifold after arithmetization — i.e., a coordinate (x) is assigned to each point P. This will sometimes be stressed by the symbol $P(x)$. Obviously $\mathfrak{R}_n(P(x)) \equiv \mathfrak{R}_n(x)$.

the *space domain* or, more simply, *space*, with respect to the point P.

In order to make this process of dividing manifold $\mathfrak{R}_n(P)$ into space $\mathfrak{S}(P)$ and time $\mathfrak{T}(P)$ more specific, we assume that physical events have meaning in some scalar field $\varphi = \varphi(P)$ that satisfies the second-order differential equation in the manifold $\mathfrak{R}_n(P)$ of the form

$$A_{\alpha\beta} \frac{\partial^2 \varphi}{\partial x_\alpha \partial x_\beta} + R = Q, \tag{3.4}$$

where the coefficients $A_{\alpha\beta}$ and R in general are functions of the field φ and its first derivatives $\varphi_\alpha = \partial\varphi/\partial x_\alpha$, Q is an external source of the field and depends only on the point P; and $\alpha, \beta = 1, 2, \ldots, n$.

We shall now consider propagation of the signal subject to Equation (3.4). By a signal, we mean the wave whose front is defined by the function

$$S = S(x_1, x_2, \ldots, x_n) = \text{const}, \tag{3.5}$$

A discontinuity of the first derivative occurs at the boundary at the front (see Figure 1). Such a discontinuity is called *weak*. From the theory

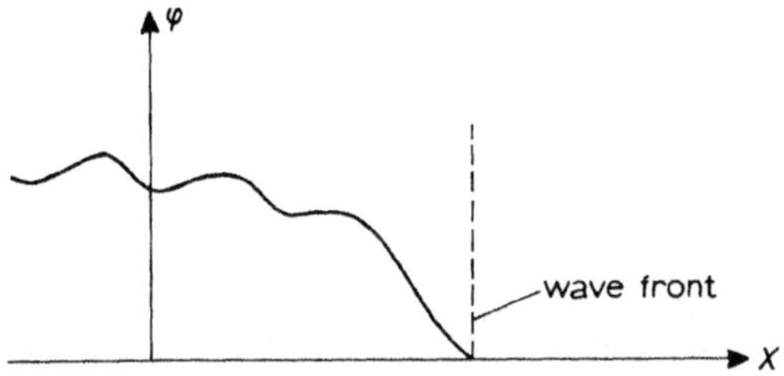

Fig. 1 A weak discontinuity on a wave front surface.

of differential equations [9, 10] it is well known that the equation for the surface of such a discontinuity is completely defined by the coefficients of the second derivatives of Equation (3.4)

$$A_{\alpha\beta} \frac{\partial S}{\partial x_\alpha} \frac{\partial S}{\partial x_\beta} = 0. \tag{3.6}$$

To an accuracy of the order of $(1/\omega)$, this equation agrees with the equation for the phase Φ of high frequency, ω, waves

$$\varphi = a e^{i\omega\Phi}, \quad \omega \to \infty. \tag{3.7}$$

Substituting (3.7) in (3.4) for $\omega \to \infty$ gives the equation

$$A_{\alpha\beta} \frac{\partial \Phi}{\partial x_\alpha} \frac{\partial \Phi}{\partial x_\beta} + O\left(\frac{1}{\omega}\right) = 0. \tag{3.6'}$$

Equation (3.6) is an equation from geometrical optics. The vector **N**, which is normal to the surface

$$= \Phi(x_1, x_2, \ldots, x_n) = \text{const} \tag{3.5'}$$

agrees with the direction of the radial lines and has components

$$N_\alpha \sim \frac{\partial \Phi}{\partial x_\alpha} \qquad (\alpha = 1, 2, \ldots, n).$$

Let us take a real random vector $\xi(\xi_1, \xi_2, \ldots, \xi_n)$ and generate the quadratic form

$$F(\xi, \xi) = A_{\alpha\beta}\xi_\alpha\xi_\beta\,. \tag{3.8}$$

If this form is definite,

$$F(\xi, \xi) > 0 \qquad \text{or} \qquad F(\xi, \xi) < 0 \tag{3.9}$$

for any ξ, then the equation for the normal vector **N**

$$F(\mathbf{N}, \mathbf{N}) = 0 \tag{3.10}$$

will have imaginary roots and, therefore, there will be no propagation of the field φ in the manifold $\mathfrak{R}_n(x)$. In this case the events form a motionless world. If we forcibly vary the field φ, for example by varying the source field $Q(x)$ in the vicinity of some point P, so that

$$\partial Q(x) = \varepsilon\delta^4\,(x - x_P)\,, \tag{3.11}$$

where ε is a small number, then the state of the field will change in all manifolds $\mathfrak{R}_n(x)$. In this case Equation (3.4) would become an elliptic function [9], and no separate manifold would arise within $\mathfrak{R}_n(x)$.

Let us consider the second case, when the form of $\mathfrak{R}_n(x)$ is indefinite, namely,

$$F(\xi, \xi) \lesssim 0 \tag{3.9'}$$

depending on the domain of values of the variables ξ. In this case, which is called hyperbolic, the actual directions N satisfy Eq. (3.6). The ensemble of these directions forms a *characteristic cone*. This cone (see Figure 2) divides the manifold $\mathfrak{R}_n(x)$ about each point P into two regions: $\mathfrak{T}(\mathfrak{T}=\mathfrak{T}_+ + \mathfrak{T}_-)$ and $\mathfrak{S}$. Region $\mathfrak{T}_+$ may be reached by any signal originating from P and having a velocity of propagation smaller than the signal velocity v_φ that is characteristic of the field φ. Similarly, the region $\mathfrak{T}_-$ contains points P', from which signals can reach the points P if their velocities $v \leqslant v_\varphi$. The region $\mathfrak{S}$ contains points which are reached by signals from P and which have a velocity $v > v_\varphi$.

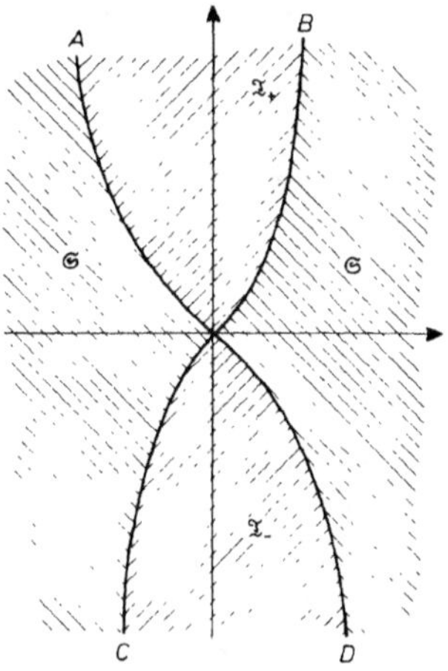

Fig. 2. The time manifolds $\mathfrak{T}_+$ and $\mathfrak{T}_-$ and the spatial manifold $\mathfrak{S}$. *ABCD* is the cross section of the characteristic cone.

We may restate this by saying that the characteristic cone divides the manifold $\mathfrak{R}_n(x)$ into manifolds $\mathfrak{T}$ and $\mathfrak{S}$ with reference to signals of the type φ. In the region $\mathfrak{T}$ (in the upper cone) lie events which may be genetically related to event at the point P; in the region $\mathfrak{T}_-$ (in the lower cone) lie events which may influence the character of event P.

We shall now consider the case when there are two types of fields, φ and ψ, which have different velocities of propagation v_φ and v_ψ. Let $|v_\varphi| < |v_\psi|$. The characteristic cones for these two fields are shown in Figure 3. We shall investigate the situations which arise when we use either the signal φ or ψ for dividing space and time. To take a specific case, we define v_φ as the velocity of sound and v_ψ as the velocity of light

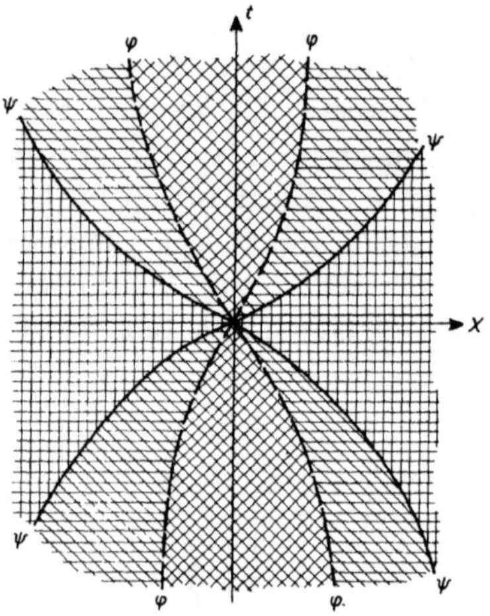

Fig. 3. Two characteristic cones for the fields φ and ψ.

$(|v_\varphi| < |v_\psi|)$. We further assume that, within the population, there is a blind group that does not accept signals ψ; their shortcomings will be the basis for their selection of the sound field as the field with which they determine the relationship between phenomena. And so these people divide the manifold $\mathfrak{R}_n(P)$ into $\mathfrak{T}_\varphi$ and $\mathfrak{S}_\varphi$, but have no idea of the division of $\mathfrak{R}_n(x)$ into $\mathfrak{T}_\psi$ and $\mathfrak{S}_\psi$. How will the interrelationship of events in the manifold $\mathfrak{R}_n(P)$ appear to them?

We shall examine this relationship using Figure 4. Let A and B be two spatially separated points. The distance between them could be measured with a sound signal v_φ, or with a rigid scale. Let the event at A be a spark, which produces both light and sound. At the point A', an observer receives sound reflected from point B (an echo); but the light from the spark, arriving at point B'', can produce noise which will reach the observer at A at the point A'', preceding in time the point A'. The blind observer will come to the conclusion that the sound phenomenon A'' violates the law of cause and effect. The signal A'' arrived earlier than it could be expected if its cause were the reflection of sound from point B.

However, this same observer could discover the constancy with which the phenomenon A'' precedes the phenomenon A', and come to the correct conclusion that, in addition to the sound field φ, there exists yet another field ψ hidden from him that propagates with a velocity v_ψ which is greater than the speed of sound v_φ. In other words, the velocity v_φ which he had assumed to be limiting is in fact not so, because there is another velocity $v_\psi > v_\varphi$. And so the blind observer would be faced with a dilemma; either to allow the existence of events which disturb causality (with reference to the dividing of $\mathfrak{R}_n(P) = \mathfrak{T}_\varphi + \mathfrak{S}_\varphi$), or to review the method by which he divided $\mathfrak{R}_n(P)$ into $\mathfrak{T}$ and $\mathfrak{S}$. We see that for the ordering of events with observance of causality, it is necessary to use the signal having the greatest velocity of all the possibilities.

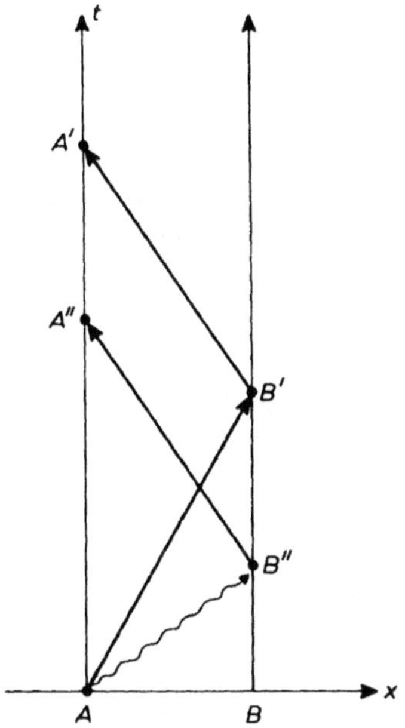

Fig. 4. The dark straight lines represent propagation of sound, the wavy line represents the propagation of light. The sound signal $(B'' - A'')$ appears to violate "acoustical" causality.

To what extent the division in space and time is closely related to the physics of the phenomena, it is possible to see from this case that if a constancy in precedence of the event A'' by the event A' were not discovered, the observer would, to the regret of science, consider the existence of phenomena that do not obey causality.

Such a situation is not difficult to imagine if one allows that the state of the medium ("ether") through which the signal ψ propagates, would be not less capricious than, say, that of the Earth's atmosphere, and this state would influence the velocity of the signal v_ψ. The time AA' would then be subject to chance variations which would at least make the interpretation of the "unforeseen" phenomenon at A more difficult.

In concluding this section we shall consider one curious example, which will serve as an additional illustration of the principle of dividing manifolds $\mathfrak{R}(P)$ into time $\mathfrak{T}$ and space $\mathfrak{S}$ manifolds [11]. Let the field φ obey the scalar Klein's equation. We first consider the case of two measurements x_1 and x_4. In this case

$$\frac{\partial^2\varphi}{\partial x_4^2} - \frac{\partial^2\varphi}{\partial x_1^2} \pm m^2\varphi = 0. \tag{3.12}$$

Let $x_4 = t$ be time and $x_1 = x$ be the spatial coordinate†: in Equation (3.12) we had two signs for m^2 (m being proportional to the mass of particles in the field φ). Assuming $\varphi \sim e^{+i(\omega t - kx)}$, we obtain

$$\omega = \pm\sqrt{k^2 \pm m^2}. \tag{3.13}$$

From this it follows that, for particles with real masses $m^2 > 0$, the group velocity of waves – and, therefore, the signal velocity – will be equal to

$$u = \frac{\partial\omega}{\partial k} = \frac{k}{\omega} \leqslant 1. \tag{3.14}$$

Conversely, for the case $m^2 < 0$,

$$u = \frac{k}{\omega} > 1. \tag{3.15}$$

Therefore, in the first case the signal would propagate inside the cone $\mathfrak{T}$ (see Figure 5) and if events were ordered with the help of the fastest signal $u = 1$, then, because $u \leqslant 1$, normal ordering of cause and effect will be fulfilled by all signals. If, however, $m^2 < 0$, then the signals that propagate with a velocity $u > 1$, will violate causality (they will propagate within the space cone $\mathfrak{S}$).

It is easy to see, however, that the interpretation we took for the variables x_1 and x_4 in the manifold $\mathfrak{R}_2(x)$, is inaccurate [11] for the case

† In Equation (3.12) the speed of light c is set equal to 1.

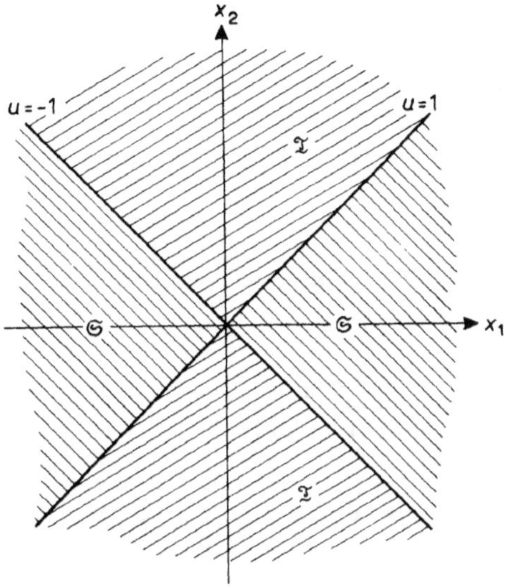

Fig. 5. In the case of $m^2>0$ $x_2=t$, $x_1=x$; for $m^2<0$ $x_2=x$, $x_1=t$, i.e., the spatial and temporal regions change places.

$m^2<0$. In fact, by definition of the time region $\mathfrak{T}$, this is a region inside of which signals propagate. Therefore, in the case where $m^2<0$, the correct interpretation requires that $x_4=x$, in the region $\mathfrak{S}$, and $x_1=t$, in the region $\mathfrak{T}$. This is immediately evident from Equation (3.12) which transforms into itself when $m^2\to -m^2, x_4\rightleftarrows x_1$.

In the four-dimensional case $\mathfrak{R}_4(x)$, the situation is more complicated. In this case, the equation takes on the form

$$\frac{\partial^2\varphi}{\partial x_4^2}-\frac{\partial^2\varphi}{\partial x_3^2}-\frac{\partial^2\varphi}{\partial x_2^2}-\frac{\partial^2\varphi}{\partial x_1^2}\pm m^2\varphi=0, \tag{3.12'}$$

By writing the equation in this form we assume that we purposely considered x_4 as time t, and x_1, x_2, x_3 as the three spatial coordinates. We now look for the Green's function for this equation. For this, we turn to the inhomogeneous equation

$$\frac{\partial^2 G}{\partial x_4^2}-\frac{\partial^2 G}{\partial x_3^2}-\frac{\partial^2 G}{\partial x_2^2}-\frac{\partial^2 G}{\partial x_1^2}\pm m^2 G=\delta(x_4)\,\delta(x_3)\,\delta(x_2)\,\delta(x_1). \tag{3.16}$$

Assuming that

$$G(x) = \int \tilde{G}(q)\, e^{iqx}\, d^4 q, \tag{3.17}$$

where $qx = \sum_{i=1}^{4} q_i x_i = \sum_{k=1}^{3} q_k x_k - q_4 x_4$, and substituting this into (3.16) we get

$$\tilde{G}(q) = \frac{1}{[q_1^2 + q_2^2 + q_3^2 \pm m^2 - q_4^2]}, \tag{3.18}$$

and, therefore,

$$G(x) = \int \frac{e^{iqx}\, d^4 q}{[\mathbf{q}^2 \pm m^2 - q_4^2]}, \tag{3.19}$$

where $\mathbf{q} = (q_1, q_2, q_3)$. In the case $m^2 > 0$ we may consider q_4 as the frequency $\omega = \pm(\mathbf{q}^2 + m^2)^{1/2}$. The group velocity of all waves emanating from the point $x = 0$, is $u = \partial\omega/\partial k \leqslant 1$ and we can interpret the coordinate x_4 as time t, and the coordinates x_1, x_2, x_3 as the three spatial coordinates. For $m^2 < 0$ [minus sign in Equation (3.19)], signals originating from the point $x = 0$ with a velocity $u = \partial\omega/\partial k \geqslant 1$, will go inside the cone D, as in

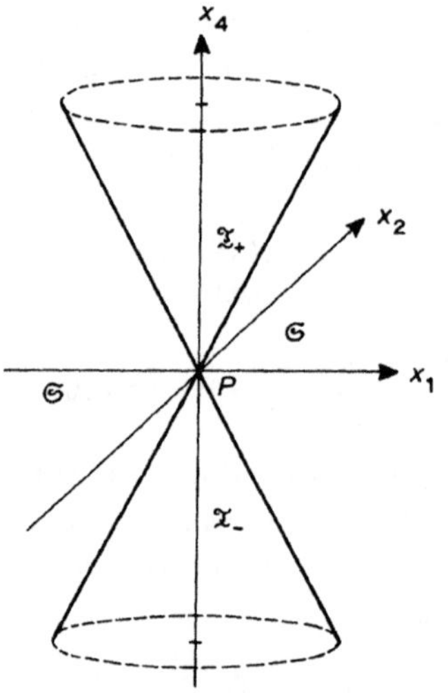

Fig. 6. For $m^2 > 0$, a signal from the vertex of the characteristic cone is concentrated inside the region $\mathfrak{T}$ which is therefore considered as temporal. For $m^2 < 0$, the signal propagates in the region $\mathfrak{S}$ and in this case $x_1 = t$, $x_2 = t_2$, $x_4 = x$.

the two-dimensional case, and if we also insist in this case on considering x_4 as time, we come to a contradiction of the concept that the zone of influence in the case of $m^2 < 0$ is the region inside the light cone (see Figure 6). We must therefore admit that we are concerned with the manifold in which there are three times, $t_1 = x_1$, $t_2 = x_2$, $t_3 = x_3$, and the one spatial dimension $x = x_4$.

4. Affine Manifold

As was discussed above, various methods of arithmetization of a manifold of events are possible in principle. Some of them will naturally reveal the deep characteristics of the space-time relationship, others will behave as a warped mirror in which reality is distorted. This caricature can go so far that reality becomes almost unrecognizable or, philosophically speaking, almost *unknowable*.

The most general characteristics of geometric manifolds which have to be reflected in the system of arithmetization that is chosen, can be listed as follows: (a) continuity, and (b) characteristic topological connectedness†. We shall interpret (a) by assuming that around each point $P(x)$ in the manifold $R_n(P(x))$, there is a region formed by neighboring points.

If this manifold is deformed without forming discontinuities, then the neighboring points remain neighbors and the structure of the manifold will be characterized exclusively by features of the topological connectedness [point (b)], i.e. by the ensemble of deformations that do not disturb the continuity of the manifolds. In Figure 7, three-dimensional surfaces are shown, which represent members of a manifold. The surfaces of the

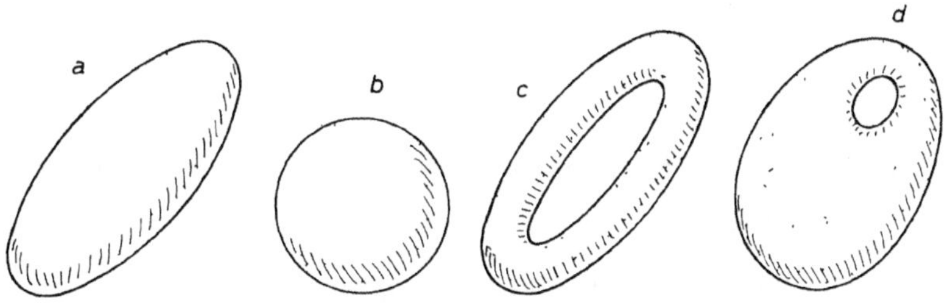

Fig. 7. Four topologically equivalent surfaces.

† The definition (a) excludes from consideration manifolds that consist of a great number of points, e.g., a spatial lattice.

bodies a and b, and c and d, taken in pairs, are topologically equivalent but the surfaces of a and b are not equivalent to those of c and d, because the former surface cannot be changed into the second type by a continuous deformation.

We shall carry out an arithmetization of the manifold such that the neighboring points in the manifold will correspond to the neighboring numbers. In view of the continuity of the manifold, these numbers will also be continuous.

In accordance with the dimension n of the manifold $\mathfrak{R}_n(P(x)) \equiv \mathfrak{R}_n(x)$, the number of numbers that characterize any point in the manifold is equal to n. Earlier, we called these n numbers the coordinates $(x) = (x_1, x_2, \ldots, x_n)$ of the point P. It is possible to select these coordinates in various ways, and, from the mathematical point of view, it is sufficient if the various possible coordinate systems have a one-to-one relationship between them, if not throughout the manifold $\mathfrak{R}_n(x)$, then at least in its separate parts that cover the entire manifold $\mathfrak{R}_n(x)$. If such parts are $\mathfrak{R}_n'(x)$, $\mathfrak{R}_n''(x)$, $\mathfrak{R}_n'''(x), \ldots$, then

$$\mathfrak{R}_n(x) \subset \mathfrak{R}_n'(x) \cup \mathfrak{R}_n'' \cup \mathfrak{R}_n''' \cdots.$$

The general coordinate transformation in any one of these regions [which may cover the whole manifold $\mathfrak{R}_n(x)$], can be written in the form [12, 14]

$$x'^{\mu} = f^{\mu}(x), \tag{4.1}$$

where $f^{\mu}(x)$ is a differentiable function and $\mu = 1, 2, \ldots, n$. We write

$$\frac{\partial x'^{\mu}}{\partial x^{\nu}} = \frac{\partial f^{\mu}}{\partial x^{\nu}} = \alpha^{\mu}_{\nu}, \tag{4.2}$$

so that

$$\mathrm{d}x'^{\mu} = \alpha^{\mu}_{\nu}\, \mathrm{d}x^{\nu}. \tag{4.3}$$

We consider Equation (4.3) as solvable for $\mathrm{d}x^{\nu}$, so that the determinant

$$|\alpha^{\mu}_{\nu}| \neq 0. \tag{4.4}$$

In the manifold $\mathfrak{R}_n(x)$, we may examine some functions of the point P: the scalar field $\varphi(P)$, the vector field $A^{\mu}(P)$ $(\mu = 1, 2, \ldots, n)$, the tensor field of various ranks $T^{\mu\nu} \cdots (P)$ $(\mu, \nu, \cdots = 1, 2, \ldots, n)$, which transform in the following manner under transformation of coordinates (4.1):

$$\varphi'(P') = \varphi(P') = \varphi(P), \tag{4.5}$$

$$A'^{\mu}(P') = \alpha^{\mu}_{\nu}(P)\, A^{\nu}(P), \tag{4.5'}$$

$$T'^{\mu\nu\cdots}(P') = \alpha^{\mu}_{\sigma}(P)\, \alpha^{\nu}_{\rho}(P) \cdots T^{\sigma\rho\cdots}(P), \tag{4.5''}$$

By P' we mean the point that is obtained by transforming point P by (4.1). Vectors and tensors which transform according to (4.3), (4.5′) and (4.5″) are called *contravariant*.

We can consider vectors $A_{\mu}(P)$ and tensors $T_{\mu\nu}\ldots(P)$ that transform according to

$$A'_{\mu}(P') = \beta^{\nu}_{\mu}(P)\, A_{\nu}(P), \tag{4.6}$$

$$T'_{\mu\nu\ldots}(P') = \beta^{\sigma}_{\mu}(P)\, \beta^{\rho}_{\nu}(P) \cdots T_{\sigma\rho\ldots}(P), \tag{4.6'}$$

where

$$\beta^{\nu}_{\mu} = \frac{\partial x^{\nu}}{\partial x'^{\mu}}. \tag{4.6''}$$

Such vectors and tensors are called *covariant* [14].

Because vectors in various points of the manifold transform in various ways [the coefficients $\alpha^{\mu}_{\nu}(P)$ and $\beta^{\nu}_{\mu}(P)$ are functions of the point P], the comparison of two vectors at different points in the manifold $A(P)$ and $A(P')$ cannot be made without further operations. In order to compare these two vectors, the vector $A(P)$ must be brought over to the point P',

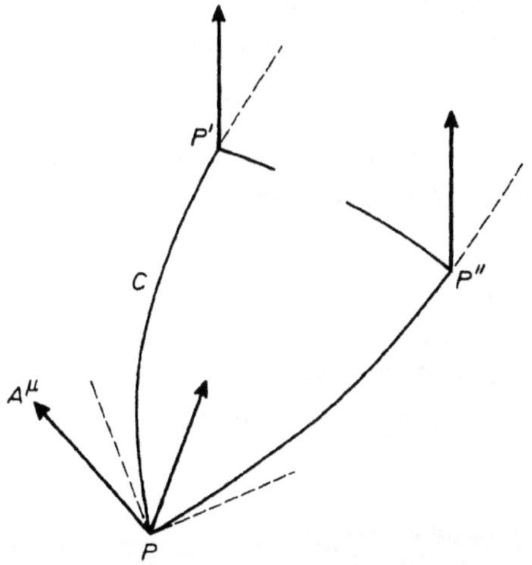

Fig. 8. A parallel translation of a vector along the closed contour $PP'P''$ which is made up of segments of geodesic lines.

keeping it *parallel to itself*. During such an operation, the components of the vector $A^\mu(P)$ undergo a change δA^μ (see Figure 8). We consider such a change for two points infinitesimally close to each other, $P(x)$ and $P' = P(x+\mathrm{d}x)$. The magnitude of δA^μ has to be a function of the components of the displaced vector ($\delta A^\mu \equiv 0$! for the nul vector), and also a linear function (the sum of the vectors must transform like a vector). Futhermore, the magnitude of δA^μ has to be proportional to the displacement $\mathrm{d}x^\mu$, for $\mathrm{d}x^\mu = 0$, $\delta A^\mu = 0$. Therefore, we may write δA^μ in the form

$$\delta A^\mu = -\,\Gamma^\mu_{\alpha\beta} A^\alpha\, \mathrm{d}x^\beta\,, \tag{4.7}$$

where $\Gamma^\mu_{\alpha\beta}$ is a function of the points P.

The coefficient $\Gamma^\mu_{\alpha\beta}$ is called the *coefficient of affine connectedness*, or *Christoffel's symbol*. The complete change of the vector $A(P)$ in going from point P to P' is equal to

$$DA^\mu = \mathrm{d}A^\mu - \delta A^\mu\,, \tag{4.8}$$

and, correspondingly, the derivative of the components of the vector A with respect to x_ν is

$$\frac{DA^\mu}{Dx_\nu} = \frac{\partial A^\mu}{\partial x_\nu} + \Gamma^\mu_{\alpha\nu} A^\alpha\,. \tag{4.9}$$

This is called the *covariant derivative* [13].

Let us now consider some line in an n-dimensional manifold

$$x^\mu = X^\mu(\tau)\,, \tag{4.10}$$

where τ is a parameter that defines the line. Amongst the various lines that pass through the points P and P', there is one whose tangent

$$u^\mu = \frac{\mathrm{d}x^\mu}{\mathrm{d}\tau} \tag{4.11}$$

remains parallel to itself. On this line the covariant derivative of the tangent vector is equal to zero; i.e.,

$$Du^\mu = \mathrm{d}u^\mu + \Gamma^\mu_{\alpha\beta} u^\alpha\, \mathrm{d}x^\beta = 0 \tag{4.12}$$

or

$$\frac{\mathrm{d}u^\mu}{\mathrm{d}\tau} + \Gamma^\mu_{\alpha\beta} u^\alpha u^\beta = 0\,. \tag{4.12'}$$

Such a line is called a *geodesic* [12–14].

By definition, the parallel translation has to conserve the "angle" between the tangent to the geodesic line and any other vector, if it is translated along the line. Therefore, for any vector A, we have

$$DA^{\mu} = 0\,. \tag{4.13}$$

This change of the components of a vector under parallel translation is a postulate affine geometry, and is closely related to the concept of curvature of an affine manifold.

Let us consider an infinitely small contour C that consists of geodesic lines (see Figure 8). We translate the vector A parallel to itself along the contour. Then the size of the component of the vector is defined as

$$\Delta A^{\mu} = \int\limits_{C} \Gamma^{\mu}_{\alpha\beta} A^{\alpha}\, \mathrm{d}x^{\beta}\,. \tag{4.14}$$

Applying Stokes' Theorem, we obtain

$$\Delta A^{\mu} = \frac{1}{2}\left[\frac{\partial(\Gamma^{\mu}_{\alpha\sigma})}{\partial x_{\rho}} - \frac{\partial(\Gamma^{\mu}_{\alpha\rho}A^{\alpha})}{\partial x^{\sigma}}\right] \Delta f^{\rho\sigma}\,, \tag{4.15}$$

where $\Delta f^{\rho\sigma}$ is the component of area drawn along the contour C. The partial derivative in (4.15) can be expressed, using (4.13), in terms of the components of the vector A. We thus obtain

$$\Delta A^{\mu} = -\tfrac{1}{2}\, R^{\mu}_{\alpha\rho\sigma} A^{\alpha}\, \Delta f^{\rho\sigma}\,, \tag{4.16}$$

where

$$R^{\mu}_{\alpha\rho\sigma} = \frac{\partial \Gamma^{\mu}_{\alpha\sigma}}{\partial x^{\rho}} - \frac{\partial \Gamma^{\mu}_{\alpha\rho}}{\partial x^{\sigma}} + \Gamma^{\mu}_{\lambda\rho}\Gamma^{\lambda}_{\alpha\sigma} - \Gamma^{\mu}_{\lambda\sigma}\Gamma^{\lambda}_{\alpha\rho} \tag{4.17}$$

is the *curvature tensor of the manifold (Riemann tensor)*. We can show that this tensor is antisymmetric with respect to the indices σ, ρ, i.e.,

$$R^{\mu}_{\alpha\rho\sigma} = -\,R^{\mu}_{\alpha\sigma\rho}\,, \tag{4.18}$$

and

$$R^{\mu}_{\alpha\rho\sigma} + R^{\mu}_{\sigma\rho\alpha} + R^{\mu}_{\rho\alpha\sigma} = 0\,. \tag{4.19}$$

5. The Riemann Manifold

The affine manifold considered above is still rather general. If affine geometry is supplemented by the concept of separation between two points, or in other words, if it brings a metric into the manifold, then we arrive at *metric geometry* or *Riemannian geometry*. Instead of the distance between two points P and P', we will use the more general concept of *interval*, and keep in mind that the manifold contains both space and time.

We shall denote the interval between two infinitely close points $P(x)$ and $P(x+\mathrm{d}x)$ by $\mathrm{d}s$. According to Riemann's basic hypothesis, $\mathrm{d}s^2$ is the squared form of the separation of the coordinates of points P and P':

$$\mathrm{d}s^2 = g_{\mu\nu}\,\mathrm{d}x^\mu\,\mathrm{d}x^\nu, \tag{5.1}$$

where $g_{\mu\nu}=g_{\mu\nu}(P)$ is the symmetric tensor $g_{\mu\nu}=g_{\nu\mu}$, known as the *metric tensor*. If we rotate the coordinate system in the neighborhood of P, and by proper choice of scale, (5.1) can be rewritten in cannonical form [14] as

$$\mathrm{d}s^2 = \varepsilon_{\mu\mu}(\mathrm{d}x^\mu)^2, \tag{5.1$'$}$$

where $\varepsilon_{\mu\mu}=\pm 1$. The quantity $f=\sum_{\mu=1}^{n}\varepsilon_{\mu\mu}$ is called the *signature of the metric*. If the signature is equal to the number of dimensions, $f=n$, then the metric form is

$$\mathrm{d}s^2 = \sum_{\mu=1}^{n}(\mathrm{d}x^\mu)^2 \tag{5.1$''$}$$

and it expresses Pythagoras' theorem for infinitely small regions in n-dimensional space.† If $f\neq n$, then the interval $\mathrm{d}s^2$ may be written in the form

$$\mathrm{d}s^2 = \sum_{\mu=1}^{m}(\mathrm{d}x^\mu)^2 - \sum_{\mu=m+1}^{n}(\mathrm{d}x^\mu)^2, \tag{5.1$'''$}$$

so that $f=2m-n+1$. In this case the Pythagoras theorem is satisfied in all manifolds $\mathfrak{R}_m(x)$ or $\mathfrak{R}_{n-m}(x)$. Such a space is called *pseudo-Euclidean*.

The choice for the quadratic form for $\mathrm{d}s^2$ has been considered many times. In fact, this hypothesis of Riemann shows that "ordinary" geometry, which is based on the concept of rigid bodies, holds in the neighborhood of each point P. For this type of geometry, the basis of the quadratic form, which defines the concept of separation, was given by Helmholtz (compare with Section 2). In this manner, the general case of Riemannian geometry reduces either to Euclidean, or to pseudo-Euclidean geometry in

† We note that here the concept of the infinitely small has a pure mathematical meaning. Physically it could be a cubic kilometer in the Galaxy, or a cubic fermi in an atomic nucleus.

the neighborhood of each point in the n-dimensional manifold $\mathfrak{R}_n(x)$.

We recall several important formula of Riemannian geometry. The metric form (5.1) enables us to define the scalar product of two vectors (AB), taken at the point P:

$$(AB) = g_{\mu\nu} A^\mu B^\nu = A^\mu B_\mu , \tag{5.2}$$

where

$$B_\mu = g_{\mu\nu} B^\nu \tag{5.3}$$

is the covariant form of the vector B.

Solving (5.3) in terms of B^ν, gives

$$B^\mu = g^{\mu\nu} B_\nu , \tag{5.4}$$

where $g^{\mu\nu} = M^{\mu\nu}/g$, and $M^{\mu\nu}$ is the minor of the determinant

$$g = |g_{\mu\nu}| . \tag{5.5}$$

From the definition of tensor transformation (see Section 4), it follows that the scalar product is *invariant* under coordinate transformation (4.3). The length of the vector A, or, to be more exact, its square must not change under parallel displacement from P to a neighboring point P':

$$L^2 = (AA), \tag{5.6}$$

and the angle between any two vectors A and B must not change

$$\cos \theta = \frac{(AB)}{\sqrt{(AA)}\sqrt{(BB)}}, \tag{5.7}$$

i.e., the covariant derivative of the scalar product (AB) must be equal to zero; and

$$D(AB) = 0 . \tag{5.8}$$

This relationship allows us to express the coefficients of affine connectedness $\Gamma^\mu_{\alpha\beta}$ in the terms of the metric tensor $g_{\mu\nu}$. We obtain

$$\Gamma^\mu_{\alpha\beta} = \frac{1}{2} g^{\mu\sigma} \left(\frac{\partial g_{\sigma\alpha}}{\partial x^\beta} + \frac{\partial g_{\beta\sigma}}{\partial x^\alpha} - \frac{\partial g_{\alpha\beta}}{\partial x^\sigma} \right). \tag{5.9}$$

From Equations (5.9) and (4.17) we may also obtain an expression for the curvature tensor from the second derivative of the metric tensor. If we introduce the covariant curvature tensor

$$R_{\mu\alpha\beta\gamma} = g_{\mu\lambda} R^{\lambda}_{\alpha\beta\gamma}, \tag{5.10}$$

then we find that

$$R_{\mu\alpha\beta\gamma} = \frac{1}{2}\left(\frac{\partial^2 g_{\mu\gamma}}{\partial x^\alpha \partial x^\beta} + \frac{\partial^2 g_{\alpha\beta}}{\partial x^\mu \partial x^\gamma} - \frac{\partial^2 g_{\mu\beta}}{\partial x^\alpha \partial x^\gamma} - \frac{\partial^2 g_{\alpha\beta}}{\partial x^\mu \partial x^\gamma}\right)$$
$$+ g_{\rho\sigma}(\Gamma^\rho_{\alpha\beta}\Gamma^\sigma_{\mu\gamma} - \Gamma^\rho_{\alpha\gamma}\Gamma^\sigma_{\mu\beta}). \tag{5.11}$$

In the general theory of relativity, the symmetric tensor

$$R_{\mu\nu} = R^{\alpha}_{\mu\alpha\nu} \tag{5.12}$$

plays an important role, as does the scalar curvature

$$R = g^{\mu\nu} R_{\mu\nu}. \tag{5.13}$$

Thus, in metric Riemannian geometry, all quantities that are important in affine geometry can be expressed in terms of the metric tensor and its derivatives.

We will also consider group properties of Riemannian geometry. For this purpose, instead of general coordinate transformation we use the infinitely small transformation

$$x^{\mu'} = x^\mu + \xi^\mu(x), \tag{5.14}$$

where $\xi(x)$ is an infinitely small vector. It is also possible to consider this transformation as an infinitely small displacement of the point P to the point $P' = P(x')$. For such a displacement, the metric tensor undergoes the transformation

$$g'_{\mu\nu} = \frac{\partial x^\rho}{\partial x'^\mu}\frac{\partial x^\sigma}{\partial x'^\nu} g_{\rho\sigma}$$
$$= g_{\mu\nu}(P) - g_{\mu\rho}(P)\frac{\partial \xi^\rho(P)}{\partial x^\nu} - g_{\rho\nu}(P)\frac{\partial \xi^\rho(P)}{\partial x^\mu}. \tag{5.15}$$

On the other hand, the original metric at the point P' may be expressed in terms of the metric at the point P by means of the expansion

$$g_{\mu\nu}(P') = g_{\mu\nu}(P) + \frac{\partial g_{\mu\nu}}{\partial x_\rho}\xi^\rho(P) + \cdots \tag{5.16}$$

From this, the change in the tensor $g_{\mu\nu}$ caused by the displacement $P \to P'$ is equal to

$$\Delta g_{\mu\nu} = - g_{\mu\rho} \frac{\partial \xi^\rho}{\partial x^\nu} - g_{\rho\sigma} \frac{\partial \xi^\rho}{\partial x^\mu} - \frac{\partial g_{\mu\nu}}{\partial x^\rho} \xi^\rho + \cdots \tag{5.17}$$

The metric does not change under this displacement if $\Delta g_{\mu\nu} = 0$, i.e., if

$$g_{\mu\rho} \frac{\partial \xi^\rho}{\partial x^\nu} + g_{\rho\nu} \frac{\partial \xi^\rho}{\partial x^\mu} + \frac{\partial g_{\mu\nu}}{\partial x^\rho} \xi^\rho = 0 . \tag{5.18}$$

These are equations of the group of infinitely small displacements. They are called the Killing equations [12.14]. If it is possible to find an infinitely small vector $\xi^\mu(P)$ for the manifold $\mathfrak{R}_n(P)$ with the metric (5.1) that satisfy Killing's equations, then transformations characterized by the vector $\xi^\mu(P)$ are allowed in the manifold. In particular, in the case of Euclidean geometry $g_{\mu\nu} = \delta_{\mu\nu}$ and Killing's equations give

$$\frac{\partial \xi^\mu}{\partial x^\nu} + \frac{\partial \xi^\nu}{\partial x^\mu} = 0 , \tag{5.19}$$

from which

$$\xi^\mu = \alpha^\mu_\nu x^\nu + a^\mu , \tag{5.20}$$

where $\alpha^\mu_\nu = -\alpha^\nu_\mu$ and α^μ are infinitely small constants. Thus the group of motions in Euclidean space consists of rotation $\hat{O}(\alpha)$ and translation $\hat{T}(\alpha)$. From the equations given above, we can see that the metric equation (5.1), which defines *geometry in the small*, fully characterizes the Riemannian manifold. It is as if an elementary cell of which the structure of *geometry as a whole* is composed.

We have already discussed the basic assumptions that dictate the choice of the quadratic form (5.1) for the squared interval. In his prominent thesis [15], Riemann first showed the connection between the mathematical and physical postulates, which acquired fundamental meaning in Einstein's general theory of relativity. Modern physics, as we shall see later, stands before the necessity of critically examining the geometrical representation that is applied to the microworld. Therefore, in the conclusion of this section, it is appropriate to recall the ingenious words of Riemann, concerning geometry in the small:

The question of whether the assumption of geometry in the infinitely small is valid is closely related to the question of the internal cause of the origin of metric relationships in space. Of course this question also refers to the studies of space, and in these studies we must remember the remarks made above about the fact that, in the case of a discrete manifold, the principle of metric relationships is already contained in the very concept of this manifold whereas, in the case of a continuous manifold, we must search for it in some other place. From this we see that the reality which establishes the idea of space forms a discrete manifold, or that we must try to explain the origin of the metric relationships by something external, by binding forces which act upon the reality.

We can hope to find the solution of these problems only when, starting from the contemporary conception which has been proved experimentally and is based on the foundations laid by Newton, we begin refining it successively being guided by facts which cannot be explained by it. The investigations like those performed in this work, namely which start from general concepts, serve only as a tool that does not allow the limitedness and long standing prejudices to hinder progress in the understanding of things.

We stand here on the threshold of a region that belongs to another science, to physics, and now we have no indication how to cross it. [15]†

6. THE PHYSICS OF ARITHMETIZATION OF THE SPACE-TIME MANIFOLD

The assignment of specific coordinates to physical events is not only a mathematical process. The issuing of a passport in itself implies definite manupulations with material objects which are necessary for defining the location of an event in space and time.

The classical method of arithmetization was that of using a *uniformly running clock* ("rigid" time) and *a rigid scale*. In this measuring system, however, the problem of agreement between the clock and the scale, which move with respect to each other, required the use of signals whose laws of propagation had the greatest generality and simplicity. This problem became particularly poignant when applied to bodies moving with high velocities. (Historically, these bodies were electrons whose velocities approached the speed of light.) The signals which comply most naturally with the conditions established above are *light signals*. The velocity of such signals proves to be a limiting velocity for all material bodies and fields. In particular, the velocity of a signal propagating in a solid is significantly smaller than the speed of light c. Therefore, we do not have the right, within the limits of accuracy required by the theory of

† Weyl, in his article dedicated to F. Klein, notes that possibly in the foundations of geometry lie discrete manifolds "... now, less than ever before, is it possible to affirm that in the foundations of geometry there do not lie any discrete manifolds" [16].

relativity, to consider a solid rod as absolutely rigid. It is necessary to keep in mind its elasticity and ability to get deformed. We may formally state that the theory of relativity excludes the possibility of the exitence of absolutely rigid rods. Therefore, according to the concept stated by Weyl [17] and developed by Martske [18], we shall not use rigid scales, but rather pulses of light (scintillations) for measuring both distances and time intervals.

A light clock may be built from two mirrors that face each other and do not absorb light. If a pulse of light is trapped between two such mirrors, it will be reflected alternately from one mirror to the other (see Figure 9).

The number of reflections N may be taken for the measure of time. Such "pulses of light" are of course optional. We could have taken any

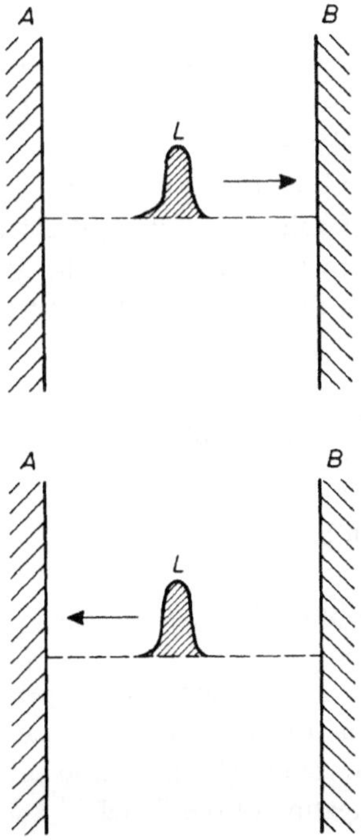

Fig. 9 A light clock. A and B are mirrors, L is a light pulse.

other periodic process, for example the reflection of sound, or the uniform rotation of a gyroscope. We note that in any case we do not have an a priori basis for being convinced of the strict uniformity of motion of any selected clock. The irregularity of motion of a clock could be discovered only in the case that an irregularity were observed in the motion of all phenomena. Therefore, it would be natural to suspect the clock of irregularity as the cause for any observed changes in the rate of change of all phenomena. By selecting other clocks, which are closer to the ideal – in other words, less dependent of the external situation – we would eventually become convinced that the previous clock "misbehaved".

The light clock with mirrors has the advantage, which has been sufficiently well substantiated experimentally and is in agreement with the principles of the theory of relativity, that the speed of light c does not depend on the motion of the source of light (in the above case, the mirrors) nor on the motion of the observer. Thus the uniformity of motion of the light clock is based on very general principles, and the nature of the mirrors and the possibility of changes in their composition is immaterial as long as the separation of the mirrors does not change and their absorption of light remains negligible.

The absence of dispersion of light waves in vacuum is an important condition for the use of light pulses in light clocks and as signals for determining the metric to be possible. Let the light signal at the time $t=0$ be given in space in the form $\varphi(x, 0)$. We will write it in the form of superposition of plane waves

$$\varphi(\mathbf{x}, 0) = \int\limits_{k>0} C(\mathbf{k})\, \mathrm{e}^{\mathrm{i}\mathbf{k}\mathbf{x}}\, \mathrm{d}\mathbf{k}\,. \tag{6.1}$$

This signal at time t will have the form

$$\varphi(\mathbf{x}, t) = \int C(\mathbf{k})\, \mathrm{e}^{\mathrm{i}(\mathbf{k}\mathbf{x}-\omega t)}\, \mathrm{d}\mathbf{k} = \varphi(|\mathbf{x}| - ct, 0) \tag{6.1'}$$

where ω is the frequency, $\omega = |k|$, and c is the speed of light. This shows that the signal has its own form and does not spread out in space. In other words, the dots and dashes of the Morse code do not run into each other in proportion to the propagation of the signal.†

† This property of light pulses holds even in the three-dimensional case (see Section 17).

We shall now turn to a conceptual experiment for the definition of the interval between phenomena. In Figure 10a, the ordinate is time and the abscissa, the spatial coordinate x. The line AA' is the *world line* of the observer A. He has two mirrors at his disposal that form a light clock, the

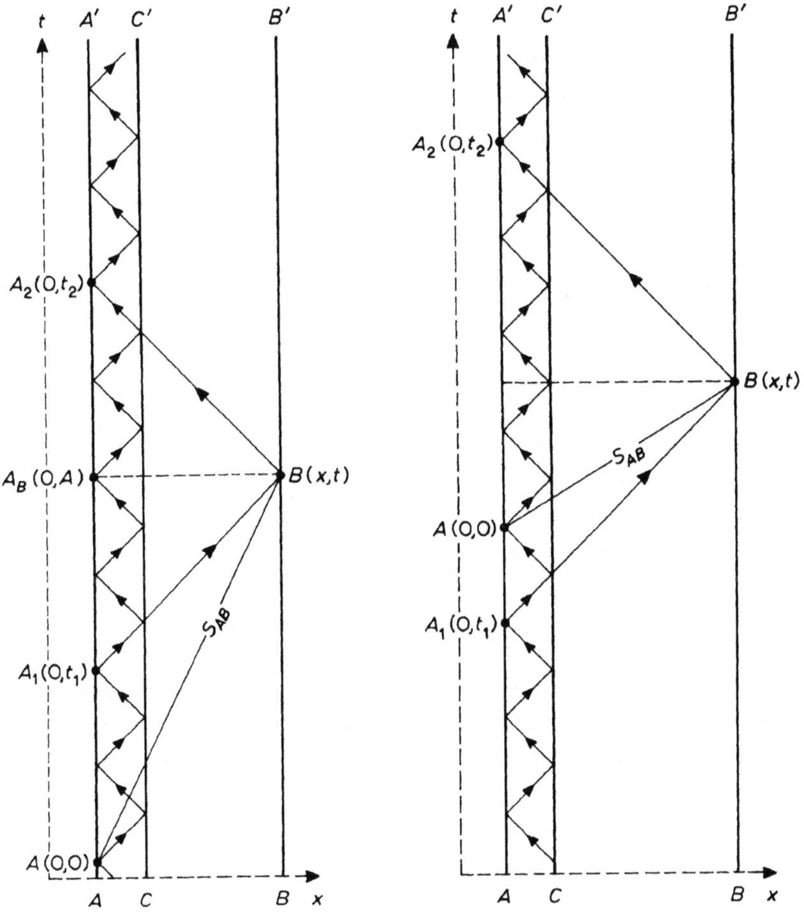

Fig. 10. A method of measuring the interval s_{AB} between events A and B using the reflection of a light pulse. (a) A time interval; (b) a spatial interval AA' is the world line of the observer A; CC' is the line of the second mirror of his clock, and BB' is the world line of event B.

first having the world line AA' and the second, a similar world line CC'. The zig-zag line between AA' and CC' is the world line of the light pulse that is alternately reflected from mirrors A and C. Each break in this line can be used as a measure of time. Let the world line of another body B be BB'. In order to define the intervals s_{AB} between the events A and B, we send a light pulse from A to B, where we place a mirror which sends the signal back to A. It will be sent at the time $t_1 = t - AB/c$ by the light clock at A, and it will return to A at the time $t_2 = t + AB/c$. Therefore, the interval s_{AB} between the events A and B will be defined as

$$s_{AB}^2 = t_1 t_2 = t^2 - \frac{AB^2}{c^2} > 0. \tag{6.2}$$

Here $t_1 = N_1\tau$ and $t_2 = N_2\tau$, where N_1 and N_2 are the number of reflections, measured by the light clock, and τ is the period of the clocks. In this case the interval s_{AB}^2 is greater than zero. Such an interval is called *time interval*.

In Figure 10b we show an example in which the light signal from A cannot get to the event B. It is not difficult to see that in this case it follows that $t_1 < 0$ [see the point $A_1(0, t_1)$ in Figure 10b], and so the interval can be written in the previous form

$$s_{AB}^2 = t_1 t_2 = t^2 - \frac{AB^2}{c^2} < 0, \tag{6.3}$$

but now it is less than zero. In this case the interval is called the *spatial interval*.

If the speed of light is bounded, then the process of arithmetization we considered above produces a division of the manifold of events $\mathfrak{R}_n(x)$ into space $\mathfrak{S}(x)$ and time $\mathfrak{T}(x)$ with the help of the light cone

$$s^2 = t^2 - \frac{r^2}{c^2} = 0, \tag{6.4}$$

where

$$r^2 = x^2 + y^2 + z^2.$$

Moreover, the definition of the form of the causal relationship of events is brought out. Events at the world point P can only influence events which are located inside the upper cone $\mathfrak{T}_+(P)$ (see Figure 6) which is defined by

$$t - \frac{r}{c} = 0. \tag{6.5}$$

All events that are located inside the cone of the absolute past $\mathfrak{T}_-(P)$,

$$t + \frac{r}{c} = 0, \tag{6.6}$$

can influence events at P. Events which are located in the spatial region $\mathfrak{S}$

$$t^2 - \frac{r^2}{c^2} < 0, \tag{6.7}$$

cannot be related causally (genetically) to events at P.

In connection with the fact that all geometrical relationships are defined with a clock and light signal, the geometry of a four-dimensional manifold with signature $f = -2$, is sometimes called *chronogeometry* (see [18]).

7. Arithmetization of Events in the Case of the Non-Linear Theory of Fields

This section should serve as an illustration of the principle of dividing the manifold of events $\mathfrak{R}(x)$ into space $\mathfrak{S}(x)$ and time $\mathfrak{T}(x)$, which was presented in Section 3. In the case of non-linear theory of fields, the velocity of the signal depends on the magnitude of the field and its derivatives. Therefore, the division itself of $\mathfrak{R}(x)$ into space and time manifolds depends to a great extent on the dynamics of the motion of the signal. In particular the characteristic cone (see Figure 2) in this case is distorted, and for some values of the field and its derivatives, it can become imaginary. So as not to make the discussion more difficult, we shall consider the field to be scalar, and restrict it to two dimensions, $x_1 = x$, $x_4 = t$.

As a basis for such a theory, we shall take the variational principle with the Lagrange function that depends on the invariants formed from the field $\varphi(x, t)$ and its first derivatives $p = \partial\varphi/\partial t$, $q = \partial\varphi/\partial x$.† Such in-

† This theory presents the scalar version of non-linear electrodynamics of Born and Infeld, which they developed in the 1930s [19, 20]. In the discussion of the scalar theory, we follow article [21].

variants are

$$I = \tfrac{1}{2}\varphi^2, \tag{7.1}$$
$$K = \tfrac{1}{2}(p^2 - q^2). \tag{7.1'}$$

The Lagrange function $\mathscr{L}$ is equal to

$$\mathscr{L} = \mathscr{L}(K, I), \tag{7.2}$$

and the principle of variations states that

$$\delta L = \int \delta\mathscr{L}\, \mathrm{d}x\, \mathrm{d}t = 0. \tag{7.3}$$

It is obvious that the theory in this form guarantees invariance with respect to the Lorentz transformations. Performing the variation, we note that

$$\delta\mathscr{L} = \frac{\partial\mathscr{L}}{\partial K}(p\,\delta p - q\,\delta q) + \frac{\partial\mathscr{L}}{\partial I}\varphi\,\delta\varphi, \tag{7.4}$$

as usual, allows the surface terms of the integral (7.3) to vanish. Equating the coefficients of $\delta\varphi$ to zero and using (7.3) we obtain

$$-\frac{\partial}{\partial t}\left(\frac{\partial\mathscr{L}}{\partial K}p\right) + \frac{\partial}{\partial x}\left(\frac{\partial\mathscr{L}}{\partial K}q\right) + \frac{\partial\mathscr{L}}{\partial I}\varphi = 0 \tag{7.5}$$

or

$$A\frac{\partial^2\psi}{\partial t^2} + 2B\frac{\partial^2\psi}{\partial t\,\partial x} + C\frac{\partial^2\psi}{\partial x^2} + R = 0, \tag{7.6}$$

where

$$A = (1 + \alpha p^2),\quad B = \alpha pq,\quad C = (1 - \alpha q^2),\quad R = \gamma - 2\beta K \tag{7.7}$$

and

$$\alpha = \frac{\partial^2\mathscr{L}}{\partial K^2}\Big/\frac{\partial\mathscr{L}}{\partial K},\quad \beta = \frac{\partial^2\mathscr{L}}{\partial K\,\partial I}\Big/\frac{\partial\mathscr{L}}{\partial K},\quad \gamma = \frac{\partial\mathscr{L}}{\partial I}\Big/\frac{\partial\mathscr{L}}{\partial K}. \tag{7.8}$$

In the case when $\mathscr{L}$ is linearly dependent on the invariants K, I, when $\alpha = \beta = 0$ and $\gamma =$ constant, we get back to the linear Klein's equations for a scalar field φ. The characteristic curves of Equation (7.6) are determined from the equation (see, e.g., [22] and [23])

$$A\xi^2 - 2B\xi + C = 0, \tag{7.7'}$$

where

$$\xi = \frac{\mathrm{d}x}{\mathrm{d}t} \tag{7.8'}$$

is the velocity of propagation of the wave front. We call this a *weak* discontinuity (as we did earlier in Section 3). From (7.7) we obtain

$$\xi = \frac{\pm\sqrt{1 + 2\alpha K} - \alpha pq}{1 + \alpha p^2} \tag{7.9}$$

or, for small $\alpha(K, I)$

$$\xi = \pm 1 \mp \tfrac{1}{2}\alpha(p \pm q)^2 + \cdots. \tag{7.10}$$

If we now suppose that, in the region $a<x<b$ for $t=0$, there is some initial state $\varphi, p, q=\partial\varphi/\partial x$, then, from (7.9) and (7.10) we see that it will propagate with a characteristic velocity ξ, that differs from the speed of light c which is set equal to ± 1 in this section. Two possible cases of field propagation are shown in Figures 11a and 11b. In (a), the velocity of propagation $|\xi|<1$ and in (b), $|\xi|>1$. In the latter case the non-linear signal violates relativistic causality and arithmetization of the manifold of events has to be reexamined.

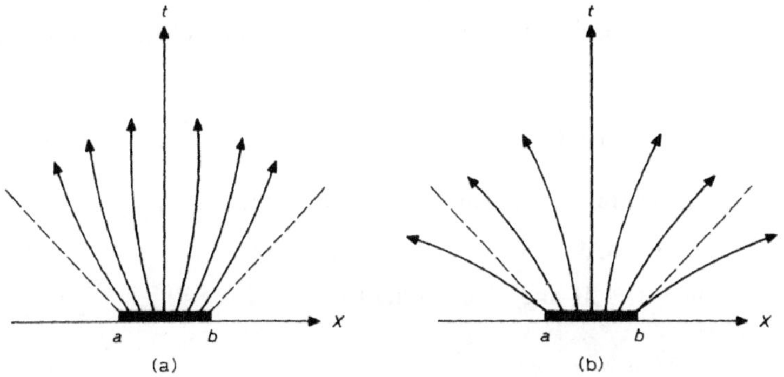

Fig. 11. Properties of a nonlinear field. (a) $|\xi| = |\mathrm{d}x/\mathrm{d}t| < 1$; (b) $|\xi| = |\mathrm{d}x/\mathrm{d}t| > 1$.

It is an interesting fact that in non-linear theory it is impossible to rule out the situation in which, for fixed values of φ, p and q, the characteristic curves become imaginary, so that equation (7.6) becomes elliptical. In this case the concept of the cause of successive events loses its meaning and we shall have to deal with a "lump" of events which mutually cause each other but *do not follow* one from the other [21]. If we assume that such situations take place somewhere inside particles in the region of strong fields, then further away from the particles, the equations should become hyperbolic. On the surface where this transition takes place, the discriminant

$$B^2 - AC = 1 + 2\alpha K \tag{7.11}$$

vanishes. The change over from elliptical to hyperbolic equations is accompanied by a discontinuity in the first derivative of the field φ (p or q or both) (see [10, 23]).

In conclusion we note that the theory of non-linear electromagnetic fields is closely related to the theory of scalar fields. In this case the invariants K and I have the form

$$K = \frac{1}{i}(\mathscr{E}^2 - \mathscr{H}^2), \qquad I = (\mathscr{E}\mathscr{H}), \tag{7.12}$$

where $\mathscr{E}$ and $\mathscr{H}$ have the same sign as the electromagnetic field strength. Varying the electromagnetic potential $\mathbf{A}$ and φ so that

$$\mathscr{E} = -\frac{\partial \mathbf{A}}{\partial t} - \operatorname{grad}\varphi, \qquad \mathscr{H} = \operatorname{rot}\mathbf{A}, \tag{7.13}$$

and using the variational principle (7.3), we obtain the equation for the electromagnetic field

$$\frac{\partial \mathbf{D}}{\partial t} = \operatorname{rot}\mathbf{B}, \qquad \operatorname{div}\mathbf{D} = 0, \tag{7.14}$$

where $\mathbf{D}$ and $\mathbf{B}$ are the electromagnetic inductions

$$\mathbf{D} = M\mathscr{E} - N\mathscr{H}, \qquad \mathbf{B} = M\mathscr{H} + N\mathscr{E}, \tag{7.15}$$

and the quantities M and N are equal to

$$M = \frac{\partial \mathscr{L}}{\partial K}, \quad N = \frac{\partial \mathscr{L}}{\partial I}. \tag{7.15'}$$

The second group of equations

$$\frac{\partial \mathscr{H}}{\partial t} = -\operatorname{rot} \mathscr{E}, \qquad \operatorname{div} \mathscr{H} = 0 \tag{7.16}$$

are obtained directly from (7.13). For the one-dimensional case, $\mathscr{E}(\mathscr{E}_x, 0, 0)$, $\mathscr{H}(0, \mathscr{H}_y, 0)$ and $\mathscr{E}_x = \mathscr{E}(t, z)$, $\mathscr{H}_y = \mathscr{H}(t, z)$ we obtain from (7.14) and (7.16)

$$\begin{gathered}(1 + \alpha\mathscr{E}^2)\frac{\partial \mathscr{E}}{\partial t} + \alpha\mathscr{E}\mathscr{H}\frac{\partial \mathscr{E}}{\partial z} - \alpha\mathscr{E}\mathscr{H}\frac{\partial \mathscr{H}}{\partial t} \\ + (1 - \alpha\mathscr{H}^2)\frac{\partial \mathscr{H}}{\partial z} = 0, \\ \frac{\partial \mathscr{E}}{\partial z} + \frac{\partial \mathscr{H}}{\partial t} = 0,\end{gathered} \tag{7.17}$$

where, as before, $\alpha = \partial^2 \mathscr{L}/\partial k^2 / \partial \mathscr{L}/\partial k$. The equation for the characteristic direction $\xi = \partial z/\partial t$ will now have the form

$$(1 + \alpha\mathscr{E}^2)\,\xi^2 + 2\alpha\mathscr{E}\mathscr{H}\xi + (1 - \alpha\mathscr{H}^2) = 0, \tag{7.18}$$

from which we obtain

$$\xi = \frac{\pm\sqrt{1 + 2\alpha K} - \alpha\mathscr{E}\mathscr{H}}{1 + \alpha\mathscr{E}^2} \tag{7.19}$$

or, for small α

$$\xi = \pm 1 \mp \tfrac{1}{2}\alpha(\mathscr{E} \pm \mathscr{H})^2 + \cdots. \tag{7.20}$$

From the above, we can see that the non-linear field theory allows the existence of signal velocities which can be either smaller or greater than the speed of light c, or even be imaginary. Thus the theory of relativity, which assumes a constant speed of light in a vacuum, implicitly assumes a linear theory [24]. The subject of signal propagation in Born's non-linear theory is considered in detail in [25]. We shall return to Born's theory in Section 43 in connection with quantization of space-time.

8. The General Theory of Relativity and Arithmetization of Space-Time

In the general theory of relativity the metric of the Riemannian manifold of events becomes dependent on the motion of matter. The coefficient $g_{\mu\nu}(x)$ in the metric form that defines the interval between close events,

$$ds^2 = g_{\mu\nu}(x)\, dx^\mu\, dx^\nu, \tag{8.1}$$

in the general theory of relativity is a functional of the energy-momentum tensor for matter $T_{\mu\nu}(x)$

$$g_{\mu\nu}(x) = G_{\mu\nu}\{T_{\mu\nu}(x)\}. \tag{8.2}$$

This functional relationship is defined in terms of the well-known Einstein equation [13, 14, 26].

$$R_{\mu\nu} - \tfrac{1}{2}\, g_{\mu\nu} R = \frac{\varkappa}{c^2}\, T_{\mu\nu}(x), \tag{8.3}$$

where $\kappa = 8\pi\gamma/c^2 = 1.86 \times 10^{-27}$ cm gm^{-1} is Einstein's constant,† $R_{\mu\nu}$ is the curvature (see Section 5).

Two additions have to be made to Equation (8.3): the first, an equation of motion of matter (e.g., the equation of relativistic hydrodynamics if the matter can be represented in the form of a fluid, or the electromagnetic field equation if the matter is a field, etc.). The second addition is a boundary and initial condition.

The general theory of relativity allows the same general coordinate transformation as the Riemannian geometry (see Section 5), namely,

$$x^{\mu'} = f^\mu(x), \tag{8.4}$$

where the function $f^\mu(x)$ maps $\mathfrak{R}_4(x)$ onto $\mathfrak{R}_4(x')$ in a continuous and unambiguous way. The great arbitrariness contained in this transformation must be used very carefully. The formal equally of various coordinate systems may, in fact, reduce to almost zero by the particular features inherent in the problem itself.

The mathematically arbitrary choice of coordinate system has to be checked against other methods of implementing the selected coordinate systems by physical methods. In other words, we must show practical methods of measuring intervals in a manifold of events.

A basic property of Riemannian geometry, which is *pseudo-euclidean in the infinitely small,* suggests that the method of defining intervals

† Where $\gamma = 6.67 \times 10^{-8}$ cm^2 gm^{-1} s^{-2} is Newton's gravitational constant.

between events A and B that were discussed earlier can be carried over into the general theory of relativity only if the events A and B are infinitely close to each other. Let the square of the interval, in a selected coordinate system, have the form

$$ds^2 = g_{44}(dx^4)^2 + 2g_{4s}\,dx^s\,dx^r + g_{rs}\,dx^r\,dx^s, \tag{8.5}$$

where $s, r = 1, 2, 3$. For a light signal $ds^2 = 0$, so that

$$dx^4 = -\frac{g_{4s}\,dx^s}{g_{44}} \pm \sqrt{\frac{(g_{4s}\,dx^s)^2}{g_{44}^2} - \frac{g_{rs}}{g_{44}}\,dx^r\,dx^s}, \tag{8.6}$$

where the two signs $\pm$ corresponds to the direct signal from A to B and the return signal from B to A.

Let us now return to Figure 10 (p. 31) and consider the events A and B as infinitely close. The coordinates of the event A are $x_A^1 = x_A^2 = x_A^3 = x_A^4 = 0$ and the coordinates of event B are dx_B^4, dx_B^3, dx_B^2, dx_B^1. We obtain the characteristic time of A from (8.5), assuming $dx^s = 0$:

$$ds = d\tau = \sqrt{g_{44}}\,(dx^4). \tag{8.7}$$

From Figure 10 we have

$$d\tau_1 = \sqrt{g_{44}}\,(dx_B^4 - dx_1^4), \qquad d\tau_2 = \sqrt{g_{44}}\,(dx_B^4 - dx_2^4), \tag{8.8}$$

where dx_1^4 is the change in dx^4 corresponding to the arrival of the signal from A_1 to B, and dx_2^4 is that for the same signal from B to A_2. These quantities are defined from (8.6).

By multiplying $d\tau_1$ and $d\tau_2$, we obtain

$$ds_{AB}^2 = d\tau_1\,d\tau_2 = g_{44}(dx_B^4)^2 + g_{rs} \tag{8.9}$$

This is a generalization of Equation (8.5), for the time interval s_{AB}. Similarly, for the spatial interval, we obtain

$$-\,ds_{AB}^2 = -\,d\tau_1\,d\tau_2 = -\,g_{44}(dx_B^4)^2 - g_{rs}\,dx^r\,dx^s. \tag{8.9'}$$

By assuming that $dx_B^4 = 0$, we obtain an expression for the interval between A and B

$$dl^2 = d\tau_1\,d\tau_2 = -\,g_{rs}\,dx_r\,dx^s, \tag{8.10}$$

or

$$dl = \sqrt{-\,g_{rs}\,dx_r\,dx^s}. \tag{8.10'}$$

The interval between the two points A and B separated by a finite interval

$$l_{AB} = \int_A^B \sqrt{-g_{rs}\, dx^r\, dx^s}, \tag{8.11}$$

depends on the integration path. We can define the "true" interval as the interval along the geodesic line

$$l^0_{AB} = \int_A^B \sqrt{-g_{rs} \frac{dx^r}{d\sigma} \frac{dx^s}{d\sigma}}\, d\sigma, \tag{8.12}$$

The integral (8.12) is taken under the condition that the coordinates of the curve $x^s(\sigma)$ satisfy the equation

$$\frac{d^2 x^s}{d\sigma^2} + \Gamma_{jk} \frac{dx^j}{d\sigma} \frac{dx^k}{d\sigma} = 0. \tag{8.12'}$$

According to the basic Einstein Equation (8.3), in the presence of matter the curvature of space differs from zero, and in the general case, because of the motion of matter, it depends on time. Therefore, the problem of arithmetization of manifolds of events in the general theory of relativity becomes very complicated.†

The most natural method of arithmetization of events seems to be a sequential arithmetization from one point to its infinitely close neighbor. By using a light signal, we can define the separation of point A and its neighbor B and synchronize the clocks at A and B. We shall, by definition, consider as simultaneous the moments $B(x, t)$ and $A_B(0, t)$ which correspond to the average of the signal leaving point A_1 and the reflected signal returning to point A_2 (see Figure 10):

$$\begin{aligned} \sqrt{g_{44}}\, dx^4 &= \tfrac{1}{2}(d\tau + \sqrt{g_{44}}\, dx^4_1 + d\tau_2 + \sqrt{g_{44}}\, dx^4_2) \\ &= \tfrac{1}{2}(d\tau_1 + d\tau_2) - g_{4s}\, dx^s / \sqrt{g_{44}}. \end{aligned} \tag{8.13}$$

If event A_B is placed at this point, then $d\tau_1 = -d\tau_2$, and the difference in time x^4 between two "simultaneous" events A and B will be given by

$$dx^4_B = -\, g_{4s}\, dx^s / g_{44} \tag{8.13'}$$

† For details, see references [14] and [18]

This is the value by which the reading of the clock B should be changed. We note that the measurement was made using the metric at point A. We could also have made the measurement using the metric at point B.

In view of the fact that the intrinsic time at points A and B flows differently, the results of the interval measurements will also be different. For small separations of A and B, this difference will only show up in second-order terms. In fact, the difference in the interval, as measured at A and B, is

$$\Delta s_{AB}^2 = [g_{\mu\nu}(B) - g_{\mu\nu}(A)]\, \mathrm{d}x^\mu\, \mathrm{d}x^\nu$$

$$\left[\frac{\partial g_{\mu\nu}}{\partial x^\alpha}\,\mathrm{d}x^\alpha + \frac{\partial^2 g_{\mu\nu}}{\partial x^\alpha\, \partial x^\beta}\,\mathrm{d}x^\alpha\, \mathrm{d}x^\beta + \cdots\right]\mathrm{d}x^\mu\, \mathrm{d}x^\nu, \tag{8.14}$$

By a suitable choice of coordinate system, the quantity in the square brackets, which contains the first-order differentials, can be made equal to zero. And thus the intrinsic difference in the interval measurements made at points A and B can be made to be quantities of the fourth order. This remaining difference can be related to the curvature of space, and so, in general cannot be removed by choice of coordinate system.

We now return to the question of arithmetization of a manifold of events. We saw that, at least in the neighborhood of a point, the arithmetization that is characteristic of the special theory of relativity is preserved in the general theory of relativity: The whole matter reduces to measuring intervals in time and to the sending of light signals.

Thus a natural expression for the interval is pseudo-euclidean

$$\mathrm{d}s^2 = c^2\, \mathrm{d}t^2 - (\mathrm{d}x_1^2 + \mathrm{d}x_2^2 + \mathrm{d}x_3^2), \tag{8.15}$$

where $c\,\mathrm{d}t = \mathrm{d}x_4$, $\mathrm{d}t$ is the proper time and $\mathrm{d}x_1$, $\mathrm{d}x_2$, $\mathrm{d}x_3$ are the directions along the axes 1, 2, 3, measured with a light signal. In the coordinate system in which the interval has the form (8.15), the value of the coordinates coincides with the numerical value of the intervals in time and space.

In principle, therefore, we have two possibilities. We can assign more or less arbitrary numbers to events (satisfying only topological continuity) and then *calibrate* the system using a clock and light signals, i.e., relate these numbers to values of intervals between infinitely close events. The other method corresponds to the above. Time and space intervals between events are obtained, and then these measurements are "*forced into*" a suitable coordinate system.

9. Chronogeometry

Although the arithmetization of events is a priori arbitrary, it requires the application of definite physical methods which themselves have to be able to encompass a widened sphere of applicability, in other words, to encompass both the generality of the method as well as the increased accuracy.

For a long time physicists used an astronomical clock, the rotation of the Earth, for measuring time, and the platinum-iridium rod, the "standard meter" as a measure of length.

In recent years, standard time and length were borrowed from atomic physics, the most accurate clocks being atomic or molecular, and the wavelength λ of the red-orange line of the inert gas krypton (^{86}Kr) is accepted as the standard length. The length of the standard meter is equal to $1{,}650{,}763.73\lambda$. Atomic and molecular clocks owe their origin to the development of technology of atomic and molecular beams and to the practical application of induced radiation [18]. A beam of atomic cesium gives a very narrow line and this guarantees the stability of the frequency, (the frequency being approximately 9000 MHz) to 10^{-12} for several months. The hydrogen atom also possesses such stability (its wavelength is about 21 cm, frequency 1420 MHz). Details about the design of atomic or molecular clocks may be found in reference [27]. The speed of light in a vacuum may be used instead of a scale for measuring length. The most accurate modern data [27] cite the value as

$$c = 299{,}792.5 \pm 0.4 \quad \text{km s}^{-1}$$

The use of clocks as standards for arithmetization of the space-time manifold must be considered as the most perfect and universal method of arithmetization that the state of modern physics offers. Thus geometry, which lies in the foundations of modern physics, is called *chronogeometry*. This name stresses the fundamental role played by the light signal, and time. There is no reason for thinking that geometrical measurements in the macroscopic world will cease to interest physicists and mathematicians. The problem of arithmetization in a non-stationary world obviously still awaits its resolution. An even deeper problem arises in the case when the variations have a stochastic character, so that the metric tensor $g_{\mu\nu}$ becomes a random quantity.

We shall not, however, go into these problems of the large world. On the contrary, we shall return to the main theme of this monograph, to the geometry of the microworld. We shall proceed from the assumption that at least in a stationary macroworld, or in the bounded regions of a non-stationary macroworld, the principles of arithmetization in physical events are sufficiently clear, and it is understood what the symbols x, y, z, t mean for macroscopic events. Chronogeometry underlies the understanding of their physical meaning.

The principle of chronogeometry will be used throughout this monograph in explicit form, or in the form of an implicit assumption, as a basis for considering geometrical measurement in the microworld.

Next, we shall proceed to the problems of the microworld, and see the gradual growth in degree of abstraction in the use of the concepts of the space-time coordinates x, y, z, t in the world of elementary particles. We must prepare ourselves for the fact that this growing abstractness is close to denying the physical meaning of these variables that we are accustomed to considering as space-time coordinates.

CHAPTER II

GEOMETRICAL MEASUREMENTS IN THE MICROWORLD

10. Some remarks on measurements in the microworld

Standard courses in quantum mechanics like to stress that the measuring instrument affects the microobject being measured. However, a more important fact in understanding the actual situation in the microworld is that the microobject, with absolute certainty, will affect the macroscopic instrument.

It is important to note that the measuring instrument has to be an *unstable macroscopic* system, because the microphenomenon could only disturb such a macroscopic system from its state of unstable equilibrium and thereby produce a macroscopic phenomenon. This macroscopic phenomenon can then be used to measure the state of the microsystem.

We can represent the situation in quantum mechanics schematically with the aid of the symbolic equation

$$\mathfrak{M} = M + (A + D), \qquad (10.1)$$

where $\mathfrak{M}$ denotes the whole macroscopic situation in which the microsystem μ "lives", and M denotes those parts of the macrosituation which determines the initial state of the microsystem μ. The recurrence of microsystems μ that are embedded in one and the same macroscopic situation M, forms the quantum ensemble $\mathfrak{R}$:

$$\mathfrak{R} = M + \mu, M + \mu, M + \mu, \ldots \qquad (10.2)$$

The part of $\mathfrak{M}$ in parentheses, $(A+D)$, is formed by the macroscopic instrument which consists of an analyzer A and a detector D.

We assume that the part $(A+D)$ of the whole arrangement does not exert a substantial influence on the formation of M [otherwise the instrument $(A+D)$ could not be separated from M]. This fact is also necessary if we are to be able to exchange one instrument $(A+D)$ for another $(A+D)'$ that is capable of measuring the state of the microobject by other measurable parameters of this microsystem. The analyzer A sorts the microsystems according to state. The purpose of the detector D is to register the presence of the microsystem in some or other state. The detector is an unstable macroscopic system capable of changing its state under the action of the microsituation. An example to illustrate this system is given in Figure 12, where S is the source of particles and C is a slit which selects particles having a certain direction. This source and slit form the part M of the situation which determine the original ensemble.

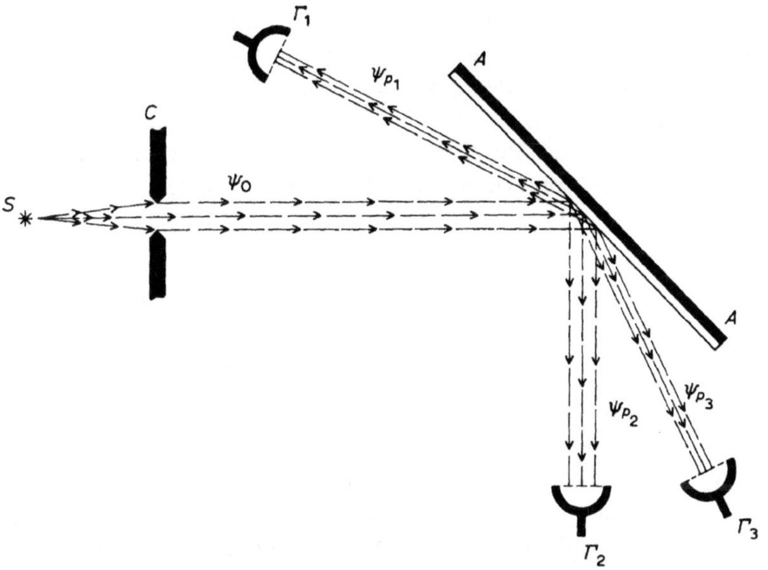

Fig. 12. Illustration of a quantum mechanical measurement. The source of particles S and the slit C form the part M of the macrosituation $\mathfrak{M}$ which defines the initial set Ψ_0, $M = S + C$. The analyzer is the diffraction grating AA'. It resolves Ψ_0 into beams with the momenta $\psi_{p_1}, \psi_{p_2}, \psi_{p_3}, \ldots$. The Geiger counters $\Gamma_1, \Gamma_2, \Gamma_3, \ldots$ form the detector D so that $D = \Gamma_1 + \Gamma_2 + \Gamma_3 + \cdots$ Finally, $\mathfrak{M} = M + (A + D)$.

For simplicity we assume that it can be expressed in the form of the wave function Ψ_0.† The diffraction grating AA' resolves the beam into a spectrum of various momenta p of the particles of the original beam. Mathematically this implies that the original wave function Ψ_0 describes the state of the original ensemble of particles that is resolved by the

† In other words, we consider this ensemble as "pure" (see [28]).

grating into the spectrum

$$\Psi_0 = \sum_s C_{p_s} \psi_{p_s}, \tag{10.3}$$

where ψ_{p_s} is the state having a certain momentum $\mathbf{p}_s$, and C_{p_s} is the amplitude of this state. Finally, the detector D is shown as the collectors of microparticles $\Gamma_1, \Gamma_2, \ldots, \Gamma_n$. Such collectors could be Geiger counters. When a charged microparticle is incident on one, it causes a cascade of electrons which excite a discharge in the gas in the appropriate counter. The Geiger counter, being charged, forms an unstable macroscopic system which registers the appearance of a microparticle in one of its possible states $\psi_{p_1}, \psi_{p_2}, \ldots$

This example illustrates the fact that all information on phenomena in the microworld comes to us through phenomena in the macroworld by the evolution of some unstable macroscopic phenomenon.†

11. The Measurement of Coordinates of the Microparticles

Non-relativistic quantum mechanics formally allows the exact measurements of the coordinates of a microparticle x, y, z to be made in any manner at a given moment of time t [28, 30].

As all physical quantities, the coordinates x, y, z are represented by the Hermitian operators $\hat{x}$, $\hat{y}$, $\hat{z}$ in quantum mechanics. These operators, together with the corresponding operators p_x, p_y, p_z, which are projections of the momentum on the Ox, Oy, Oz axes, are subject to the well-known laws of commutation‡

$$[\hat{x}, \hat{p}_x] = i\hbar, \quad [\hat{y}, \hat{p}_y] = i\hbar, \quad [\hat{z}, \hat{p}_z] = i\hbar, \tag{11.1}$$

$$[\hat{x}, \hat{y}] = [\hat{y}, \hat{z}] = [\hat{z}, \hat{x}] = 0, \tag{11.1'}$$

$$[\hat{p}_x, \hat{p}_y] = [\hat{p}_y, \hat{p}_z] = [\hat{p}_z, \hat{p}_x] = 0. \tag{11.1''}$$

In contrast to the spatial coordinates, the fourth coordinate, time t, is not described by any operator in quantum mechanics (non-relativistic or relativistic), and remains a classical quantity (parameter): Therefore, in

† This point of view of measurements in the microworld was successively carried out in monograph [29].

‡ Here, as usual, $[A, B] = AB - BA$.

quantum mechanics the time t is not treated in the same way as the spatial coordinates x, y, z. This disparity in time and space is preserved in relativistic quantum mechanics. This fact should not be surprising because the characteristic "equality" of time t and coordinates x, y, z in the theory of relativity does not surpass the property which is expressed by the equation

$$s^2 = c^2t^2 - (x^2 + y^2 + z^2). \tag{11.2}$$

As soon as we turn to real physical events the role of time becomes evident in a perfectly clear manner. In fact, the more important problems in physics are formulated thus: From the initial value of dynamic variables, determine their value at later times. In other words, given the values of the variables at $t=0$, we need to find values for $t>0$ (or $t<0$). This problem in mechanics arises from the very essence of the laws of dynamics. In the theory of fields the corresponding problem is that of Cauchy. By the very nature of the laws of dynamics, we cannot, for example, correctly formulate the problem "given t, y and z for $x=0$, find their values for $x\neq 0$". This important circumstance is expressed in quantum mechanics by the fact that in the space of the variables x, y, z, t [the space $\mathfrak{R}_4(x)$] there are no connections between these variables, whereas in the space of conjugate momenta $p_x, p_y, p_z, p_t = E$† [the space $\mathfrak{R}_4(p)$], there are connections arising from dynamics.

Thus in the non-relativistic theory, in the absence of external fields

$$E = \frac{\mathbf{p}^2}{2m}, \quad \mathbf{p} = (p_x, p_y, p_z), \tag{11.3}$$

where m is the mass of particles. Correspondingly, in the relativistic theory

$$E = +\sqrt{\mathbf{p}^2 + m^2}. \tag{11.3'}$$

By virtue of the relationship (11.3) or (11.3′), the space $\mathfrak{R}_4(p)$ reduces to a space of fewer dimensions, $\mathfrak{R}_3(p)$. In the relativistic case (11.3′) the space is that of Lobachevski (see Section 31).

In quantum mechanics, the relationships (11.3) and (11.3′) can also

† The fourth component of the momentum p_t is the energy E.

be considered as corollaries of Schrödinger's equation

$$i\hbar \frac{\partial \Psi}{\partial t} = \hat{H}\Psi, \tag{11.4}$$

where Ψ is the wave function and $\hat{H}$ the Hamiltonian operator. This operator is a function of the momentum operators p_x, p_y, p_z. The form of the Hamiltonian $\hat{H}$ defines the dynamics in quantum mechanics. In particular, we note that the relativistic Dirac equation, in which time is absolutely symmetric with the spatial coordinates (see Section 14), can also be written in the nonsymmetric "Schrödinger" form (11.4).

For stationary states, the time dependence of the wave function Ψ is contained in the factor $e^{-iEt/\hbar}$. In this case (11.4) gives

$$E\Psi = \hat{H}\Psi, \tag{11.5}$$

That is, the possible values of energy E (or of the fourth component of the momentum $P_t = E$) are the eigenvalues of the Hamiltonian operator $\hat{H}$. They can be discrete whereas the "conjugate" fourth coordinate t is continuous. Because of this, there does not exist a relationship [31, 32]

$$[\hat{p}_t, \hat{t}] = i\hbar \quad (\hat{p}_t = \hat{E}), \tag{11.6}$$

similar to (11.1), (11.1′), (11.1″). And, therefore, the uncertainty relation

$$\overline{\Delta E^2}\ \overline{\Delta t^2} \geqslant \frac{\hbar^2}{4}, \tag{11.6′}$$

corresponding to that for x and p_x

$$\overline{\Delta x^2}\ \overline{\Delta p_x^2} \geqslant \frac{\hbar^2}{4}, \tag{11.1‴}$$

does not exist which results from (11.1). Instead of (11.6′), we have the relationship

$$\overline{\Delta E^2}\ \tau^2 \geqslant \frac{\hbar^2}{4}, \tag{11.7}$$

where τ is not the statistical dispersion of time, but the interval of time (fully defined) in which the average value of a dynamic variable L (which is not an integral of motion) changes by $\overline{\Delta L}$, compared to its statistical deviation $(\overline{\Delta L^2})^{1/2}$ [29, 33].

Let us now turn to measuring the characteristics of the operator of the coordinates x. It is sufficient for our purpose here to limit ourselves to one dimension x. According to the basic principles of quantum mechanics, the observed value x' of the variable x is the characteristic value of the operator $\hat{x}$ which is defined by the equation

$$\hat{x}\psi_{x'} = x'\psi_{x'}, \tag{11.8}$$

where $\psi_{x'}$ is the characteristic function of the operator x belonging to the characteristic value x'. The explicit form of $\psi_{x'}$ depends on the representation selected. In the momentum representation, $\psi_{x'}$ is a function of p_x and the operator $\hat{x}$ is equal to

$$\hat{x} = + i\hbar \frac{\partial}{\partial p_x}. \tag{11.9}$$

In the coordinate representation, $\psi_{x'}$ is a function of x and the operator $\hat{x}$ is equal to

$$\hat{x} \equiv x. \tag{11.10}$$

From Equation (11.8) it follows that, in the momentum representation

$$\psi_{x'}(p_x) = \frac{\exp(ip_x x/\hbar)}{\sqrt{2\pi\hbar}}, \tag{11.11}$$

and in the coordinate representation

$$\psi_{x'}(x) = \delta(x - x'). \tag{11.12}$$

If we select the normalized functions $\psi_{x''}$ they are related by the unitary transformation

$$\psi_{x'}(x) = \int_{-\infty}^{\infty} S(x \mid p_x)\, \psi_{x'}(p_x)\, \mathrm{d}p_x, \tag{11.13}$$

where the matrix elements $S(x \mid p_x)$ are equal to

$$S(x \mid p_x) = \frac{\exp(-ip_x/\hbar)}{\sqrt{2\pi\hbar}}. \tag{11.14}$$

It follows from the explicit form of the eigenfunctions of the operator

$\hat{x}$ that, in the state with a certain value $x = x'$, there coexist states with any momentum p_x, from $-\infty$ to $+\infty$, having equal importance, equal to $|S(x|p_x)|^2 = \frac{1}{2}\pi\hbar$. Therefore, when making an exact measurement of the coordinates of a microparticle x, an infinitely large energy must be involved in this measurement.

This necessary energy may be borrowed from the energy of the particle itself, or from another physical system that takes part in the measurement (e.g., from any part of the measuring instrument). Thus for measuring coordinates, the interference of the instrument with the microparticle state is both inevitable and essential. When considering this interference, it is better to speak, not of the measuring the microparticle coordinate x, but of *localizing* [30] the microparticle about the point $x = x'$ as a result of its interaction with the measuring device.

The state $\psi_{x'}$ which arises from the localization of the particle quickly diffuses, so that the *average value* of the square of the deviation $\overline{\Delta x^2} = \overline{(x - \bar{x})^2}$ will grow more rapidly with the passage of time.

In order to investigate the diffusion of the state obtained by localization at time $t = 0$ and point $x = x'$, we shall consider the function $\psi_{x'}(x)$ as the limit of the sequence of the function $\psi_{x'}(x, a)$ as $a \to 0$:

$$\psi_{x'}(x, a) \equiv \frac{\exp(-(x - x')^2/a^2)}{\sqrt{\pi}\, a}$$

$$= \frac{1}{2\pi\hbar} \int_{-\infty}^{\infty} \exp\left(-\frac{a^2}{4} p_x^2 + \frac{ip_x(x - x')}{\hbar}\right) dp_x \quad (11.15)$$

The quantity $C(p_x) = (\frac{1}{2}\pi\hbar) \exp[-(ap_x/4)^2 - (ip_x x'/\hbar)]$ is the same wave function, but written in the p_x representation. For $a \to 0$, we have

$$\lim_{a \to 0} \psi_{x'}(x, a) = \delta(x - x'). \quad (11.16)$$

From Schrödinger's equation

$$i\hbar \frac{\partial \psi}{\partial t} = \hat{H}\Psi \quad (11.17)$$

($\hat{H} = \hat{p}_x^2/2m$ is the Hamiltonian operator for a free particle, and m is the mass of that particle); each plane wave $\psi_{p_x}(x) = \exp(ip_x x/\hbar)/(2\pi\hbar)^{1/2}$ in the superposition of waves (11.15), for $t>0$, becomes

$$\psi_{p_x}(x, t) = \frac{\exp[\mathrm{i}(p_x x - Et)/\hbar]}{\sqrt{2\pi\hbar}}, \tag{11.18}$$

where $E(p_x)=p_x^2/2m$ is the kinetic energy of the microparticle having momentum p_x and mass m. Therefore, at the time t the wave function (11.15) transform into the function

$$\psi_{x'}(x, a, t) = \frac{1}{2\pi\hbar} \int_{-\infty}^{\infty} \exp\left(-\frac{a^2 p_x^2}{4} + \frac{\mathrm{i}p_x(x-x')}{\hbar} - \frac{\mathrm{i}p_x^2 t}{2m\hbar}\right) \mathrm{d}p_x. \tag{11.19}$$

This integral may be evaluated when

$$Z = (Ap_x + B), \qquad A^2 = \left(\frac{a^2}{4} + \frac{\mathrm{i}t}{2m\hbar}\right), \qquad B = \frac{\mathrm{i}(x-x')}{2\hbar A}. \tag{11.20}$$

Evaluating this integral gives

$$|\psi_{x'}(x, a, t)|^2 = \frac{1}{2\sqrt{\pi}\,\hbar A}\,\mathrm{e}^{-B^2}. \tag{11.21}$$

From this form, it is easy to obtain the width of the packet

$$\overline{\Delta x^2(t)} = a^2 + \frac{\hbar^2}{a^2 m^2}t^2 \tag{11.22}$$

and note its growth in time. For $a\to 0$, when $\psi_{x'}(x, a)\to\delta(x-x')$, the velocity with which the packet diffuses becomes of the order of $v\sim\hbar/ma\to\infty$, which is formally postulated in the non-relativistic theory.

We have considered the eigenfunctions of the operator $\hat{x}$ which are quadratically not integrable. We may now consider a second class of functions $\psi_{x'}(x, a)$ which remain quadratically integrable, namely,

$$\psi_{x'}(x, a) = \frac{\exp[-(x-x')^2/2a^2]}{\sqrt[4]{\pi}\sqrt{a}}. \tag{11.23}$$

This type of function tends to $(\delta(x-x'))^{1/2}$ as $a\to 0$.

For such a function

$$\int |\psi_{x'}(x, a)|^2 \, dx = 1 \tag{11.24}$$

for any a. We may readily see that, for the state described by the function (11.23), the quantity $\overline{\Delta x^2}$ is equal to

$$\overline{\Delta x^2} = a^2/2 \tag{11.25}$$

and for $a \to 0$, $x^2 = 0$. As regards to the operator x, it operates on the function $\psi_{x'}(x, a)$ thus:

$$\hat{x}\psi_{x'}(x, a) = x'\psi_{x'}(x, a) + \Delta(x', x, a), \tag{11.26}$$

where

$$\Delta(x', x, a) = \sqrt[4]{\pi}\sqrt{a}\left[(x - x')\frac{\psi(x', a)}{\sqrt[4]{\pi}\sqrt{a}}\right] \tag{11.26'}$$

and for $a \to 0$

$$\Delta(x', x, a) \to \sqrt[4]{\pi}\sqrt{a}(x - x')\,\delta(x - x') = 0. \tag{11.27}$$

Therefore, for $a \to 0$, the function (11.23) satisfies Equation (11.3) to any degree of accuracy.

12. The Mechanics of Measuring Coordinates of Microparticles

Thus far, the discussion concerning measurements of the coordinates of a microparticle x has been rather formal. From this formal point of view, the coordinate measurements indicate that some initial states described by the wave function $\psi_0(x)$ "reduce" to the state

$$\psi_0(x) \to \psi_{x'}(x). \tag{12.1}$$

However the realization of the process of localization of particles near some point $x = x'$ is not simple, even in a *gedanken* experiment.

Thus it is important to divide measurements into two groups, the *direct* and *indirect*, as was done by Mandelshtam [34]. In the case of the direct measurements, the measurement itself is realized in the first link joining the microobject with the macroscopic instrument. In the second case the measurements have one or more intermediate links.

For example, if a beam of free microparticles is incident on a photographic plate and it produces a "spot" on the plate, such a phenomenon is called a *direct measurement* of the coordinates of the free particles. However, the dimensions of the spot are not smaller than the dimensions of the atoms in the plate. They are in fact larger, because the smallest spot has the dimension of the grain size of the plate. Thus the initial wave function that describes the incident beam $\psi_0(x)$ does not tend to $\psi_{x'}(x) = \delta(x-x')$ in such a measurement, but to some function $\psi_{x'}(x, a)$ which has the dimension a which is greater than or equal to the atomic size of the photographic plate. However, theory does not limit the size of atoms, which may be very small, and so there is no inconsistency within the theory.

As an example of an indirect measurement, we may consider the measurement of the coordinates of an electron within an atom. In principle such a measurement can be made by observing the scattering of a narrow beam of microparticles from a bound atomic electron. The scattering of these particles can be registered directly on a screen, and be related to direct measurements. The narrow beam of particles which is used for the probe forms an *intermediate* link. Therefore such a measurement is called indirect.

A. *Direct Measurement of Coordinates of a Microparticle*

We shall now turn to the mathematical side of the problem and consider the interaction of the microparticle and the atoms in the photographic plate [29]. Insofar as we are interested in the basic points we will make several simplifying assumptions which are compatible with the principles of the theory. We consider the photographic plate as a system of infinitely heavy atoms whose location in space is known exactly. We assume that the primary microphenomenon which "initiates" the photochemical reaction consists of the excitation of the atom in such a way that its electron goes from the state $\psi_0(x)$ to the state $\psi_n(x)$ where x is the coordinate of the electron in the atom. If a is the dimension of the atom, then ψ_0 and ψ_n disappear rapidly for $|x| \gg a$. We denote the coordinates of the microparticle by Q and assume that the energy of interaction of the microparticle with the electron has the form of a δ function

$$W = g\delta(x - Q), \tag{12.2}$$

where g is the interaction constant. Let the excitation energy of the atom be $\varepsilon = \varepsilon_n - \varepsilon_0$, and the energy of the microparticle be $E = p^2/2m$ where m is the mass of the particle, p its momentum before the interaction and p' its momentum after the interaction with the atom. We assume that $E \gg \varepsilon$. Therefore the microparticle is the main source of energy which is necessary for its localizing. The change in excitation $E_0 \to E_n$, which is localized exactly in the region around the atom plays the part of the analyzer A which sorts the microparticles according to their coordinates Q to an accuracy of a, the atomic dimension. In further excitations the atom can initiate hidden transformations in the sensitive grain of photoemulsion. These, or similar processes, play the part of the detector D which destroys the interference of the various states of the microparticles. We shall not go into details on the design of detector, and limit ourselves to the design of the primary function of the measuring device, that of analyzing particles by coordinates. In our case, the macroscopic nature of the analyzer A is expressed in the assumption that the mass of the atoms in the photographic plate is infinitely large so that their location in space may be defined exactly.

At the initial time $t = 0$, the wave function of this system (microparticle μ and analyzer A) has the form

$$\Phi_0(x, Q) = \psi_0(x)\, e^{ipQ/\hbar}. \tag{12.3}$$

This function describes the atom in the unexcited state and the microparticle in the state of a plane wave having momentum p. For $t > 0$, we write the wave function in the system as

$$\Phi(x, Q, t) = \Phi_0(x, Q)\, e^{i\omega_0 t} + \varphi(x, Q, t), \tag{12.4}$$

where the frequency $\omega_0 = (E + \varepsilon_0)/\hbar$. We shall consider the function $\varphi(x, Q, t)$ as small and compute its contribution using perturbation theory. According to this theory (see, e.g., [28], Chapter XI),

$$= \int d^3p' \sum_n \frac{1}{i\hbar} \int_0^t d\tau\, e^{i\Omega_n \tau} (np' |W| 0p)\, \psi_n(x)\, e^{iQp}, \tag{12.5}$$

where $(np'|W|\,Op)$ is the matrix element of the energy of interaction, taken between the initial state $\Phi_0(x, Q)$ (12.3) and the excited state

$$\Phi_n(x, Q) = \psi_n(x)\, e^{ip'Q/\hbar}. \tag{12.3'}$$

This matrix element is equal to

$$\begin{aligned}(np'|W|0p) &= \int \Phi_n^*(x, Q)\, W(x, Q)\, \Phi_0^*(x, Q)\, dx\, dQ \\ &= g \int \Phi_n(Q, Q)\, \Phi_0(Q, Q)\, dQ \\ &= g \int \rho_{n0}(Q)\, e^{i(p-p', Q)}\, dQ = g\tilde{\rho}_{n0}(q);\; q = p - p'.\end{aligned} \tag{12.6}$$

Here $\rho_{n0}(Q) = \psi_n(Q)\, \psi_0^*(Q)$ and $\tilde{\rho}_{n0}(q)$ is the Fourier component of $\rho_{n0}(Q)$. Further, Ω_n is the frequency of the excitation

$$\Omega_n = \frac{(p^2/2M + \varepsilon_0 - p'^2/2M - \varepsilon_n)}{\hbar}.$$

If we consider small t: $0 \ll t \ll 1/\Omega_n$, then

$$\begin{aligned}\varphi(x, Q, t) &= \sum_n \int d^3p'\, \frac{g}{i\hbar}\, \tilde{\rho}_{n0}(q)\, e^{iQp'} \psi_n(x)\, t \\ &= \sum_n \frac{g}{i\hbar}\, \psi_n(x)\, \rho_{n0}(Q)\, e^{iQp}\, t.\end{aligned} \tag{12.7}$$

In order to make this result clearer, we assume that in the sum of excited states only one term is important and for the others $\rho_{m0} \approx 0$ where $m \neq n$. Then we have

$$\varphi(x, Q, t) = t\, \frac{g}{i\hbar}\, \psi_n(x)\, \rho_{n0}(Q)\, e^{iQp}. \tag{12.8}$$

From this we may easily see that, for small t, soon after the interaction the wave function of the microparticle is

$$\psi_n(Q) = \rho_{n0}(Q)\, e^{iQp} \tag{12.9}$$

and accurately reproduces the transitional density of the electron in the atom $\rho_{n0}(x)$. This function is the wave packet with a probability distribution for the coordinates of the microparticle

$$W_n(Q) = |\rho_{n0}(Q)|^2 \tag{12.10}$$

and with an average momentum of

$$\bar{\mathbf{p}} = \mathrm{i}\hbar \int \Phi_n(Q) \frac{\partial}{\partial Q} \Phi_n(Q)\, \mathrm{d}Q \approx |\mathbf{p}|\,. \tag{12.11}$$

(where the symbol $\approx$ indicates that equality is satisfied to an accuracy of the electron momentum, which has the order of magnitude of $\hbar/a \ll \mathbf{p}$).

In this manner, the act of exciting an atom is accompanied by the localization of the microparticle in the wave packet (12.9), and so the excitation of the atom can be considered as the transition of the particle from the state of a plane wave $e^{ipQ/\hbar}$ for $t=0$ to a localized state $\psi_n(Q)$ which is defined in (12.9) for $0 \ll t \ll 1/\Omega_n$. For $t \to \infty$, the wave packet diffuses. However, if the excitation of the atom initiates a chemical or some other reaction, then it must be included in the examination of the chain of events leading up to the macroscopic event. In this case the state will not tend to a diffused packet for $t \to \infty$, but will be described by a new state which results from the process of atomic excitation followed by the chemical reaction.

B. *Indirect Coordinate Measurements of a Microparticle*

We shall now turn to the indirect measurements of the microparticle coordinates. We first consider ray theory which we shall use as a probe for measuring the coordinates of a microparticle that is in a bound state, as for example an interatomic electron.

A screen $3-3$ with aperture D are shown in Figure 13. The aperture forms a packet of microparticles that have an energy $E = p^2/2m = \hbar\omega$ and momentum $p = \hbar k$. Let us superimpose a plane wave that has a certain total energy but a different direction of propagation. It has the form

$$\psi = \int A(\alpha) \exp\{ikr[\cos \upsilon \cos \alpha + \sin \upsilon \sin \alpha \cos(\varphi - \beta)] - i\omega t\} \sin \alpha \, \mathrm{d}\alpha \, \mathrm{d}\beta\,, \tag{12.12}$$

where r, υ, φ are the polar coordinates of the radius vector r and α, β are the polar angles of the wave vector **k**. Integrating over the angle $d\beta$ gives

$$\psi = \int A(\alpha) \sin\alpha \, d\alpha \exp\{ikr \cos \upsilon \cos\alpha\} \, J_0(kr \sin \upsilon \sin\alpha) \tag{12.13}$$

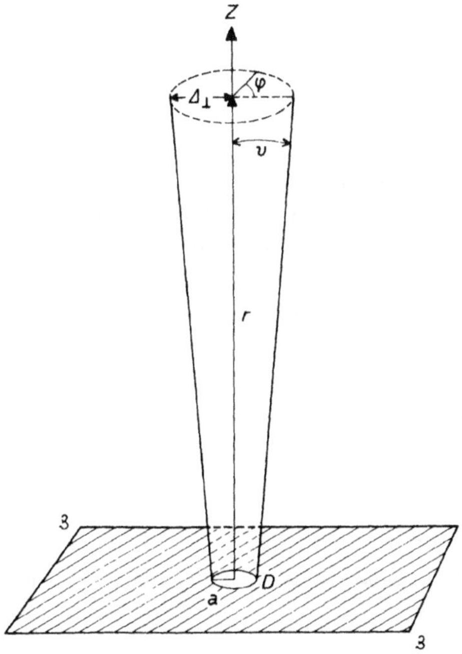

Fig. 13. Diffraction of a narrow beam. 3-3 is an absorbing screen; D is an aperture with diameter a; $\Delta_\perp$ is the width of the beam at a distance r from the screen; $\Delta_\perp^2 = a^2 + (\lambda^2/a^2 \, r^2$.

[$J_0(x)$ is the zero-order Bessel function]. For $\upsilon = \pi/2$ we obtain the value of the field at the surface of the screen. This field must vanish outside the aperture. We can set $f(r) = B \exp(-r^2/a^2)$ in a simple approximation to express the opacity of the screen for $r > a$. In the integral (12.13) only small angles give a significant contribution and, therefore, we may approximate the integral by

$$\psi = e^{ikz} \int_0^\infty A(\alpha) \, \alpha \, d\alpha \exp(-\tfrac{1}{2} ikr \, \alpha^2) \, J_0(kr \sin \upsilon \, \alpha). \tag{12.14}$$

For $z=0$ we have

$$Be^{-r^2/a^2} = \int_0^\infty J_0(kr\alpha)\, A(\alpha)\, \alpha\, d\alpha\,, \qquad (12.15)$$

from which†

$$A(\alpha) = \frac{Bk^2a^2}{4\pi} \exp\left(-\frac{k^2a^2\alpha^2}{4}\right). \qquad (12.16)$$

Substituting this expression for $A(\alpha)$ into (12.14) and using the inverted equation (12.15), for large r and small υ we obtain the intensity of the wave in the beam $|\psi|^2$ which is equal to

$$|\psi|^2 = \left(\frac{kBa^2}{2r}\right)^2 \exp\left(-\frac{1}{2}\frac{a^2}{\lambda\!\!\!^{-2}}\theta^2\right). \qquad (12.17)$$

From this equation it follows that a beam of particles originating from an aperture of radius r, spreads due to diffraction in such a way that the square of its transverse dimension $\rho^2 = r^2\theta^2$ is defined in the quantity

$$\Delta_\perp^2 = \frac{\lambda\!\!\!^{-2} r^2}{a^2}. \qquad (12.18)$$

If we consider the time evolution of this process assuming $r = vt = pt/m$, we find that

$$\Delta_\perp^2 = \frac{\hbar^2}{m^2a^2}t^2, \qquad (12.19)$$

which is in agreement with the result we obtained earlier for the transverse dispersion of a wave packet. In order that the beam would form a good probe, its width $\Delta_\perp$ must be substantially smaller than the dimensions of the system d within which the location of the microparticle is to be determined. That is, taking into account the initial dimensions, we have the condition that

$$a^2 + \frac{\lambda\!\!\!^{-2} r^2}{a^2} \ll d^2 \qquad (12.20)$$

† See [35] for an example.

or, for $a \ll d$,

$$\frac{d}{\lambda} \gg \frac{r}{a}. \tag{12.20'}$$

These conditions are extremely rigid and if applied to some real object, they are practically unrealizable. In fact there is no way of forming a slit or any other device for producing a beam whose width $a<d$, if d is the atomic dimension. However the theory formally allows the slit to be of any width and thus is not internally inconsistent. Keeping this in mind, let us consider the interaction of a microparticles to be used as a probe to determine the location of the first bound particle. Let the mass of the first particle be m, its coordinate $\mathbf{x}$, its eigenfunctions $\psi_n(\mathbf{x})$ and E_n the corresponding energy level. These functions satisfy the equation

$$\hat{H}_m(\mathbf{x})\,\psi_n(\mathbf{x}) = E_n\psi_n(\mathbf{x}), \tag{12.21}$$

where $\hat{H}_m(\mathbf{x})$ is the Hamiltonian operator describing the dynamics of the bound particle. Let the mass of the particles in the beam probe be μ, their coordinates $\mathbf{Q}$ and the momentum operator $\hat{\mathbf{P}}$. Thus their Hamiltonian, written more simply, becomes

$$\hat{H}_\mu(\mathbf{Q}) = \frac{\hat{\mathbf{p}}^2}{2\mu}. \tag{12.22}$$

As in Section 11, we assume that the interaction of the m-particles and μ-particles occurs at one point, i.e.,

$$W(\mathbf{x}, \mathbf{Q}) = g\delta(\mathbf{x} - \mathbf{Q}), \tag{12.23}$$

where g is the interaction constant. The wave function of the m-particles together with the beam-probe $\psi(\mathbf{x}, \mathbf{Q})$ satisfy Schrödinger's equation

$$[\hat{H}_m(\mathbf{x}) + \hat{H}_\mu(\mathbf{Q}) + W(\mathbf{x} - \mathbf{Q})]\,\psi(\mathbf{x}, \mathbf{Q}) = E\psi(\mathbf{x}, \mathbf{Q}). \tag{12.24}$$

We write $\psi(\mathbf{x}, \mathbf{Q})$ in the form

$$\psi(\mathbf{x}, \mathbf{Q}) = \psi_0(\mathbf{x})\,u_0(\mathbf{Q}) + \sum_{n \neq 0} \psi_n(\mathbf{x})\,u_n(\mathbf{Q}), \tag{12.25}$$

where $\psi_0(\mathbf{x})$ describes the state of the m-particles before the interaction, and $u_0(\mathbf{Q})$ describes the beam of μ-particles together with the elastically scattered wave. Substituting this function into (12.24) and multiplying scalarly by $\psi_m^*(\mathbf{x})$, we obtain the system of equations for the function $u_m(\mathbf{Q})$:

$$\left[\frac{\hat{\mathbf{p}}^2}{2\mu} + (E_m - E)\right] u_m(\mathbf{Q}) = -g\psi_0(\mathbf{Q})\,\psi_m^*(\mathbf{Q})\,u_0(\mathbf{Q}), \qquad (12.26)$$

We neglected the product $W\sum_{n\neq 0}\psi_n(\mathbf{x})\,u_n(\mathbf{Q})$ as being of the second order of smallness (for small g). For elastic scattering which is described by $u_0(\mathbf{Q})$, we have according to (12.26)

$$(\nabla^2 + k^2)\,u_0(\mathbf{Q}) = \frac{-2\mu g}{\hbar^2}\,|\psi_2(\mathbf{Q})|^2\,u_2(\mathbf{Q}) = \rho(\mathbf{Q}), \qquad (12.27)$$

where $k^2 = 2\mu E/\hbar^2$. Using well-known methods[†] we obtain

$$u_0(\mathbf{Q}) = \varphi_0(\mathbf{Q}) + \int \mathfrak{G}(\mathbf{Q} - \mathbf{Q}')\,\rho(\mathbf{Q}')\,\mathrm{d}\mathbf{Q}', \qquad (12.28)$$

where the function $\varphi_0(\mathbf{Q})$ describes the initial narrow beam [see Equation (12.12)], and the function $u_0(\mathbf{Q})$ describes this beam together with the scattered wave. $\mathfrak{G}(\mathbf{Q}-\mathbf{Q}')$ is the Green's function of the homogeneous equation (12.27)

$$\mathfrak{G}(\mathbf{Q} - \mathbf{Q}') = \frac{-\,\mathrm{e}^{\mathrm{i}k|Q-Q'|}}{4\pi\,|\mathbf{Q} - \mathbf{Q}'|}, \qquad (12.29)$$

and the quantity $\rho(\mathbf{Q})$ is equal to

$$\rho(\mathbf{Q}) = -\frac{2\mu g}{\hbar^2}\,|\psi_0(\mathbf{Q})|^2\,\varphi_0(\mathbf{Q}). \qquad (12.30)$$

If we consider large separations between the scattering regions $|\mathbf{Q}| = r \to \infty$, then equation (12.28) reduces to the well-known result[†]

$$u_0(\mathbf{Q}) = \varphi_0(\mathbf{Q}) + \frac{1}{4\pi}\,\frac{\mathrm{e}^{ikr}}{r}\,A(\mathbf{q}), \qquad (12.28')$$

where the amplitude of the scattered wave $A(\mathbf{q})$ is equal to

$$A(\mathbf{q}) = \int \mathrm{e}^{\mathrm{i}qQ}\rho(\mathbf{Q})\,\mathrm{d}\mathbf{Q}\,; \qquad (12.31)$$

[†] See, e.g., [28], Chapter 13.
[†] See, e.g., [28].

Here the vector $\mathbf{q}=\mathbf{p}'-\mathbf{p}$ is transmitted momentum due to scattering.

From Equation (12.28) we can see that the source of the wave scattering in the region the density $\rho(\mathbf{Q})$ differs substantially from zero. This region is the region of intersection of the beam-probe of μ-particle $\varphi_0(\mathbf{Q}, t)$ with the region of localization of the m-particle. This region is defined by the behavior of the function $|\psi_0(\mathbf{Q})|^2$.

Consequently, the region in which $\rho(\mathbf{Q})$ differs substantially from zero has definite coordinates in the (x, y) plane to the accuracy of the beam width a and an undefined coordinate z which is the beam direction (see Figure 14). If we can now record the scattering of the μ-particles off the

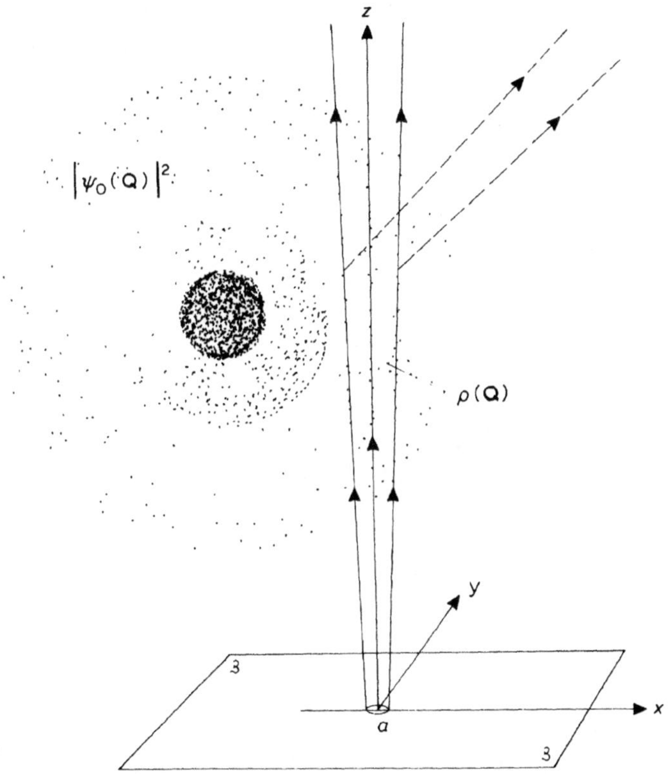

Fig. 14. Coordinate measurements in an indirect experiment. 3-3 is an absorbing screen; a is the diameter of the aperture which produces a narrow beam-probe.

atom with some detector, then we will have shown that at the moment of scattering, the m-particle had the coordinates x, y which are equal to the coordinates of the aperture that defined the beam of μ-particles to an accuracy of the beam width Δx, $\Delta y \approx a$. We note however that this assertion would be exact only if the beam had sharp boundaries. In fact such boundaries do not exist and there is a finite probability that the μ-particle is scattered on an m-particle that is located at a distance greater than a from the axis of the beam. Furthermore, because of the diffuseness of the boundary of the atom, the m-particle may be extremely far from the center of the atom, in the region where the beam-probe diffused to such an extent that its dimensions $\Delta_\perp \gg a$.

And so, even in this *gedanken* experiment in which we assumed a narrow slit and an interaction at one point, the inference about the actual coordinates related to the m-particle atom can only be made in terms of probabilities.

The question can be formulated more strictly in the following manner: Given a scattered μ-particle, what is the probability that it was scattered by a m-particle located in some sharply defined region $\Omega(\mathbf{x})$? In answering this question we note that the quantity which defines the full probability of some elastic scattering of μ-particles off of bound m-particles is the integral

$$I = \int |A(\mathbf{q})|^2 \, \mathrm{d}^3 q\,, \tag{12.32}$$

where $A(\mathbf{q})$ is the amplitude of the scattered wave (12.31). The quantity I (12.32) measures in relative units the flow of elastically scattered μ-particles.

Using (12.31), this integral can be rewritten in the form

$$I = \frac{1}{(4\pi)^2} \int \rho(\mathbf{x}')\, \rho(\mathbf{x}'')\, \Delta(\mathbf{q}, \mathbf{x}' - \mathbf{x}'')\, \mathrm{d}^3x'\, \mathrm{d}^3x''\,, \tag{12.33}$$

where the function $\Delta(\mathbf{q}, \mathbf{x})$ $(\mathbf{x} = \mathbf{x}' - \mathbf{x}'')$ is defined by the equation

$$\Delta(\mathbf{q}, \mathbf{x}) = \frac{\sin Q_1 x}{x} \frac{\sin Q_2 y}{y} \frac{\sin Q_3 z}{z}\,, \tag{12.34}$$

and Q_1, Q_2, Q_3 are the maximum values of the projections of the vector $\mathbf{q}$ on the coordinate axes. From (12.33) we can see that the probability for elastic scattering I cannot be represented as a sum of contributions from various spatial regions but is defined as the sum of correlations.

However, if the wavelength λ of the beam-probe is small in comparison to the region in which $\rho(\mathbf{x})$ varies, then the limits of integration for q_x, q_y, q_z can be changed to $\pm\infty$. This is equivalent to exchanging the function $\Delta(\mathbf{q}, \mathbf{x})$ for the function $\delta(\mathbf{x})$. Then in place of (12.33) we obtain

$$I \approx \frac{1}{(4\pi)^2} \int \rho^2(\mathbf{x})\,\mathrm{d}^3x \tag{12.35}$$

and the probability that the scattering took place in the region Ω can be written in the form

$$I(\Omega) = \frac{1}{(4\pi)^2} \int_{\Omega} \rho^2(\mathbf{x})\,\mathrm{d}^3x. \tag{12.36}$$

In these calculations we assumed that the interaction of the μ-particles and m-particles occurred at one point [see (12.23)]. If it did not occur at one point then it is, in general, impossible to define the probability $I(\Omega)$ in regions Ω which are smaller than the sphere of interactions. In fact, substituting $\delta(\mathbf{x}-\mathbf{Q})$ for the distribution function of the interaction $V(\mathbf{x}-\mathbf{Q})$ in (12.23) gives

$$W(\mathbf{x}-\mathbf{Q}) = gV(\mathbf{x}-\mathbf{Q}), \tag{12.23'}$$

and repeating the steps used to obtain Equation (12.27) we obtain

$$(\nabla^2 + k^2)\,u_0(\mathbf{Q}) = \rho_V(\mathbf{Q}), \tag{12.27'}$$

in place of (12.27), where

$$\rho_V(\mathbf{Q}) = \frac{2\mu g}{\hbar^2} \int V(\mathbf{x}-\mathbf{Q})\,|\psi_0(\mathbf{x})|^2\,\varphi_0(\mathbf{Q})\,\mathrm{d}^3x. \tag{12.30'}$$

From this it follows that

$$I_V = \frac{1}{4\pi} \int \rho_V(\mathbf{x}')\,\rho_V(\mathbf{x}'')\,\Delta(\mathbf{q}, \mathbf{x}'-\mathbf{x}'')\,\mathrm{d}^3x'\,\mathrm{d}^3x'', \tag{12.33'}$$

or for $\lambda \to 0$, $|\mathbf{q}| \to \infty$ we obtain, analogously to (12.35),

$$I_V = \frac{1}{(4\pi)^2} \int \rho_V^2(\mathbf{x})\,\mathrm{d}^3x. \tag{12.34'}$$

As before, we may formally introduce the quantity

$$I_V(\Omega) = \frac{1}{(4\pi)^2} \int\limits_{\Omega} \rho_V^2(\mathbf{x})\, \mathrm{d}^3 x\,. \tag{12.35'}$$

However, from (12.30) we can see that the quantity $\rho_V^2(\mathbf{x})$ is non-locally related to the function $|\psi_0(\mathbf{x})|^2$, so that it does contribute to the integral (12.35) even in the regions of space where the quantity $|\psi_0(\mathbf{x})|^2$ is equal to zero. The interpretation of integral (12.35) can be retained in the case when $|\psi_0(\mathbf{x})|^2$ is very evenly distributed inside the scale which defines the sphere of interaction, that is the region where interaction potential $V(\mathbf{x}-\mathbf{Q})$ is negligibly small.

13. Indirect Measurement of a Microparticle Coordinates at a Given Instant of Time

So far, the measurement of the coordinate x of the microparticle was considered for an undefined moment in time. In order to measure the coordinate x in some specific small interval of time Δt the function of the aperture in the *gedanken* experiment of forming a beam-probe necessarily becomes more complicated. We must assume that it is possible to open the aperture at some moment in time, say $t=0$, and to close it at some moment $t=\Delta t$. Then, in place of a beam shown in Figure 14, we obtain a packet, defined not only along axes Ox and Oy, but also in the direction of its motion, along the axis Oz.

This packet will diffuse according to the equations obtained in Section 11. Now, in place of a beam-probe of μ-particles we shall have a packet-probe of μ-particles. Such a wave packet-probe can be made up of a superposition of plane waves which diffuse primarily in the Oz direction (see Figure 14). In accordance with the above, we shall write the wave function of the μ-particles which we shall use to probe the location of the m-particles which are bound inside the microsystem, in the form

$$\varphi_0(\mathbf{Q}, t) = \int C(\mathbf{p}) \exp\left(-i\frac{(Et - \mathbf{pQ})}{\hbar}\right)\frac{\mathrm{d}^3 p}{(2\pi\hbar)^{3/2}}, \tag{13.1}$$

where

$$C(\mathbf{p}) = \frac{\exp\{-b^2[p_x^2 - p_y^2 - (p_z - p)^2]\}}{b^{3/2}\pi^{3/4}}. \tag{13.2}$$

Here p is the average momentum in the direction of motion of the packet along the Oz axis, and the quantity $1/b^2$ is a measure of the rms deviation of the momentum $\overline{\Delta P_x^2}$, $\overline{\Delta P_y^2}$, $\overline{\Delta P_z^2}$. Using the calculations described in Section 11, we find that for sufficiently large values of the parameter b, the function $\varphi_0(\mathbf{Q}, t)$ may be written in the form

$$\varphi_0(\mathbf{Q}, t) = \frac{e^{i(\omega^2 - k\zeta)}}{(2\pi\hbar)^{3/2}} f(\mathbf{Q}, t), \tag{13.3}$$

where the function $f(\mathbf{Q}, t)$ is equal to

$$f(\mathbf{Q}, t) = \exp\left[-\frac{\xi^2 + \eta^2}{\Delta_1^2} - \frac{(\zeta - vt)}{\Delta_\parallel^2}\right]. \tag{13.4}$$

In these equations $\omega = E/\hbar$; $E = p^2/2\mu$ is the average energy of the μ-particles; $\mathbf{P}$ is their average momentum directed along the $O\xi$ axis, $\mathbf{k} = \mathbf{p}/\hbar$ is the wave vector, $v = p/\mu$ is the average velocity along the $O\xi$ axis and $\Delta_\perp^2$, $\Delta_\parallel^2$ are the average widths of the wave packet in the transverse direction and that corresponding to the direction of the μ-packet. In agreement with (11.22) we have

$$\Delta_\perp^2 = \Delta_\parallel^2 = a^2 + \frac{\hbar^2}{\mu^2 a^2} t^2, \tag{13.5}$$

where $a^2 \approx \hbar^2/b^2$. As in Section 12, let the Hamiltonian of the m-particle be $\hat{H}_m(\mathbf{x})$, and the Hamiltonian of the μ-particles be $\hat{H}_\mu(\mathbf{Q})$ [see Equations (12.21) and (12.22)]. Then the Schrödinger equation for the wave function $\Psi(\mathbf{x}, \mathbf{Q}, t)$ of the system of "m-particles plus the wave packet of μ-particles" may be written as

$$i\hbar \frac{}{\partial t} = [\hat{H}_m(\mathbf{x}) + \hat{H}_\mu(\mathbf{Q}) + W(\mathbf{x} - \mathbf{Q})]\ \Psi, \tag{13.6}$$

where W is the energy of interaction of the m-particles and μ-particles. We shall consider that this interaction occurs at one point [see Equations (12.23)]. We now write the function $\psi(\mathbf{x}, \mathbf{Q}, t)$ in the form

$$\Psi = \Phi(\mathbf{x}, \mathbf{Q}, t) \exp\left(-i\frac{E_0 t}{\hbar}\right), \tag{13.7}$$

where E_0 is the energy of m-particles in the state $\psi_0(\mathbf{x})$. The function $\Phi(\mathbf{x}, \mathbf{Q}, t)$ may be written as the sum

$$\Phi(\mathbf{x}, \mathbf{Q}, t) = \psi_0(\mathbf{x})\, \varphi_0(\mathbf{Q}, t) + u(\mathbf{x}, \mathbf{Q}, t), \tag{13.8}$$

where $u(\mathbf{x}, \mathbf{Q}, t)$ is the scattered wave, and $\psi_0(\mathbf{x})\, \varphi_0(\mathbf{Q}, t)$ is the initial wave. Let us further write $u(\mathbf{x}, \mathbf{Q}, t)$ as the series

$$u(\mathbf{x}, \mathbf{Q}, t) = \sum_n u_n(\mathbf{Q}, t) \bar{\bar{\Psi}}_n(\mathbf{x}) \tag{13.9}$$

in the eigenfunctions $\psi_n(\mathbf{x})$ of the Hamiltonian operator $\hat{H}_m$ of the m-particles. In the usual manner (cf. Section 12) we find the system of equations for the function u_n

$$i\hbar \frac{\partial u_m}{\partial t} = \frac{\hat{\mathbf{P}}^2}{2\mu} u_m + g\psi_0(\mathbf{Q})\, \psi_m^*(\mathbf{Q})\, \varphi_0(\mathbf{Q}, t), \tag{13.10}$$

which are analogous to the system (12.26).† In particular, for elastic scattering, we obtain

$$i\hbar \left(\frac{\partial}{\partial t} - \frac{\hat{\mathbf{P}}^2}{2\mu} \right) u_0 = g\, |\psi_0(\mathbf{Q})|^2\, \varphi_0(\mathbf{Q}, t). \tag{13.11}$$

In finding the solution to this equation, we note that the Green's function $\mathfrak{G}$ of the homogeneous Equation (13.11) is

$$i\hbar \left(\frac{\partial}{\partial t} - \frac{\hat{\mathbf{P}}^2}{2\mu} \right) \mathfrak{G} = -\, \delta(\mathbf{Q} - \mathbf{Q}')\, \delta(t - t'), \tag{13.12}$$

and has the well-known solution [29,36]

$$\mathfrak{G} = \frac{1}{(t - t')^{3/2}} \exp\left(-\frac{i}{\hbar} \frac{(\mathbf{Q} - \mathbf{Q}')^2}{2(t - t')^2} \right) \quad \text{for} \quad t > t', \tag{13.13}$$

$$\mathfrak{G} = 0 \quad \text{for} \quad t < t'. \tag{13.13'}$$

Thus the solution for the non-homogeneous Equation (13.11) can be written in the form

$$u_0(\mathbf{Q}, t) = \varphi_0(\mathbf{Q}, t) + \alpha \int_t^{\infty} (t - t')^{-3/2} \exp\left(-\frac{i}{\hbar} \frac{m}{2} \frac{(\mathbf{Q} - \mathbf{Q}')^2}{(t - t')} \right) \times |\psi_0(\mathbf{Q}')|^2\, \varphi_0(\mathbf{Q}', t')\, \mathrm{d}^3 Q'\, \mathrm{d}t', \tag{13.14}$$

where the coefficient α contains all constant factors.

The function $u_0(\mathbf{Q}, t)$ describes elastic scattering of μ-particles. We shall now consider the case when the dimensions of the μ-packet, $\Delta_{\parallel}$

† In doing this we again assume that the interaction constant g is small.

longitudinal and $\Delta_\perp$ transverse, are large in comparison to the fundamental wavelength $\lambda\!\!\!^{-}=\hbar/p$. In this case the variation of the index of the exponential of the function $f(\mathbf{Q}, t)$ (13.4) in the period of the variations $\approx\hbar/E$ is very small, and we can perform the integration over dt' in (13.4), considering the function $\varphi_0(\mathbf{Q}, t)$ to depend on t' only through the factor $e^{-iEt'/\hbar}=e^{-i\omega t'}$. Introducing the symbol $\tau=t'-t$ in (13.4), we apply the method of steepest descent to the index of the exponent in (13.4). The integral that we must take has the form

$$I=\int_0^\infty \frac{e^{-iF(\tau)}}{\tau^{3/2}}\,d\tau \tag{13.15}$$

where

$$F(\tau)=\frac{\Delta^2}{4D\tau}+\omega\tau \tag{13.16}$$

and $4D=2\hbar/\mu$. $\Delta^2=|\mathbf{Q}'-\mathbf{Q}|^2$. Equating $dF(\tau)/d\tau$ to zero, we obtain

$$\tau_m=\pm\sqrt{\frac{\Delta^2}{4D\omega}}=\pm\frac{|\mathbf{Q}'-\mathbf{Q}|}{v}, \tag{13.17}$$

The minus sign corresponds to the retarded, and the plus sign to the advanced action. This relationship shows that the scattered wave travels from the point $\mathbf{Q}'$ to the point $\mathbf{Q}$ in a time τ_m, which is defined by the velocity of the μ-particle $v=\hbar k/\mu$. Noting that $d^2F(\tau)/d\tau^2$ is equal to

$$F''(\tau_m)=\frac{\Delta^2}{2D\tau_m^3}=\frac{v^3}{2D\,|\mathbf{Q}'-\mathbf{Q}|}, \tag{13.18}$$

we find that

$$I=\beta\frac{e^{+iK|\mathbf{Q}'-\mathbf{Q}|}}{|\mathbf{Q}-\mathbf{Q}'|}, \tag{13.19}$$

where β is a constant coefficient. Returning now to Equation (13.14) we obtain

$$u_0(\mathbf{Q}, t)=\varphi_0(\mathbf{Q}, t)+\alpha'\int\frac{e^{ik|\mathbf{Q}-\mathbf{Q}'|-i\omega t}}{|\mathbf{Q}-\mathbf{Q}'|}\times\rho\left(\mathbf{Q}', t-\frac{|\mathbf{Q}-\mathbf{Q}'|}{v}\right)d^3Q', \tag{13.20}$$

where α' is some new coefficient, and the function $\rho(\mathbf{Q}, t)$ is equal to

$$\rho(\mathbf{Q}, t) = \frac{2\mu g}{\hbar^2} |\psi_0(\mathbf{Q})|^2 \varphi_0(\mathbf{Q}, t) e^{ik\zeta} \tag{13.21}$$

It differs from the earlier result (12.28) only in the time dependence of the wave packet $\varphi_0(\mathbf{Q}, t)$. This dependence allows us to speak of that moment in time when, by the act of scattering, the μ-particle measures the coordinates of the m-particle.

If we consider larger separations as we did in Section 12, i.e., $|\mathbf{Q}| = r \to \infty$, then from (13.20) we obtain

$$u_0(\mathbf{Q}, t) = \varphi_0(\mathbf{Q}, t) + \frac{e^{ikr}}{4\pi r} A(\mathbf{q}, \mathbf{Q}, t), \tag{13.22}$$

where the amplitude of the scattered wave $A(\mathbf{q}, \mathbf{Q}, t)$ is equal to

$$A(\mathbf{q}, \mathbf{Q}, t) = \int e^{i\mathbf{q}\mathbf{Q}'} \rho\left(\mathbf{Q}', t - \frac{|\mathbf{Q}' - \mathbf{Q}|}{v}\right) d^3Q' \tag{13.23}$$

and differs from (12.31) by the retarding time dependence.

We now calculate the full intensity I by integrating $|A(\mathbf{q}, \mathbf{Q}, t)|^2$ over d^3q, assuming very short waves ($\lambda \to 0$), and instead of (12.36) we obtain

$$I(\Omega, \mathbf{Q}, t) = \frac{1}{(4\pi)^2} \int_{\Omega} \rho^2\left(\mathbf{Q}', t - \frac{|\mathbf{Q}' - \mathbf{Q}|}{v}\right) d^3Q'. \tag{13.24}$$

This equation differs from (12.36) in that the moment of time t' when the scattering takes place is $t' = t - |\mathbf{Q}' - \mathbf{Q}|/v$. Here $\mathbf{Q}$ is the radius vector of the point of observation, and $\mathbf{Q}'$ is the radius vector of the region Ω inside of which the scattering took place.

The quantity $I(\Omega, \mathbf{Q}, t)$ is a measure of the probability that the scattering of a μ-particle observed at a point in the moment t, originated as a result of scattering in the region Ω in the moment of time $t' = t - |\mathbf{Q} - \mathbf{Q}'|/v$.

CHAPTER III

GEOMETRICAL MEASUREMENTS IN THE MICROWORLD IN THE RELATIVISTIC CASE

14. The Fermion Field

A field of free fermions can be described by a four-component spinor operator $\hat{\psi}(x)$. This operator and its conjugate $\bar{\hat{\psi}}(x)$ satisfy Dirac's equations

$$\left(\gamma_\mu \frac{\partial}{\partial x_\mu} + m\right)\hat{\psi} = 0, \tag{14.1}$$

$$\bar{\hat{\psi}}\left(\gamma_\mu \frac{\partial}{\partial x_\mu} - m\right) = 0; \tag{14.1'}$$

$x_\mu (\mu = 1, 2, 3, 4)$ are the components of the four-dimensional radius vector of an arbitrary point P. They are defined by $\mathbf{x} = (x_1, x_2, x_3)$, $x_4 = ict$, where, as usual, t is present time, c is the velocity of light, and γ_μ is the well-known Dirac matrix

$$\gamma_\mu \gamma_\nu + \gamma_\nu \gamma_\mu = 2\delta_{\mu\nu}. \tag{14.2}$$

The components of the spinor $\hat{\psi}$ and $\bar{\hat{\psi}}$ satisfy the laws of commutation

$$\{\hat{\psi}_\alpha(x), \hat{\psi}_\beta(x')\} = 0, \quad \{\bar{\hat{\psi}}_\alpha(x), \bar{\hat{\psi}}_\beta(x')\} = 0, \tag{14.3}$$

$$\{\hat{\psi}_\alpha(x), \bar{\hat{\psi}}_\beta(x')\} \equiv iS_{\alpha\beta}(x - x'), \tag{14.3'}$$

$$\alpha, \beta = 1, 2, 3, 4,$$

where the brackets $\{AB\}$ denote the anticommutation $AB + BA$.

The function $S_{\alpha\beta}(x - x')$ is the matrix element of the commutation function $S(x - x')$, and is equal to†

$$S_{\alpha\beta}(x - x') = -\left(\gamma_\mu \frac{\partial}{\partial x_\mu} - m\right)_{\alpha\beta} D(x - x'), \tag{14.4}$$

† See Appendix 1, Equations (I.3) and (I.15).

where

$$D(x) = \frac{1}{(2\pi)^3} \int e^{i\mathbf{px}} \frac{\sin Et}{E} \, d^3p. \tag{14.5}$$

In this equation $\mathbf{p}$ is the momentum vector and E is the energy which is equal to $E = +(\mathbf{p}^2 + m^2)^{1/2}$.

If the plane waves e^{ipx} ($px = \mathbf{px} - Et$) are taken as the basis functions for expansion, then the spinors $\hat{\psi}$ and $\hat{\bar{\psi}}$ may be written in the form

$$\hat{\psi} = \sum_{r=1,2} \int d^3p \, \{\hat{a}_r(\mathbf{p}) \, u^r(\mathbf{p}) \, e^{ipx} + \hat{b}_r^+(\mathbf{p}) \, v^r(-\mathbf{p}) \, e^{-ipx}\}, \tag{14.6}$$

$$\hat{\bar{\psi}} = \sum_{r=1,2} \int d^3p \, \{\hat{a}_r^+ \bar{u}^r(\mathbf{p}) \, e^{-ipx} + \hat{b}_r(\mathbf{p}) \, \bar{v}^r(-\mathbf{p}) \, e^{+ipx}\}. \tag{14.6'}$$

Here $u^r(\mathbf{p})$, $v^r(-\mathbf{p})$ are the spinor amplitudes of the plane waves. They satisfy Dirac's equations

$$(i\hat{p} + m) \, u(\mathbf{p}) = 0, \tag{14.7}$$

$$(-i\hat{p} + m) \, v(-\mathbf{p}) = 0 \tag{14.7'}$$

[the operators $\hat{p} = \gamma_\mu p_\mu$ and $p_\mu = (\mathbf{p}, E/c)$] and are normalized by the equations

$$\sum_{\alpha=1}^{4} u_\alpha^r(\mathbf{p}) \, u_\alpha^{r'}(\mathbf{p})^* = \delta'_{rr'}, \tag{14.8}$$

$$\sum_{\alpha=1}^{4} v_\alpha^r(-\mathbf{p}) \, v_\alpha^{r'}(-\mathbf{p})^* = \delta_{rr'}. \tag{14.8'}$$

The four-component spinor $u^r(\mathbf{p}) \, e^{ipx}$ denotes waves that describe negatively charged fermions (electrons) with momentum $\mathbf{p}$, energy $E = +(p^2 + m^2)^{1/2}$ and with orientation spinors defined by the index $r(=1, 2)$. The wave $v^r(-\mathbf{p}) \, e^{-ipx}$ describes positively charged fermions (positrons) whose momentum is $\mathbf{p}$, energy is $E = +(\mathbf{p}^2 + m^2)^{1/2}$ and whose spinor is r. The operator properties of the spinors $\hat{\psi}$ and $\hat{\bar{\psi}}$ carried over to the amplitudes $\hat{a}$, $\hat{a}^+$, $\hat{b}$, $\hat{b}^+$ which are the operators that satisfy the anticommutation conditions

$$\{\hat{a}_r(\mathbf{p}), \hat{a}_{r'}^+(\mathbf{p}')\} = \delta_{rr'} \delta^3(\mathbf{p} - \mathbf{p}'), \tag{14.9}$$

$$\{\hat{b}_r(\mathbf{p}), \hat{b}_{r'}^+(\mathbf{p}')\} = \delta_{rr'} \delta^3(\mathbf{p} - \mathbf{p}'); \tag{14.9'}$$

All the remaining anticommutator operators, $\hat{a}$, $\hat{a}^+$, $\hat{b}$, $\hat{b}^+$ are equal to zero.

The laws of commutation ensure that the laws of commutation for operators $\hat{\psi}(\mathbf{x})$ and $\hat{\bar{\psi}}(\mathbf{x})$, (14.3), (14.3′) will be satisfied. The physical meaning of the operators $\hat{a}, \hat{a}^+, \hat{b}, \hat{b}^+$ is the following: the operator $a_r(\mathbf{p})$ which acts on the functional Ψ which describes the state of a quantized fermion field, decreases the number of electrons $\hat{n}_r(\mathbf{p})$ in the state $(\mathbf{p}, r)$ by unity

$$\hat{a}_r(\mathbf{p})\,\Psi(\ldots, n_r(\mathbf{p}), \ldots) = \Psi(\ldots, n_r(\mathbf{p}) - 1, \ldots), \tag{14.10}$$

and the operator $\hat{a}_r^+(\mathbf{p})$ generates an electron in the state $(\mathbf{p}, r)$

$$\hat{a}_r^+(\mathbf{p})\,\Psi(\ldots, n_r(\mathbf{p}), \ldots) = \Psi(\ldots, n_r(\mathbf{p}) + 1, \ldots). \tag{14.10′}$$

The operators $\hat{b}, \hat{b}^+$ have the corresponding value for positrons

$$\hat{b}_r(\mathbf{p})\,\Psi(\ldots, n_r^+(\mathbf{p}), \ldots) = \Psi(\ldots, n_r(\mathbf{p}) - 1, \ldots), \tag{14.10″}$$

$$\hat{b}_r^+(\mathbf{p})\,\Psi(\ldots, n_r^+(\mathbf{p}), \ldots) = \Psi(\ldots, n_r^+(\mathbf{p}) + 1, \ldots). \tag{14.10‴}$$

We note that according to Fermi-Dirac statistics, the population of a state $n_r(\mathbf{p})$ is limited to 0 or 1.†

The law of conservation of electric current follows from Equations (14.1) and (14.1′)

$$\frac{\partial J_\mu}{\partial x_\mu} = 0, \tag{14.11}$$

The components of this current are defined by the equation

$$J_\mu = ie\hat{\bar{\psi}}\gamma_\mu\hat{\psi}, \tag{14.12}$$

where e is the electron charge. The fourth component of this current

$$J_4 = i\rho = ie\hat{\bar{\psi}}\gamma_4\hat{\psi} \tag{14.13}$$

can have either sign depending on whether the charge density of electrons is greater or less than that of positrons at a given point.

However, if we consider a single particle state, then the quantity ρ has a definite sign and if we divide it by the charge of a particle e, we can con-

† The details of quantization of a Dirac field may be found in books on quantum theory of fields. Here we are following the notation used in [37, 38].

sider it as the probability density $\omega(x)$ of finding a particle in a place **x** at a given time t.

To prove this important fact, let us consider a single-particle state, for example an electron. In general this state will have an undefined momentum **p** and an undefined spin orientation r. The functional that describes such a state has the form

$$\Psi(1) = \sum_{r'=1,2} \int c_{r'}(\mathbf{p}') \, \delta(n_{p'}^{r'} - 1) \prod_{\substack{p \neq p' \\ r \neq r'}} \delta(n_{\mathbf{p}}^{r}) \, \mathrm{d}^3 p' . \tag{14.14}$$

On the basis of (14.6), (14.10)–(14.10''') and (14.11), the matrix element of the operator of a fermion field $\hat{\psi}$ that corresponds to the transition from the vacuum state $\Psi(0)$ to a state $\Psi(1)$ is equal to

$$\psi(x) = \langle \Psi(0) | \, \hat{\psi}(x) \, | \Psi(1) \rangle = \int c_r(\mathbf{p}) \, \mathrm{d}^3 p \, u^r(\mathbf{p}) \, \mathrm{e}^{ipx} \tag{14.15}$$

It describes the wave packet of a fermion with a spin orientation r, with momenta defined by the amplitude $c(\mathbf{p})$ and a positive energy $E = + +(p^2 + m^2)^{1/2}$.

A normalization of the functional of the field

$$\langle \Psi(1), \Psi(1) \rangle = 1 , \tag{14.16}$$

leads to the condition normalizing the amplitude $c(\mathbf{p})$,

$$\int c^*(\mathbf{p}) \, c(\mathbf{p}) \, \mathrm{d}^3 p = 1 . \tag{14.17}$$

This matrix element of the field operator $\hat{\psi}(x)$ (14.15) has the meaning of the electron wave function $\psi(x)$ in Dirac's one-electron theory.

Let us now calculate the average value of the fourth component of the current J_4, or, as it is more convenient, the quantity $\omega = J_4/ie$ in a single particle state $\psi(1)$ (14.11). From (14.13) and (14.15) we have

$$\begin{aligned} \omega(x) &= \langle \Psi(1) | \, \bar{\hat{\psi}} \gamma_4 \hat{\psi} \, | \Psi(1) \rangle \\ &= \bar{\psi}(x) \, \gamma_4 \psi(x) \equiv \psi^*(x) \, \psi(x) \equiv \sum_{\alpha=1}^{4} \psi_\alpha^*(x) \, \psi_\alpha(x) \geqslant 0 . \end{aligned} \tag{14.18}$$

The positive definiteness of the quantity $\omega(x)$ allows us to consider it as the probability density for single particle states. By use of the conservation law (14.11), the total probability

$$W = (\psi, \psi) = \int \bar{\psi}\gamma_4\psi \, \mathrm{d}^3x = \int \omega(x) \, \mathrm{d}^3x \tag{14.19}$$

remains constant in time namely,

$$\frac{\mathrm{d}W}{\mathrm{d}t} = \int \frac{\partial\omega}{\partial t} \, \mathrm{d}^3x = - \int \operatorname{div} \mathbf{J} \, \mathrm{d}^3x = - \int J_N \, \mathrm{d}\sigma = 0 \, . \tag{14.20}$$

In conclusion, we note that the basic functions

$$\psi_{p,r}(x, \alpha) = \frac{\exp \mathrm{i}(\mathbf{px} - Et)}{(2\pi)^{3/2}} u_\alpha^r(\mathbf{p}) , \tag{14.21}$$

$$\psi_{p,r}(x, \alpha) = \frac{\exp \mathrm{i}(\mathbf{px} - Et)}{(2\pi)^{3/2}} v_\alpha^r(\mathbf{p}) \tag{14.21'}$$

$$(r = 1, 2) ,$$

which appear in Equations (14.6) and (14.6′) form a complete system of functions. The condition of completeness for $t' = t$ gives

$$\sum_{r=1}^{2} \sum_{u,\, v} \int \mathrm{d}^3p \psi_{p,r}^*(x, \alpha) \, \psi_{p,r}(x', \beta) = \delta^3(\mathbf{x} - \mathbf{x}') \, \delta_{\alpha\beta} \, . \tag{14.22}$$

where the sum is taken over both spin orientations $r = 1, 2$ and both signs for energy (spinors u and v). If we limit ourselves to using only one of the energy signs, for example $E > 0$, and correspondingly to one spinor u, then in place of (14.22) we obtain

$$\sum_{r=1}^{2} \int \mathrm{d}^3p \psi_{p,r}^*(x, \alpha) \, \psi_{p,r}(x', \alpha') = - \mathrm{i}(\gamma_4 S_{\alpha\beta}^+(x - x')) , \tag{14.23}$$

where

$$S_{\alpha\beta}^+(x - x') = - \left(\gamma_\mu \frac{\partial}{\partial x_\mu} - m\right) \Delta^+(x - x') \tag{14.24}$$

and

$$\Delta^+(x) = \frac{i}{(2\pi)^3} \int \frac{\mathrm{e}^{-\mathrm{i}px}}{2E} \, \mathrm{d}^3p \, . \tag{14.25}$$

From this we can see the $\delta^3(\mathbf{x} - \mathbf{x}')$ cannot be expanded in terms of a fundamental system of functions $\psi_{p,r}(x, \alpha)$ if we limit ourselves to states with only positive frequencies (electrons) or only negative frequencies (positrons).

This fact, that the function $S^{+}_{\alpha\beta}(x-x')$ is spread in the region $\Delta x \sim \hbar/mc$, reminds us that Dirac particles cannot in general be localized more closely than within limits of its Compton wavelength. In the following section we shall consider this problem in detail.

15. The Uncertainty Relation for Fermions

We consider the single particle state $\Psi(1)$. For such a state, we can use the ordinary, non-quantized Dirac equation for one fermion with the important restriction that only one sign for the energy state $E = +(\mathbf{p}^2 + m^2)^{1/2} > 0$ be used.

The wave function that describes such a state has the form

$$\psi(\mathbf{x}, t) = \sum_{r=1}^{2} \int \frac{d^3p}{(2\pi)^{3/2}} c_r(\mathbf{p})\, u^r(\mathbf{p})\, e^{ipx}. \tag{15.1}$$

We shall consider it to be quadratically integrable. We calculate the rms deviation of any coordinate, e.g., z, in this state at some instant in time.

Using no restriction other than a choice of the frame of reference, we can assume that the average value of $\bar{z}=0$, $t=0$ and the average momentum $\bar{\mathbf{p}}$ is also equal to zero. For simplicity it is sufficient to consider the state with one specific spin orientation $r=1$ or $r=2$. In the selected frame of reference, the value of the rms deviation $\overline{\Delta z^2} \equiv \overline{(z-\bar{z})^2} = \overline{z^2}$ may, using (14.18), be written in the form

$$\overline{\Delta z^2} = \int \omega(\mathbf{x}, t)\, z^2\, d^3x = \int \psi^*(\mathbf{x}, t)\, z^2 \psi(\mathbf{x}, t)\, d^3x \tag{15.2}$$

Normalization gives

$$\int \omega(\mathbf{x}, t)\, d^3x = 1. \tag{15.3}$$

The expression (15.2) can be transformed using the relationship

$$\begin{aligned} z\psi(\mathbf{x}) &= \frac{1}{i} \int \frac{d^3p}{(2\pi)^{3/2}} c(\mathbf{p})\, u(\mathbf{p}) \frac{\partial e^{i\mathbf{px}}}{\partial p_z} \\ &= i \int \frac{d^3p}{(2\pi)^{3/2}} \frac{\partial [c(\mathbf{p})\, u(\mathbf{p})]}{\partial p_z} e^{i\mathbf{px}}. \end{aligned} \tag{15.4}$$

Similarly we may obtain the product $z\psi^*(x)$ and by carrying out the integration over d^3x, we find that

$$\overline{\Delta z^2} = \int \mathrm{d}^3p \, \frac{\partial\,[c\,(\mathbf{p})\,u\,(\mathbf{p})]}{\partial p_z} \frac{\partial\,[c\,(\mathbf{p})\,u\,(\mathbf{p})]}{\partial p_z} \tag{15.5}$$

or

$$\overline{\Delta z^2} = \int \mathrm{d}^3p \left|\frac{\partial c\,(\mathbf{p})}{\partial p_z}\right|^2 |u|^2 + \int \mathrm{d}^3p\,|c\,(\mathbf{p})|^2 \left|\frac{\partial u}{\partial p_z}\right|^2 + \int \mathrm{d}^3p \left\{\frac{\partial c^*}{\partial p_z}\, c\, \frac{\partial u}{\partial p_z}\, u + c^*\, \frac{\partial c}{\partial p_z} \frac{\partial u}{\partial p_z}\, u + \frac{\partial c^*}{\partial p_z}\, c u^*\, \frac{\partial u}{\partial p_z}\right\}. \tag{15.5'}$$

As we shall see later, the first integral on the right leads to the uncertainty relation; the second two integrals are characteristic of the theory of relativity and require special investigation.

In the frame of reference we selected we can, without any important restrictions, consider that the amplitude $c(\mathbf{p})$ has the form

$$c\,(\mathbf{p}) = f\left(\frac{|\mathbf{p}|}{p_0}\right) \frac{1}{p_0^{3/2}}, \tag{15.6}$$

where the function $f\,(|\mathbf{p}|/|p_0|)$ is real and decreases sufficiently rapidly for $|\mathbf{p}|/p_0 \gg 1$. This indicates that the quantity p_0 characterizes the dispersion momentum $\overline{\Delta p_z^2}$ in the state we are considering. In fact,

$$\overline{\Delta p_z^2} = \overline{(p_z - \bar{p}_z)^2} = \overline{p_z^2} = \int \frac{\mathrm{d}^3p}{p_0^3} f^2\left(\frac{|\mathbf{p}|}{p_0}\right) p_z^2 = p_0^2 \int \xi^4\, \mathrm{d}\xi\, \cos^2\vartheta\, \mathrm{d}\Omega\, f^2(\xi) = \alpha' p_0^2. \tag{15.7}$$

where $p_z = |\mathbf{p}| \cos\vartheta$, $\mathrm{d}\Omega = \sin\vartheta\, \mathrm{d}\vartheta\, \mathrm{d}\varphi$, $\xi = |\mathbf{p}|/p_0$, and α' is a coefficient of order 1.

For a real amplitude $c(\mathbf{p})$, Eq. (15.5′) becomes simplified because of the normalization of the spinor $u^+u = 1$ [see (14.8)]. Using this normalization, we obtain

$$\overline{\Delta z^2} = \int \mathrm{d}^3p \left(\frac{\partial c\,(\mathbf{p})}{\partial p_z}\right)^2 + \int \mathrm{d}^3p c^2\,(\mathbf{p}) \left|\frac{\partial u\,(\mathbf{p})}{\partial p_z}\right|^2. \tag{15.8}$$

We now introduce the variable $\xi = |\mathbf{p}|/p_0$ in place of $|\mathbf{p}|$ and rewrite the first integral in (15.8) in the form

$$I_1 = \frac{1}{p_0^2}\frac{4\pi}{3}\int_0^\infty \left(\frac{\mathrm{d}f}{\mathrm{d}\xi}\right)^2 \xi^2\,\mathrm{d}\xi = \frac{\alpha''}{p_0^2} \tag{15.9}$$

(the coefficient α'' is of the order 1) and the second integral in the form

$$I_2 = \frac{1}{p_0^2}\frac{4\pi}{3}\int f^2(\xi)\,\xi^2\,\mathrm{d}\xi M\left(\xi, \frac{m}{p_0}\right), \tag{15.10}$$

where the function $M(\xi, m/p_0)$ is defined by the integral

$$M\left(\xi, \frac{m}{p_0}\right) = \frac{1}{4\pi}\int \frac{\partial u^*}{\partial p_z}\frac{\partial u}{\partial p_z}\,\mathrm{d}\Omega. \tag{15.11}$$

This function is calculated in Appendix II and is equal to

$$M\left(\xi, \frac{m}{p_0}\right) = \begin{cases} \dfrac{3p_0^2}{4m^2} & \text{for } \xi \ll \dfrac{m}{p_0}, & (15.12) \\[2ex] \dfrac{3p_0^2}{4m^2}\,4\dfrac{m^2}{p_0^2} & \text{for } \xi \gg \dfrac{m}{p_0}. & (15.12') \end{cases}$$

From the properties of the function $M(\xi, m/p_0)$ we have

$$I_2 = \begin{cases} \dfrac{\pi}{m^2}\beta' & \text{for } \dfrac{m}{p_0} \ll 1, & (15.13) \\[2ex] \dfrac{4\pi}{p_0^2}\beta'' & \text{for } \dfrac{m}{p_0} \gg 1, & (15.13') \end{cases}$$

where β' and β'' are coefficients of order 1.

Returning now to Eq. (15.8) and reducing it to a form that shows its dimensions more clearly, we obtain

$$\overline{\Delta z^2} = \alpha\frac{\hbar^2}{\overline{\Delta p_z^2}} + \beta\frac{\hbar^2}{m^2c^2} \quad \text{for } \overline{\Delta p_z^2} \ll m^2c^2, \tag{15.14}$$

$$\overline{\Delta z^2} = \gamma\frac{\hbar^2}{\overline{\Delta p_z^2}} \quad \text{for } \overline{\Delta p_z^2} \gg m^2c^2. \tag{15.14'}$$

From these equations in which α, β, γ are numerical coefficients of order

1, we see that the additional relativistic terms which result from the dependence of a spinor momentum are not important because from (15.14) the additional term is small in comparison to the first, and for $\overline{\Delta p_z^2} \gg m^2c^2$, Equation (15.14′) should be used which differs from the term in (15.14). From (15.14) it follows that, for $\overline{\Delta p_z^2} \to \infty$, the magnitude of the variance of the coordinates $\overline{\Delta z^2}$ may be made arbitrarily small.

Thus, although the function $\delta(\mathbf{x}-\mathbf{x}')$ is not in the class of functions that are allowed for the single-particle state, the theory nevertheless allows the possibility of localizing to any degree of accuracy the single-particle state with a specific sign for the energy that is in the class of quadratically integrable functions.†

16. The Boson Field

The field of spinless bosons is described by the scalar or pseudoscalar operator $\hat{\varphi}(x)$ which has specific symmetry properties in an isotopic (charged) space.

We shall consider a complex scalar field which describes charged particles. Such a field satisfies Klein's equation‡

$$(\Box^2 - m^2)\,\hat{\varphi} = 0, \tag{16.1}$$

$$(\Box^2 - m^2)\,\hat{\varphi}^+ = 0, \tag{16.1'}$$

and the commutation laws

$$[\varphi(x), \varphi^+(x')] = \mathrm{i}D(x-x'), \tag{16.2}$$

where $[A, B] = AB - BA$ is the ordinary commutator, and the function $D(x-x')$ is defined by Equation (14.5):

$$D(x) = \frac{1}{(2\pi)^3}\int e^{\mathbf{ikx}}\,\frac{\sin\omega t}{\omega}\,\mathrm{d}^3k; \tag{16.3}$$

Here $\mathbf{k}$ is the momentum of the boson, ω is its energy in the system of identity elements where $\hbar = 1$. The remaining commutators

$$[\hat{\varphi}(x), \hat{\varphi}(x')] = [\hat{\varphi}^*(x), \hat{\varphi}^*(x')] = 0. \tag{16.2'}$$

† In this section we followed the work of [30].

‡ See [37, 38].

The boson field $\hat{\varphi}(x)$, in analogy to the fermion field, can be written in the form of an expansion of a fundamental system of orthogonal functions

$$\hat{\varphi}(x) = \int d^3k \,\{\hat{a}(\mathbf{k})\, U(\mathbf{k})\, e^{ikx} + \hat{b}^+(\mathbf{k})\, U^*(\mathbf{k})\, e^{-ikx}\}, \quad (16.4)$$

$$\hat{\varphi}^+(x) = \int d^3k \,\{\hat{a}^+(\mathbf{k})\, U^*(\mathbf{k})\, e^{-ikx} + \hat{b}(-\mathbf{k})\, U(-\mathbf{k})\, e^{ikx}\}, \quad (16.4')$$

where

$$U(\mathbf{k}) = \sqrt{\frac{1}{2\omega}}, \quad (16.5)$$

$\omega = +(\mathbf{k}^2 + m^2)^{1/2}$, m is the boson mass. The operators $\hat{a}$, $\hat{a}^+$, $\hat{b}$, $\hat{b}^+$ have meanings analogous to the operators in Equations (14.6) and (14.6′) for the fermion field. That is, they satisfy the commutation laws

$$[\hat{a}(\mathbf{k}), a^+(\mathbf{k})] = \delta^3(\mathbf{k} - \mathbf{k}'), \quad (16.6)$$

$$[\hat{b}(\mathbf{k}), \hat{b}^+(\mathbf{k})] = \delta^3(\mathbf{k} - \mathbf{k}'); \quad (16.6')$$

and the remaining commutators, $[\hat{a}, \hat{a}']$, $[\hat{a}^+, \hat{a}^{+\prime}]$, $[\hat{b}, \hat{b}']$, $[\hat{b}, \hat{b}^{+\prime}]$, $[\hat{a}, \hat{b}']$, are equal to zero. The physical meanings of the operators $\hat{a}$, $\hat{a}^+$, $\hat{b}$, $\hat{b}^+$ are the following: The operator $\hat{a}(\mathbf{k})$ operates on the functional Ψ which describes the state of a boson in a quantized field, and it decreases the number of positively charged bosons $n^+(\mathbf{k})$ in the state with momentum $\mathbf{k}$ and energy $\omega = +(\mathbf{k}^2 + m^2)^{1/2}$ to unity:

$$\hat{a}(\mathbf{k})\, \Psi(\ldots, n^+(\mathbf{k}), \ldots) = \Psi(\ldots, n^+(\mathbf{k}) - 1, \ldots); \quad (16.7)$$

The operator $\hat{a}^+(\mathbf{k})$ increases the number of positively charged particles in the state $(\mathbf{k}, \omega)$:

$$\hat{a}^+(\mathbf{k})\, \Psi(\ldots, n^+(\mathbf{k}), \ldots) = \Psi(\ldots, n^+(\mathbf{k}) + 1, \ldots). \quad (16.7')$$

The operators $b(\mathbf{k})$ and $\hat{b}^+(\mathbf{k})$ have the same meanings for negatively charged bosons:

$$\hat{b}(\mathbf{k})\, \Psi(\ldots, n^-(\mathbf{k}), \ldots) = \Psi(\ldots, n^-(\mathbf{k}) - 1, \ldots), \quad (16.8)$$

$$\hat{b}^+(\mathbf{k})\, \Psi(\ldots, n^-(\mathbf{k}), \ldots) = \Psi(\ldots, n^-(\mathbf{k}) + 1, \ldots). \quad (16.8')$$

From Equations (16.1) and (16.1′) we obtain the conservation law

$$\frac{\partial \hat{J}_\mu}{\partial x_\mu} = 0, \tag{16.9}$$

where

$$\hat{J}_\mu = \mathrm{i}e\left\{\hat{\varphi}^* \frac{\partial \hat{\varphi}}{\partial x_\mu} \quad \frac{\partial \hat{\varphi}^*}{\partial x_\mu}\hat{\varphi}\right\}, \tag{16.10}$$

The fourth component has the meaning of an operator of the electric charge density

$$\hat{J}_4 = \mathrm{i}\hat{\rho} = \mathrm{i}e\left\{\hat{\varphi}^* \frac{\partial \hat{\varphi}}{\partial t} - \frac{\partial \hat{\varphi}^*}{\partial t}\hat{\varphi}\right\}. \tag{16.11}$$

However, in contrast to the fermion case, we shall now see from the following that one cannot ascribe a positive-definite probability density $\omega(x) = \rho/e$ to bosons, nor even to single-particle states with a specific sign for the energy. We shall now consider such a state. Let $\Psi(1)$ be a functional of the field which describes the single particle state with an undefined boson momentum $\mathbf{k}$ and with a defined sign for the energy $\omega > 0$. This functional may be written in the form, absolutely analogous to that for the fermion field (14.14),

$$\Psi(1) = \int A(\mathbf{k})\, \mathrm{d}^3k\, \delta(n^+(\mathbf{k}) - 1) \times \delta[n^-(\mathbf{k})] \prod_{\mathbf{k} \neq \mathbf{k}'} \delta[n^+(\mathbf{k}')]\, \delta[n^-(\mathbf{k}')]. \tag{16.12}$$

The matrix element of the operator of a boson field corresponding to the transition from the vacuum state $\Psi(0)$ to the state $\Psi(1)$, which is described by (16.12), may be calculated from Equations (16.4), (16.4′), (16.7)–(16.8′) and is equal to

$$\varphi(x) = \langle \Psi(0)|\, \hat{\varphi}(x)\, |\Psi(1)\rangle = \int A(\mathbf{k})\, U(\mathbf{k})\, e^{ikx}\, \mathrm{d}^3k \tag{16.13}$$

It describes the wave packet of bosons which have a momentum whose amplitude is $A(\mathbf{k})$ and which have a positive energy $\omega = +(\mathbf{k}^2 + m^2)^{1/2}$.

Normalizing the functional

$$\langle \Psi(1), \Psi(1)\rangle = 1 \tag{16.14}$$

leads to the normalizing condition for the amplitude $A(\mathbf{k})$

$$\int A^*(\mathbf{k})\, A(\mathbf{k})\, \mathrm{d}^3k = 1\,. \tag{16.15}$$

This matrix element $\varphi(x)$ of the operator of the field $\hat{\varphi}(x)$ (16.13) that we have considered has the meaning of a boson wave function in the single-particle theory in a scalar field.

We shall now compute the average value of the fourth component of the current J_4 or, as it is more convenient, the quantity $\omega(x) = J_4/ie$ in a single particle state $\Psi(1)$ (16.12). From (16.11) and (16.13) we obtain

$$\begin{aligned} w(x) &= \langle \Psi(1)|\, \mathrm{i}\left(\hat{\varphi}^* \frac{\partial\varphi}{\partial t} - \frac{\partial\hat{\varphi}^*}{\partial t}\,\hat{\varphi}\right)|\Psi(1)\rangle \\ &= \mathrm{i}\left(\varphi^*(x)\frac{\partial\varphi}{\partial t} - \frac{\partial\varphi^*}{\partial t}\,\varphi(x)\right), \end{aligned} \tag{16.16}$$

where $\varphi(x)$ is defined by Equation (16.13), the integration over d^3k is taken for only positive energies and $kx = \mathbf{kx} - \omega t$, $\omega = +(\mathbf{k}^2 + m^2)^{1/2}$. The quantity W is defined as

$$W \equiv (\varphi, \psi) \equiv \int w(x)\, \mathrm{d}^3x = i \int \left(\varphi^* \frac{\partial\varphi}{\partial t} - \frac{\partial\varphi^*}{\partial t}\,\varphi\right) \mathrm{d}^3x, \tag{16.17}$$

and is conserved due to the continuity equation (16.9):

$$\frac{\mathrm{d}W}{\mathrm{d}t} = \int \frac{\partial w}{\partial t}\, \mathrm{d}^3x = -\int \operatorname{div}\mathbf{J}\, \mathrm{d}^3x = \int J_N\, \mathrm{d}\sigma = 0\,. \tag{16.18}$$

The generalized quantity

$$(\varphi, \psi) = \mathrm{i}\int \left(\varphi^* \frac{\partial\psi}{\partial t} - \frac{\partial\varphi^*}{\partial t}\,\psi\right) \mathrm{d}^3x \tag{16.17'}$$

may be considered as the scalar product of two single-particle functions φ and ψ. If we take for these two functions the basic solution

$$\varphi_k(x) = \frac{e^{\mathrm{i}(\mathbf{kx} - \omega t)}}{(2\pi)^{3/2}}\, U_k\,, \tag{16.19}$$

then they describe a complete system of functions. The condition of completeness for $t=t'$ gives

$$\int \varphi_k^*(x)\,\varphi_k(x')\,\mathrm{d}^3k = \Delta^+(x-x'), \tag{16.20}$$

where the function $\Delta^+(x)$ is equal to

$$\Delta^+(x) = \frac{i}{(2\pi)^3}\int \frac{e^{i(\mathbf{kx}-\omega t)}}{\omega}\,\mathrm{d}^3k. \tag{16.21}$$

Because of this fact, the function $\delta^3(\mathbf{x}-\mathbf{x}')$ which corresponds to the "exact" localization of bosons does not belong to the class of functions represented by the expansion (16.13). Using the fact that the single-particle wave function $\varphi(x)$, which has a certain sign for energy, satisfies the equations

$$i\frac{\partial\varphi}{\partial t} = \hat{\omega}\varphi, \tag{16.22}$$

$$-i\frac{\partial\varphi^*}{\partial t} = \hat{\omega}\varphi^*, \tag{16.22'}$$

where the operator ω' is equal to

$$\hat{\omega} = +\sqrt{m^2-\nabla^2}; \tag{16.23}$$

and using Eq. (16.11), the density $\omega(x)$ may be written in the form

$$w(x) = \tfrac{1}{2}(\hat{\omega}\varphi^*\varphi + \varphi^*\hat{\omega}\varphi). \tag{16.24}$$

From (16.24) we may see that, by itself, it does not guarantee positive definitess. In Appendix III we give an example of a state of a boson field which is written in the form of the superposition of two waves φ_2 and φ_1 which have distinctly different momenta $\mathbf{k}_2$ and $\mathbf{k}_1$ and whose energies have the same sign $(\omega_2 \gg \omega_1 > 0)$:

$$\varphi(x) = \frac{c_1}{\sqrt{2\omega_1}}\varphi_1 + \frac{c_2}{\sqrt{2\omega_2}}\varphi_2. \tag{16.25}$$

In this case

$$w(\mathbf{x},t) \approx \frac{|c_1|^2}{\omega_1}\left\{1 + \left|\frac{c_2}{c_1}\right| \cos\left[(\mathbf{k}_2-\mathbf{k}_1,\mathbf{x}) - (\omega_2-\omega_1)t + \alpha\right]\right\}, \tag{16.26}$$

and if $|c_2/c_1|>1$, then this density oscillates in space with a period that is of the order of $1/|\mathbf{k}_2-\mathbf{k}_1|$ and in time with period $1/|\omega_2-\omega_1|$.

The energy density of this field $\varepsilon(x)$ is both physically observable and a positive-definite quantity in the case of a boson field. The operator of this quantity $\hat{\varepsilon}(x)$ may be derived from the energy-momentum tensor of the boson field

$$\hat{T}_{\mu\nu}(x)=\frac{\partial\hat{\varphi}^+}{\partial x_\mu}\frac{\partial\hat{\varphi}}{\partial x_\nu}-\frac{1}{2}\delta_{\mu\nu}\left(\frac{\partial\hat{\varphi}^+}{\partial x_\mu}\frac{\partial\hat{\varphi}}{\partial x_\mu}-m^2\hat{\varphi}^+\hat{\varphi}\right). \tag{16.27}$$

From this we obtain

$$\hat{\varepsilon}(x)=-\hat{T}_{44}(x)=\frac{1}{2}\left(\frac{\partial\hat{\varphi}^+}{\partial t}\frac{\partial\hat{\varphi}}{\partial t}+\nabla\hat{\varphi}^+\nabla\hat{\varphi}+m^2\hat{\varphi}^+\hat{\varphi}\right). \tag{16.28}$$

Therefore the average value of $\varepsilon(x)$ in the state $\Psi(1)$ is

$$\begin{aligned}\varepsilon(x)&=\langle\Psi(1)|\,\hat{\varepsilon}(x)\,|\Psi(1)\rangle\\&=\frac{1}{2}\left(\frac{\partial\varphi^*}{\partial t}\frac{\partial\varphi}{\partial t}+\nabla\varphi^*\,\nabla\varphi+m^2\varphi^*\varphi\right)\geqslant 0.\end{aligned} \tag{16.29}$$

17. The Localization of Photons

The electromagnetic field belongs to the class of boson fields, but unlike the scalar meson fields, it is described by a vector operator — the vector potential operator $\hat{A}_\mu(x)$, where $\mu=1, 2, 3, 4$ with components $\hat{A}(\hat{A}_1, \hat{A}_2, \hat{A}_3)$, $\hat{\varphi}=\hat{A}_4/\mathrm{i}$.

This potential is defined to within the accuracy of a gauge transformation

$$\hat{A}_\mu(x)=\hat{A}_\mu(x)+\frac{\partial f(x)}{\partial x_\mu}, \tag{17.1}$$

where $f(x)$ is an arbitrary function that satisfies d'Alembert's equation

$$\Box^2 f(x)=0. \tag{17.1'}$$

Because of this arbitrariness, the characteristic values of the vector potential are not physically measureable quantities. The measureable quantities are the components of the antisymmetric tensor of the electromagnetic field

$$\hat{F}_{\mu\nu} = \frac{\partial \hat{A}_\mu}{\partial x_\nu} - \frac{\partial \hat{A}_\nu}{\partial x_\mu} \tag{17.2}$$

or, in vector notation

$$\hat{\mathscr{E}} = -\frac{1}{c}\frac{\partial \hat{\mathbf{A}}}{\partial t} - \nabla\hat{\varphi}, \tag{17.2'}$$

$$\hat{\mathscr{H}} = \operatorname{rot}\hat{\mathbf{A}}, \tag{17.2''}$$

where $\hat{\mathscr{E}}$ is the electric field potential operator, and $\hat{\mathscr{H}}$ is the magnetic field potential operator.

A distinctive feature of a free electromagnetic field is its transverse character. Therefore only three of the four components of the vector A_μ are independent. This places a restriction on the choice of the wave functional Ψ

$$\left\langle \Psi \left| \frac{\partial A_\mu}{\partial x_\mu} \right| \Psi \right\rangle = 0. \tag{17.3}$$

If this condition is satisfied at the initial instant in time, then it is possible to show that it will be satisfied for all time from the equation of motion.†
When the condition (17.3) is satisfied, then the equation for the vector potential of the electromagnetic field becomes

$$\Box^2 \hat{A}_\mu = 0. \tag{17.4}$$

The commutation laws for the operator $\hat{A}$ may be written in the form

$$[\hat{A}_\mu(x), \hat{A}_\nu(x')] = -iD(x-x')\,\delta_{\mu\nu}, \tag{17.5}$$

For $m=0$, $D(x)$ has the form

$$D(x) = \frac{i}{(2\pi)^2}\int e^{i\mathbf{kx}}\frac{\sin\omega t}{\omega}\,d^3k, \tag{17.6}$$

where $\omega = |\mathbf{k}|$. The expansion of the vector potential in a fundamental system of functions has the form

$$\hat{A}_\mu(x) = \sum_{\lambda=1}^{4} d^3k\, e_\mu^\lambda [\hat{A}_\lambda(\mathbf{k})\, U(\mathbf{k})\, e^{ikx} + \hat{A}_\lambda^+(\mathbf{k})\, U^*(\mathbf{k})\, e^{-ikx}], \tag{17.7}$$

† For details on the quantization of the electromagnetic field, see [37] or [38].

where $U(\mathbf{k})=1/(2\omega)^{1/2}$, $kx=\mathbf{kx}-\omega t$, e_μ^λ are the components of the unit vector which define the polarization and $\hat{A}_\lambda(\mathbf{k})$ and $\hat{A}_\lambda^+(\mathbf{k})$ are the operators of the amplitude of partial waves. We note that all four polarizations appear in this expansion; two are transverse, one longitudinal and one of time. Using Eqs. (17.2) and (17.2′) for the observed quantities $\hat{\mathscr{E}}$ and $\hat{\mathscr{H}}$ with condition (17.3) satisfied, we obtain the expansions

$$\hat{\mathscr{E}}(x)=\sum_{\lambda=1}^{2}\int \mathrm{d}^3k\,\omega\,(\hat{A}_\lambda(\mathbf{k})\,U_k(x)-\hat{A}_\lambda^+(\mathbf{k})\,U_k^*(x))\,\mathrm{e}^\lambda, \tag{17.8}$$

$$\hat{\mathscr{H}}(x)=\mathrm{i}\sum_{\lambda=1}^{2}\int \mathrm{d}^3k\,[\mathbf{k}\mathrm{e}^\lambda]\,(\hat{A}_\lambda(\mathbf{k})\,U_k(x)-\hat{A}_\lambda^+(\mathbf{k})\,U_k^*(x)), \tag{17.8′}$$

Only the two transverse polarizations for $\lambda=1, 2$ appear in these equations. From the commutation law (17.5) we obtain the commutation condition for the amplitude $A_\lambda(\mathbf{k})$

$$[\hat{A}_\lambda(\mathbf{k}),\hat{A}_{\lambda'}^+(\mathbf{k}')]=\delta^3(\mathbf{k}-\mathbf{k}')\,\delta_{\lambda\lambda'}, \tag{17.9}$$
$$[\hat{A}_\lambda(\mathbf{k}),\hat{A}_{\lambda'}(\mathbf{k}')]=[\hat{A}_\lambda^+(\mathbf{k}),\hat{A}_{\lambda'}(\mathbf{k}')]=0. \tag{17.9′}$$

The only physically observable quantity which satisfies the conservation laws, in the case of a vector field, is the energy-momentum field tensor $T_{\mu\nu}(x)$. The operator which represents this tensor has the form

$$\hat{T}_{\mu\nu}(x)=\hat{F}_{\mu\rho}\hat{F}_{\nu\rho}-\tfrac{1}{4}\,\delta_{\mu\nu}\hat{F}_{\sigma\rho}^2\,; \tag{17.10}$$

from which we have

$$\begin{aligned}\hat{T}_{44}(x)&=-\tfrac{1}{2}(\hat{\mathscr{E}}^2+\hat{\mathscr{H}}^2),\\ \hat{T}_{k4}(x)&=-\mathrm{i}(\hat{\mathscr{E}}\times\hat{\mathscr{H}})_k=-\mathrm{i}\hat{S}_k.\end{aligned} \tag{17.11}$$

The conservation law for the tensor $\hat{T}_{\mu\nu}(x)$ states that

$$\frac{\partial\hat{T}_{\mu\nu}(x)}{\partial x_\nu}=0, \tag{17.12}$$

and, in particular, for the energy-density operator

$$\varepsilon(x)=-\hat{T}_{44}(x) \tag{17.13}$$

we obtain from (17.12)

$$\frac{\partial \hat{\varepsilon}(x)}{\partial t} + \operatorname{div} \hat{\mathbf{S}}(x) = 0, \tag{17.14}$$

where $\hat{\mathbf{S}}(x)$ is the vector energy flow operator whose components are $\hat{S}(x) = i\hat{T}_{4k}(x)$. We shall now return to the single photon state with an undefined momentum $\mathbf{k}$.

For this case we note that, by virtue of the commutation laws (17.9) and (17.9′), the amplitude operators $\hat{A}_\lambda(\mathbf{k})$, $\hat{A}_\lambda^+(\mathbf{k})$ operate on the wave functional Ψ in a manner identical to that of the scalar field operator that we discussed in Section 16 [see Eqs. (16.7), (16.7′)]. For a single photon state, this wave functional has the form

$$\Psi(1) = \int A_\lambda(\mathbf{k})\, \mathrm{d}^3 k\, \delta(n_\lambda(\mathbf{k}) - 1) \prod_{\lambda = 1, 2} \prod_{\mathbf{k}' \neq \mathbf{k}} \delta[n_\lambda(\mathbf{k}')], \tag{17.15}$$

where $A_\lambda(\mathbf{k})$ is an arbitrary amplitude of the wave whose wave vector is $\mathbf{k}$ and polarization is λ ($\lambda = 1, 2$).

Using the law of operation of the operators $\hat{A}_\lambda(\mathbf{k})$ $\hat{A}_\lambda^+(\mathbf{k})$ on the wave functional [the laws are expressed by equations which correspond exactly to Equations (16.7) and (16.7′) for scalar field operators $\hat{a}(\mathbf{k})$ and $\hat{a}^+(\mathbf{k})$], we find that the matrix element of the operator $\hat{A}_\lambda(x)$ is equal to

$$A_\lambda(x) = \langle \Psi(0) |\, \hat{A}_\lambda(x) \,| \Psi(1) \rangle = \int A_\lambda(\mathbf{k})\, U(\mathbf{k})\, \mathrm{e}^{ikx}\, \mathrm{d}^3 k. \tag{17.16}$$

The function $A_\lambda(x)$ contains only positive frequencies and, therefore, it satisfies Schrödinger's equation

$$i \frac{\partial A_\lambda(x)}{\partial t} = \hat{\omega} A_\lambda(x), \tag{17.17}$$

where $\hat{\omega} = +(-\Delta^2)^{1/2}$. The function $A_\lambda^*(x)$ satisfies a similar equation

$$-i \frac{\partial A_\lambda^*(x)}{\partial t} = \hat{\omega} A_\lambda^+(x). \tag{17.17′}$$

However, because the operator $\hat{\omega}$ is not self-conjugate, the quantity $A_\lambda^* A_\lambda$ does not satisfy just any conservation law.

As we already remarked, the quantity which is conserved and is composed of real, measurable (observable) quantities, is the magnitude of the energy density $\varepsilon(x) = -T_{44}(x)$ which is expressed in terms of the electric $\mathscr{E}(x)$ and magnetic $\mathscr{H}(x)$ fields.

It is easy to show that the energy density $\varepsilon(x)$ is a single photon state is equal to

$$\begin{aligned}\varepsilon(x) &= \langle \Psi(1)| \hat{\varepsilon}(x) |\Psi(1)\rangle \\ &= \tfrac{1}{2}\left[\mathscr{E}(x)\,\mathscr{E}^*(x) + \mathscr{H}(x)\,\mathscr{H}^*(x)\right], \end{aligned} \tag{17.18}$$

where $\mathscr{E}(x)$ and $\mathscr{H}(x)$ are the matrix elements of the operators $\hat{\mathscr{E}}(x)$ and $\hat{\mathscr{H}}(x)$ for the quantum transition $\Psi(0) \to \Psi(1)$.

Thus in the case of a free electromagnetic field, one cannot construct a quantity for the probability density of finding a photon in one or another part of space at a given moment in time. However, there does exist an energy density that can be focused into any specific region of space.

18. The Diffusion of Relativistic Packets

In the preceding section we considered the state corresponding to a strong localization of quantum fields (localized packets). Such a state is not stationary, but changes with the passage of time in such a way that the degree of localization of the field decreases. An initially well defined wave packet diffuses so that the quantities $\overline{\Delta x^2}$, $\overline{\Delta y^2}$, $\overline{\Delta z^2}$ increase in time. Characteristic features of this diffusion are common to all fields and therefore it is sufficient to examine the case of a scalar field. Let the wave packet have the form†

$$\varphi(\mathbf{x}, t) = \int \frac{c(\mathbf{k})}{\omega}\, \mathrm{e}^{\mathrm{i}(\mathbf{kx} - \omega t)}\, \mathrm{d}^3 k \tag{18.1}$$

where $\omega = +(\mathbf{k}^2 + \mu^2)^{1/2}$. We assume

$$c(\mathbf{k}) = N \exp - \frac{(\mathbf{k} - \mathbf{k}_1)^2}{2b^2}, \tag{18.2}$$

where N is a normalizing factor, $N = 1/(b^3(\pi^3)^{1/2})^{1/2}$. Because the density

† See [30].

$\rho(\mathbf{x}, t)$ is equal to

$$\rho(\mathbf{x}, t) = \frac{\mathrm{i}}{2}\left(\varphi \frac{\partial \varphi^*}{\partial t} - \varphi^* \frac{\partial \varphi}{\partial t}\right), \tag{18.3}$$

the localization for $t=0$ will be strong if $\varphi(\mathbf{x}, 0)$ or $\mathrm{d}\varphi^*(\mathbf{x}, 0)/\mathrm{d}t$, or both of these quantities are strongly localized.

Placing (18.2) into (18.1) we obtain

$$\begin{aligned}\frac{\partial \varphi(\mathbf{x}, 0)}{\partial t} &= -\,\mathrm{i}N \int \exp\left[-\frac{(\mathbf{k} - \mathbf{k}_1)^2}{2b^2} + \mathrm{i}\mathbf{k}\mathbf{x}\right] \mathrm{d}^3 k \\ &= -\,\mathrm{i}\mathrm{e}^{\mathrm{i}\mathbf{k}_1\mathbf{x}} \frac{\mathrm{e}(-\mathbf{x}^2/2a^2)}{(a^3(\pi^3)^{1/2})^{1/2}}, \ldots\ldots\end{aligned} \tag{18.4}$$

where $a = 1/b$ and $a \to 0$ for $b \to \infty$, so that the state (18.1) belongs to the class of states which are strongly localized for $t=0$.

For $t>0$, we have

$$\frac{\partial \varphi(\mathbf{x}, t)}{\partial t} = -\,\mathrm{i}N\mathrm{e}^{\mathrm{i}(\mathbf{k}_1\mathbf{x} - \omega_1 t)}\, I(\mathbf{x}, t), \tag{18.5}$$

$$\varphi(\mathbf{x}, t) = \frac{N}{\omega_1} \mathrm{e}^{\mathrm{i}(\mathbf{k}_1\mathbf{x} - \omega_1 t)}\, I(\mathbf{x}, t), \tag{18.5'}$$

where

$$\begin{aligned} I(\mathbf{x}, t) = \int \mathrm{d}^3 k \exp\Bigg[&-\frac{(\mathbf{k} - \mathbf{k}_1)^2}{2b^2} \\ &+ \mathrm{i}(\mathbf{k} - \mathbf{k}_1)\,\mathbf{x} - \mathrm{i}(\omega - \omega_1)\, t\Bigg], \end{aligned} \tag{18.6}$$

It is assumed that in (18.5′) that $|\mathbf{k}_1| \gg b$.

For definiteness, let the vector $\mathbf{k}_1$ be directed along the Ox axis so that $\mathbf{k}_1 = (k_{1x}, 0, 0)$. Then

$$\omega - \omega_1 = \frac{k_{1_x}}{\omega_1} q_x + \frac{1}{2\omega_1}(q_x^2 + q_y^2 + q_z^2) - \frac{1}{2}\frac{k_{1x}^2}{\omega_1^2} q_x^2 + \cdots \tag{18.7}$$

and $\mathbf{q} = \mathbf{k} - \mathbf{k}_1$. Simple calculations yield

$$I(\mathbf{x}, t) = A(t) \exp\left[\mathrm{i}\alpha(\mathbf{x}, t) - \frac{(x - v_x t)^2}{2\Delta_\parallel^2} - \frac{y^2 + z^2}{2\Delta_\perp^2}\right], \tag{18.8}$$

where $A(t)$ is a slowly varying quantity, α is the real phase, and the quantities $\Delta_{\parallel}^2$ and $\Delta_{\perp}^2$ are given by

$$\Delta_{\parallel}^2(t) = \frac{1}{b^2} + \frac{b^2 m^4}{\omega^6} t^2, \tag{18.9}$$

$$\Delta_{\perp}^2(t) = \frac{1}{b^2} + \frac{b^2}{\omega^2} t^2. \tag{18.9'}$$

Assuming $\Delta^2(0) = 1/b^2$, these equations may be rewritten in the form

$$\Delta_{\parallel}^2(t) = \Delta^2(0) + \frac{\lambdabar^2}{\Delta^2(0)} \frac{m^4}{E^4} v^2 t^2, \tag{18.10}$$

$$\Delta_{\perp}^2(t) = \Delta^2(0) + \frac{\lambdabar^2}{\Delta^2(0)} v^2 t^2, \tag{18.10'}$$

where $\lambdabar = \hbar/p$ is the wavelength, p the particle momentum and $v = \partial E/\partial p = p/E$ is its velocity.

From (18.10) we can see that, for $m=0$, the wave-packet does not diffuse in the longitudinal direction as it ought to for particles whose rest mass is of the order $m=0$. In this case, de Broglie waves do not diffuse in vacuum so that $\partial\omega/\partial k = c = \text{constant}$. Equation (18.10′) agrees with Equation (12.19) which was derived from diffraction theory for a beam which was formed by a narrow slit of width $\Delta(0) = \text{a}$.

However, there are important differences in the behavior of relativistic and non-relativistic beams. The width of the packet increases in proportion to $\lambdabar^2 t^2 v^2$ [see (12.19)]. In the non-relativistic case $\lambdabar^2 v^2 = \hbar^2 v^2/p^2 = \hbar^2/m^2 = \text{constant}$, and in the relativistic case $\lambdabar^2 v^2 = \hbar^2/E^2$ for $E \to \infty$ and $\lambdabar^2 v^2 \to 0$. In the non-relativistic case, the dispersion is the same in the longitudinal and transverse directions, $\Delta_{\parallel} = \Delta_{\perp}$. In the relativistic case an extra factor of $(m/E)^4$ appears in the dispersion in the longitudinal direction. The factor tends to zero for $E \to \infty$ and it significantly decreases the growth $\Delta_{\parallel}^2(t)$ in time. This shows the tendency to a behavior similar to that of a zero rest mass particle for which $\Delta_{\parallel}^2(t) = \text{constant}$.

By rewriting (18.10) in the form

$$\Delta_{\perp}^2(t) = \Delta^2(0) = \frac{\Lambda_0^2}{\Delta^2(0)} \left(\frac{m}{E}\right)^2 c^2 t^2, \tag{18.10''}$$

where $\Lambda_0 = \hbar/mc$, we may treat the decrease in the rate of growth $\Delta_{\perp}^2(t)$

as if were due to a retarded time $t' = mt/E$ and the motion of the system as if it were due to the "twin effect" – the life time of a moving packet increases by E/m.

Thus relativistic effects have a stabilizing influence on a wave packet. The characteristics of relativistic packets that we have discussed are general for various covariant fields. This is evident from the fact that all of these characteristics are related not so much to the form of the amplitude of the packet $c(\mathbf{k})$ as to the law of dispersion for de Broglie waves in vacuum

$$\frac{d\omega}{dk} = \frac{k}{\omega} = \frac{k}{\sqrt{k^2 + m^2}}, \tag{18.11}$$

which behave in an absolutely different way for $k \gg m$ and $k \ll m$.

19. The Coordinates of Newton and Wigner

As we showed earlier, in the relativistic case $\delta(x)$ cannot be the eigenfunction of the coordinate operator for an elementary particle because it cannot be formed from a state with a positive energy.

Therefore Newton and Wigner [39] introduced a new operator for positive particles $\hat{q}$, whose characteristics are derived from the concept of a localized state ψ. They postulated the following characteristics of the state which represent a particle which is localized at a moment of time t at the point $x = y = z = 0$:

(a) these states form a linear set, i.e. the superposition of two localized states is also a localized state;
(b) this set transforms into itself under rotation about the origin and under spatial and temporal reflection;
(c) if ψ is a localized state, then a spatial displacement makes it orthogonal to states which are localized at the origin;
(d) they must satisfy some requirements of regularity which allow operators of the Lorentz group to be applied to localized states.

It is particularly easy to find such states for spinless particles. In the momentum representation the wave function for a spinless particle ψ is a function of the momentum $\mathbf{k}$ and $k_0 = \omega = (\mathbf{k}^2 + m^2)^{1/2}$. In the case of scalar particles, from (b) it follows that ψ is a function of $\mathbf{k}^2$, or of ω, i.e., $\psi = \psi(\omega)$.

We now use requirement (c). The operator of spatial displacement $\hat{T}_3(\mathbf{a})$ reduces to a factor $e^{i\mathbf{ka}}$ in the momentum representation

$$\varphi(\mathbf{k}) = \hat{T}_3(\mathbf{a})\,\psi(\omega) = e^{i\mathbf{ka}}\psi(\omega). \tag{19.1}$$

This state must be orthogonal to the state which is localized at the origin, i.e., to $\psi(\omega)$. We take for $\varphi(\mathbf{k})$ and $\psi(\omega)$ the covariant amplitude in the form

$$\tilde{\varphi}(x) = \int \varphi(\mathbf{k})\, e^{i\mathbf{kx}} \frac{d^3k}{\omega}, \tag{19.2}$$

so that $\varphi(\mathbf{k}) = A(\mathbf{k})\,(\omega)^{1/2}$ and $\psi(\omega) = B(\omega)\,(\omega)^{1/2}$. According to (16.17′) the scalar product of φ and ψ has the form

$$(\varphi\psi) = \int \varphi^* \psi \frac{d^3k}{\omega}; \tag{19.3}$$

Substituting the values of φ and ψ into this equation gives

$$\int \frac{|\psi(\omega)|^2}{\omega} e^{-i\mathbf{ka}}\, d^3k = 0, \tag{19.4}$$

where $a \neq 0$. From this it follows that $|\psi(\omega)|^2/\omega = \text{constant}$, that is $\psi(\omega) \approx \omega^{1/2}$. We may, for example, set

$$\psi(\omega) = \frac{\omega^{1/2}}{(2\pi)^{3/2}}. \tag{19.5}$$

Applying the displacement operator to (19.5) we obtain a function which represents a state localized at the point $\mathbf{x}$

$$\psi_x(\mathbf{k}) = \hat{T}(\mathbf{x})\,\psi(\omega) = \frac{\omega^{1/2}}{(2\pi)^{3/2}} e^{i\mathbf{kx}}. \tag{19.6}$$

It is easy to see that this function is the eigenfunction of the operator $\hat{q}_x$

$$\hat{q}_x = -i\left(\frac{\partial}{\partial k_x} + \frac{k_x}{2\omega^2}\right), \tag{19.7}$$

which is just the *coordinate operator of an elementary particle* according to Newton and Wigner.

Applying the operator $\hat{q}_x$ to (19.6) gives

$$\hat{q}_x \psi_x(\mathbf{k}) = x \psi_x(\mathbf{k}), \tag{19.8}$$

so that x is the characteristic value of the operator $\hat{q}_x$ in the state $\psi_x(\mathbf{k})$ which is localized at some point $\mathbf{x}$ at $t=0$.

We now turn to the coordinate representation of a state which is localized at the point $\mathbf{x}=0$ at the moment in time $t=0$. According to (19.2) and (19.5) we have

$$\tilde{\varphi}(r) = \frac{1}{(2\pi)^{3/2}} \int \frac{\mathrm{e}^{\mathrm{i}\mathbf{kx}}}{\omega^{1/2}} \mathrm{d}^3 k = \left(\frac{m}{r}\right)^{5/4} H_{5/4}^{(1)}(\mathrm{i}mr), \tag{19.9}$$

where $H_{5/4}^{(1)}$ is the Hankel function of the first kind and of the order $\frac{5}{4}$. For $r \to 0$ $\tilde{\varphi}(r) \approx r^{-5/2}$ and for $r \to \infty$ $(r \gg \hbar/mc)$ $\tilde{\varphi}(r) \approx \mathrm{e}^{-mr}$. From this it is obvious that the region of localization of the state is defined to an order of magnitude by the Compton wavelength of the particle $\hbar/mc$ [in (19.9) $c=1$ and $\hbar=1$].

From the behavior of the function $\hat{\varphi}(r)$ near zero it follows that the localized state belongs to the class of unnormalized functions. The "localized" state $\psi(\omega)$ and the new coordinate operator $\hat{q}_x$ that we have considered in this section follow naturally from the principles postulated in (a), (b), (c) and (d).†

However the physical interpretation of these states and operators q_x, q_y, q_z has a number of difficulties and therefore these operators are not as useful as the general coordinates x, y, z.

The first of these difficulties is contained in the fact that a state which is localized at $t=0$, $x=0$ in one coordinate system, will not be localized, in the sense of fulfilling conditions (a), (b), (c) and (d), in moving coordinate system, even if the points $x=0$ and $x'=0$ coincide at $t=t'=0$. In view of this fact the operators $\hat{q}_x$, $\hat{q}_y$, $\hat{q}_z$ do not possess the characteristics required for relativistic covariance.

The second difficulty is contained in the fact that it is not possible to construct a *gedanken* experiment to measure the operators $\hat{q}_x$, $\hat{q}_y$, $\hat{q}_z$ that is, to single out the localized state $\psi(x)$. The Newton-Wigner coordinate operator seems, therefore, to be the coordinate operator of the center of

† Note that because we restricted ourselves to considering spin-zero particles, we did not use all the requirements listed from (a)-(d).

gravity of the wave packet $\hat{\varphi}(p)$ which, *by definition* represents *the localized state*.

In view of these difficulties we prefer to use the "old" coordinates x, y, z and t which are measurable, at least in a *gedanken* experiment. Such an experiment is based on the assumption that it is possible to produce an arbitrarily narrow beam, an arbitrarily small wavelength $\lambda\!\!\!^{-} \to 0$, and therefore on the principal assumption that there exist arbitrarily small apertures $a \to 0$. Although these assumptions are rather formal, they have the advantage of being definable.

20. The Measurement of a Microparticle's Coordinates in the Relativistic Case

The formulation of the problem of measuring the coordinates of a microparticle in the relativistic case is very similar to the non-relativistic case that we discussed in detail in Sections 11 and 12. As in the non-relativistic case, we must distinguish between direct and indirect measurements.

In the case of a direct measurement, we consider the measuring of the coordinates of a free particle. In Section 14 we showed that for particles with spin $\frac{1}{2}$ which satisfy Dirac's equation, there are allowed single-particle states with a positive energy which can be localized arbitrarily close to the point $x = x'$ at the moment in time $t = t'$. From the laws of diffusion of such states, it follows that this extremely localized wave packet of a state will diffuse with the speed of light if it does not itself move with a very high speed, so that the average momentum of the particles in such a packet $\bar{p} \gg mc$ where m is the mass of the particles.

In this case (see Section 18) the longitudinal diffusion of the packet ceases and the velocity of transverse diffusion decreases by a factor of $m/E\,[E = (p^2 + m^2)^{1/2}]$.

The localized single-particle state for the case of a particle with spin zero (which satisfies the Klein-Gordon equation) and with spin one (photons which satisfy Maxwell's equations) exist only as a special exception.

As we showed in Sections 16 and 17, in these cases it is more reasonable to speak of the localization of the meson field rather than of the mesons or photons. The second feature of the relativistic case is the significant decrease of the arbitrariness in the choice of possible interactions. The interactions must satisfy certain requirements of relativistic invariance. However these requirements do not greatly restrict *gedanken* experiments because the theory of relativity does not impose any restrictions on the mass of the charged particles, except for the conditions relating to the existence of the particles in bound states. The binding energy of the particles has to be significantly smaller than their proper energy

$$\Delta E \ll mc^2 . \tag{20.1}$$

Let us now turn to considering direct measurement of coordinates, which for the non-relativistic case was considered in Section 12. As in Section 12, we consider the atoms in a photographic plate as infinitely heavy ($Mc^2 \gg E$, where E is the energy of the particle) and we assume that the excitation energy of the atom $\varepsilon = \varepsilon_n - \varepsilon_0$ is small in comparison to the energy of the particle ($E \gg \varepsilon$) whose coordinates we are measuring. In order to write the δ-function interaction (12.2) in a relativistically invariant form, we consider that it is due to an exchange of very heavy mesons. In this case

$$W = g \frac{e^{-r/a}}{r}, \tag{20.2}$$

where a is the Compton length of a hypothetically heavy meson, g is a coupling constant and $r = ((x - Q)^2)^{1/2}$. In the limit of an infinitely heavy meson $a \to 0$, the function (20.2) is concentrated in a volume of the order of $a^3 \to 0$. Under the conditions discussed above, the atoms of a photographic plate remain a non-relativistic system and relativistic considerations extend only to the description of the motion of the free particles whose coordinate is defined by observing the consequences of excitation of the atoms of the photographic plate, and its interaction with these atoms.

In the Born approximation, which we used in Section 12, relativity does not change the main character of the conclusions which were drawn in Section 12 regarding the physical processes involved in such measurements. In particular, replacing the δ-function interaction (12.2) by the interaction of ultra-heavy mesons (20.2) and replacing the Schrödinger wave function by a four-component spinor which satisfies Dirac's equations does not change the conclusion that, for small intervals of time [see (12.8)], the excitation of an atom is accompanied by the localization of a microparticle in the vicinity of the atom. This result also holds true for mesons which satisfy Klein's wave equation.

We now turn to indirect measurements. In this case we mean coordinate measurements of particles that are in bound states using observations of scattering of a narrow beam-probe. We first note that, insofar as we assume that bound microparticles retain their individuality, we have predetermined the smallness of the mass defect ΔE in comparison to the characteristic energy of the particle mc^2. Therefore, relativistic effects will not affect the basic conclusion drawn in Section 12 from considerations of non-relativistic indirect measurements. A more important fact is that the narrow beam, which is necessary for finding the location of a bound microparticle in the relativistic case, due to the reasons discussed in Section 18 can be made significantly narrower, and to have a much smaller aperture angle than that allowed in the non-relativistic packet.

In concluding this discussion we shall make a few comments regarding the possibility of obtaining such narrow beam-probes. We are, of course, not discussing their practical realization but only the theoretical possibility of this. We may consider that the necessary narrow aperture whose width is $a \gg \lambda\!\!\!^{-}$ (for $\lambda\!\!\!^{-} \to 0$) may be made from a slit in a strongly absorbing material. If the necessary absorption is due to the production of new, very heavy and correspondingly slow particles, then their generation will not "spoil" our beam-probe. The existence of apertures with such properties does not contradict the principles of quantum theory of fields and, therefore, they may serve as a formal basis for discussion of extremely narrow beams and small wave packets (see Appendix IV).

CHAPTER IV

THE ROLE OF FINITE DIMENSIONS OF ELEMENTARY PARTICLES

21. THE POLARIZATION OF VACUUM. THE DIMENSIONS OF AN ELECTRON

When analyzing the possibility of measuring the space-time coordinates of an elementary particle, they were assumed to be point particles, and thus the accuracy to which they could be determined was limited by the possibility of producing an arbitrarily narrow beam-probe which interacts by contact, interaction with the microparticles whose location in spacetime is thus measured. The latter requirement is necessary for the elimination of the effects related to the existence of a finite, or even infinite radius of interaction.

As shown before, this possibility is allowed in the form of a *gedanken* experiment. However, it also follows that the physical picture of the *point particle* is actually an approximation. In higher approximations in the powers of the interaction constant, effects arise which may be considered as the manifestation of the structure of the particle which is dynamic in nature and cannot be reduced to a distribution of some material in space.

In our discussion of the structure we will begin with the simple case of an infinitely heavy, electrically charged point particle surrounded by a Coulomb field, the potential of which is equal to

$$V(r) = \frac{e}{r}, \tag{21.1}$$

where e is the charge of the particle and r is the distance from the particle to the point of observation.

The interaction of two charged particles, where one is scattered on the other, may be represented by a Feynman diagram (see Figure 15a). M denotes the heavy point particle, line 1–2 the path of the scattered particle, and the dotted line q the path of the virtual photon which transfers momentum from one particle to the other. In addition to this simple diagram, there are other diagrams which give a contribution to this process. One of these is shown in Figure 15b. This diagram describes the production of a positron electron pair, the e^+, e^- loop, by a virtual photon q and the annihilation of this pair in the subsequent emission of a photon q which in turn causes the momentum exchange between the particles.

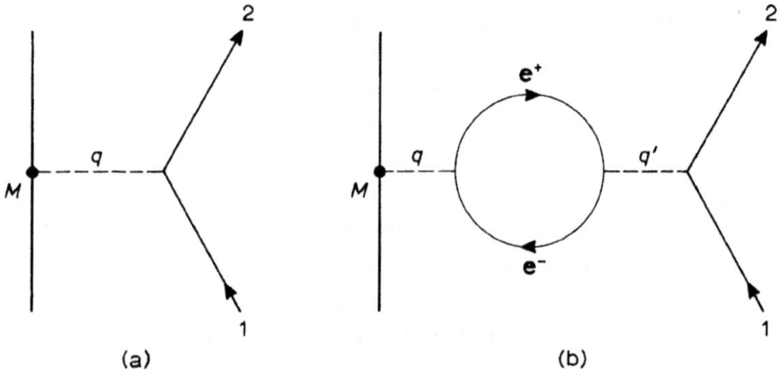

Fig. 15. (a) An illustration of the scattering of a charged particle (line 1-2) on an infinitely heavy (fixed) charged particle M. q is a virtual photon. (b) The polarization of vacuum is taken into account in this diagram. It includes the corrections to Coulomb's law that are needed to account for the polarization of vacuum. q and q' are virtual photons; e^+, e^- are a positron and electron of a virtual pair.

This rather complicated diagram takes into account in the first non-vanishing approximation the effect of the polarization of the electron-positron field by the Coulomb field of the massive particle M. The computation of this effect is very tedious and we refer the reader to books on quantum electrodynamics for the necessary details.†

The calculations show that the taking into account the polarization correction leads to a new expression for the potential of a point charge. In place of (21.1) we have

$$V(r) = \frac{e(r)}{r}, \tag{21.2}$$

where the quantity $e(r)$ is called the *effective charge*. From the calculations the effective charge is equal to

† See, e.g., Chapter VIII in [37], Chapter VI in [38].

$$e(r) = \begin{cases} e\left\{1 + \frac{\alpha}{3\pi}\ln\left(\frac{\Lambda}{r}\right)^2 + \cdots\right\} & \text{for } r \ll \Lambda, \quad (21.3) \\ e\left\{1 + \frac{\alpha}{\sqrt{\pi^4}} e^{-2r/\Lambda} \Big/ \left(\frac{r}{\Lambda}\right)^{3/2} + \cdots\right\} & \text{for } r \gg \Lambda. \quad (21.3') \end{cases}$$

where $\alpha = e^2/\hbar c$, $\Lambda = \hbar/mc$, where m is the mass of the electron. From these equations we see that, for small r, the effective charge is large. In proportion to the increase in the spacing r, this charge is screened by the vacuum polarization more and more strongly, so that for $r \gg \Lambda$ it becomes equal to the charge of the particle e that is macroscopically observable.† From Equation (21.3′) we also see that the effective charge is distributed in a region whose dimensions are of the order of magnitude of the Compton length of an electron ($\hbar/mc = 386 \times 10^{-11}$ cm). The existence of a vacuum polarization leads to the important possibility of the scattering of light off an infinitely heavy charged particle.

It is well known that, Maxwell's equations being linear, the static field and wave field may be superimposed on each other without distortion. The vacuum polarization destroys this linearity. In the first non-vanishing approximation of the interaction of a quantum of light with the field of a heavy particle, the polarization of vacuum may be described by the diagram shown in Figure 16. The lines 1 and 2 show as before the incoming and outgoing paths of the scattered photon. The loop $e^{\pm}$ illustrates the formation of virtual positron-electron pairs which are produced by the vacuum polarization. The lines q, q' are the paths of the virtual photons which transfer momentum to the heavy particle M.

We must, therefore, keep in mind that, when a photon with a frequency $\omega \gg 2mc^2/\hbar$ interacts with a heavy charged particle (e.g., with an atomic nucleus), the dominant effect will be the positron-electron pair formation (e^+e^-) and, therefore, absorption of the photon. This absorption results in the elastic diffraction scattering of photons. The total elastic scattering cross section σ_{el} consists of the elastic diffraction scattering cross section σ'_{el}, and the elastic scattering cross section that is related to the vacuum polarization σ''_{el} and which is defined by the process shown in Figure 16.

† We must remember, however, that Equations (21.3) and (21.3′) are approximations and valid only as long as the corrections $|e(r) - e|$ are small in comparison to e.

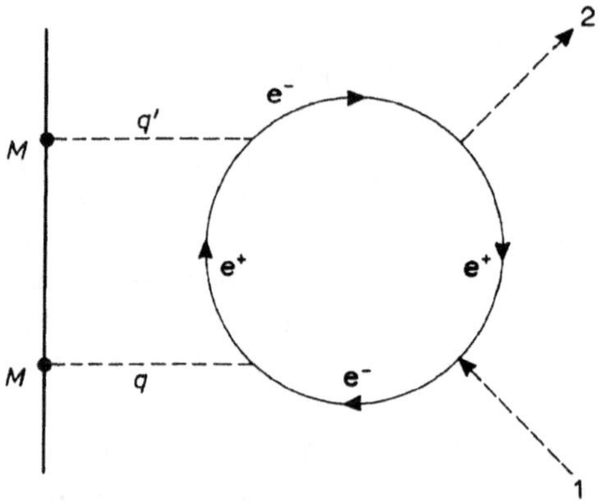

Fig. 16. An illustration of the scattering of light on a fixed charged particle M. The scattering actually takes place on the electrons and positrons ($e^{\pm}$) that result from the polarization of vacuum by the particle M. Here, as in Figure 15, q and q' are virtual photons and $e^{\pm}$ are a virtual pair.

Because the cross section for production of a $\sigma_{\pm}$ pair is comparatively large, the ratio of σ''_{el} and σ'_{el} is small:

$$\frac{\sigma''_{\text{el}}}{\sigma'_{\text{el}}} = \frac{3}{\pi^2}\left(\frac{mc^2}{\hbar\omega}\right)^2 \ln \frac{\hbar\omega}{mc^2} \tag{21.4}$$

(see [37], Section 55).

Note that for frequencies $\omega \ll 2mc^2$ for which real pairs of electrons and positions cannot be produced, the effect of vacuum polarization may be described in terms of the dielectric constant tensor for vacuum ε_{ik} and the magnetic permeability tensor μ_{ik} as

$$\varepsilon_{ik} = \delta_{ik} + \frac{1}{\mathscr{E}_0^2}\{2(\mathscr{E}^2 + \mathscr{H}^2)\,\delta_{ik} + 7\mathscr{H}_i\mathscr{H}_k\}, \tag{21.5}$$

$$\mu_{ik} = \delta_{ik} + \frac{1}{\mathscr{E}_0^2}\{2(\mathscr{E}^2 - \mathscr{H}^2)\,\delta_{ik} - 7\mathscr{H}_i\mathscr{H}_k\}, \tag{21.5'}$$

where $\mathscr{E}$, $\mathscr{H}$ are the electric and magnetic field intensities, $\mathscr{E}_i$ and $\mathscr{H}_i$ ($i = 1, 2, 3$) are the projections of these vectors on the coordinate axes and $\mathscr{E}_0$ is some critical field which is defined to an order of magnitude by the equation

$$e\mathscr{E}_0 \frac{\hbar}{mc} = 2mc^2, \tag{21.6}$$

implying that the potential difference in such a field over a separation $\hbar/mc$ is equal to the energy required for pair formation. In this manner, as a result of the polarization of space around a charged particle, a charged "atmosphere" is formed around the particle, the extent of which is defined by the electron Compton wavelength $\hbar/mc$. Because of this fact, particles lose their point nature.

In particular, if we take the particle as being localized in a very narrow wave packet whose dimensions are smaller than $\hbar/mc$, then this situation will correspond to the case of the stationary, heavy particles that we considered above. A charged region of the order of $\hbar/mc$ is formed around these particles and because of this the interaction of these particles with the particles of the probe can take place. And, therefore, an uncertainty which cannot be removed arises in the definition of the location of the particle, which is equal in order of magnitude to the quantity

$$\Delta x \sim \frac{\hbar}{mc} = 3.86 \times 10^{-11}\ \text{cm}. \tag{21.7}$$

In the case of electromagnetic interactions, this atmosphere is very tenuous. This follows from the smallness of the electromagnetic interaction constant $\alpha = e^2/\hbar c = 1/137$. Therefore, the number of interactions in the atmosphere of particles in regions of polarized vacuum will be very small in comparison to the number of point interactions. In the overwhelming majority of cases the uncertainty in definition of particle coordinates will not be related to the dimensions of the particle atmospheres.

In particular, corrections to electron scattering, to the Compton effect, to *bremsstrahlung* and to other electromagnetic effects that result from the vacuum polarization in the vicinity of an electron (or positron) are very small (see [37]).

22. The electromagnetic structure of nucleons

An entirely different situation exists for the cases in which the dimensions of the atmospheres of the particles are formed from strong interactions. Because of the significant magnitude of the strong interaction constant g ($g^2/4\pi\hbar c = 16$) the atmosphere that forms around strongly interacting particles is not, under any circumstance, rare.

This assertion may be illustrated using the electromagnetic "form factor"† which define the charge and current distributions in nucleons. The Feynman diagram shown in Figure 17a represents the scattering process on a proton in which the electron is considered as a point. Lines 3 and 4 represent the path of the scattered electron; lines 1 and 2 represent the proton path; the vertex Γ which is indicated by the shaded triangle emphasizes the extended character of the nucleon, and q is the path of the virtual photon which transfers momentum between the proton and electron. The diagram in Figure 17b illustrates the origin of the extended nucleon. At the point A a proton emits a π^+-meson by means of a strong pion-nucleon interaction. The π^+-meson at point B emits (or absorbs) a

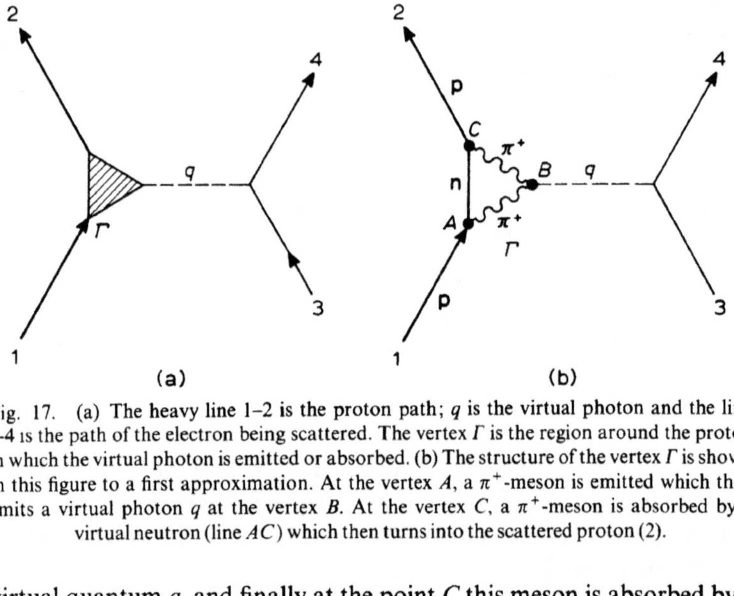

Fig. 17. (a) The heavy line 1–2 is the proton path; q is the virtual photon and the line 3-4 is the path of the electron being scattered. The vertex Γ is the region around the proton in which the virtual photon is emitted or absorbed. (b) The structure of the vertex Γ is shown in this figure to a first approximation. At the vertex A, a π^+-meson is emitted which then emits a virtual photon q at the vertex B. At the vertex C, a π^+-meson is absorbed by a virtual neutron (line AC) which then turns into the scattered proton (2).

virtual quantum q, and finally at the point C this meson is absorbed by a neutron which is transformed into a scattered proton (line 2). By this process a pion atmosphere is formed around the proton and this also creates the effect of an extended nucleon. The dimensions of this atmosphere is defined by the meson Compton wavelength $\hbar/mc = 1.4 \times 10^{-13}$ cm.

† This term is borrowed from X-ray scattering theory. The "form factor" takes into account the point character of charge distributions and currents inside atoms in this theory.

The diagram we considered above as an illustration of how an extended nucleon is formed has a conditional meaning. Due to the rather large value of the interaction constants of the pion and the nucleon field, the contribution of the other diagrams that describe processes with a large number of virtual mesons cannot be small. There are also significant resonances in the interactions of these mesons with each other. Therefore, the computation of these effects, due to the existence of the meson atmosphere of the nucleon, present great and as yet unsurmountable difficulties.

However, it is possible to construct a phenomenological theory of a nucleon's meson atmosphere in terms of the form factor. In order to do this, we compute the matrix element of the vertex particles $\Gamma_\mu(x_2, x_1, y)$, and we do not restrict ourselves to diagrams of the type shown in Figure 17b, because in the case of strong interactions such limitations are not justified.†

The quantity $\Gamma_\mu(x_2, x_1, y)$ represents a nucleon current with all interaction effects in vacuum taken into account. It is more convenient to compute the matrix element of the interaction energy of this current with the field of a virtual photon, which we write in terms of the vector potential

$$A_\mu(y) = a_\mu(q)\, \mathrm{e}^{-\mathrm{i}qy}. \tag{22.1}$$

This energy of interaction is equal to

$$W(x_2 - x_1) = \int \Gamma_\mu(x_2, x_1, y)\, A_\mu(y)\, \mathrm{d}^4 y. \tag{22.2}$$

We note that, because of the homogeneity of the vacuum, the quantity Γ_μ, the translational invariant, correspondingly depends only on the difference of the coordinates x_2, x_1, y. Therefore, in place of the coordinates, it is more convenient to use the quantities $x = x_2 - x_1$, and $X = y - \frac{1}{2}(x_1 + x_2)$. Therefore,

$$\Gamma_\mu(x_2, x_1, y) \equiv \Gamma_\mu(X, x) = \frac{1}{(2\pi)^8} \int \tilde{\Gamma}_\mu(\alpha, \beta)\, \mathrm{e}^{\mathrm{i}(\alpha X + \beta x)}\, \mathrm{d}^4\alpha\, \mathrm{d}^4\beta, \tag{22.3}$$

† In Appendix VA, the calculations of the vertex $\Gamma_\mu(x_2, x_1, y)$ are explained on the basis of perturbation theory (diagram in Figure 17b). This simple calculation is suited for the case of weak coupling between nucleons and mesons.

so that $\tilde{\Gamma}_\mu(\alpha, \beta)$ is the Fourier transform of $\Gamma_\mu(X, x)$.

Conservation of current, which is equivalent to the gauge invariance, requires that

$$\frac{\partial \Gamma_\mu}{\partial y_\mu} \equiv \frac{\partial \Gamma_\mu}{\partial X_\mu} = 0 \tag{22.4}$$

or

$$\tilde{\Gamma}_\mu(\alpha, \beta)\, \alpha_\mu = 0. \tag{22.4'}$$

We now calculate the energy $W(x_2 - x_1)$ matrix element for a transition of the nucleon from state with momentum $p_1 = p_i$ to a state with momentum $p_2 = p_f$. These states are described by the spinors

$$\bar{\psi}_f(x_2) = u_f(p_f)\, e^{ip_f x_2}, \quad \psi_i(x_1) = u_i(p_i)\, e^{-ip_i x_1}. \tag{22.5}$$

We have

$$(f|W|i) = \int \bar{\psi}_f(x_2)\, W(x_2 - x_1)\, \psi_i(x_1)\, d^4x_2\, d^4x_1. \tag{22.6}$$

Placing $W(x_2 - x_1)$ from (22.2) into this equation and using (22.3) we obtain

$$(f|W|i) = (2\pi)^4\, \delta^4(p_f - p_i + q)\, \bar{u}_f(p_f)\, \tilde{\Gamma}_\mu(q, P)\, u_i(p_i)\, a_\mu(q). \tag{22.7}$$

Therefore,

$$(f|\Gamma_\mu|i) = (2\pi)^4\, \delta^4(p_f - p_i - q)\, \bar{u}_f(p_f)\, \tilde{\Gamma}_\mu(q, P)\, u_i(p_i), \tag{22.8}$$

where $\alpha \equiv q = p_f - p_i$ is the momentum transferred and $\beta \equiv P \equiv p_f + p_i$ is the sum of the initial and final momenta.

We now define the structure of the current vector

$$J_\mu = \bar{u}_f(p_f)\, \tilde{\Gamma}_\mu(q, P)\, u_i(p_i). \tag{22.9}$$

Note that in constructing this vector we used the Dirac matrices γ_μ, $\sigma_{\mu\gamma}$ and the vectors $q = p_f - p_i$, $P = p_f + p_i$, the Dirac matrices γ_5 and $\gamma_5\gamma_\mu$ being pseudovectors. Therefore, the current is of the form

$$J_\mu = \bar{u}_f(p_f)\, \{a q_\mu + b P_\mu + c\gamma_\mu + d\sigma_{\mu\nu} q_\nu + e\sigma_{\mu\nu} P_\nu\}\, u_i(p_i), \tag{22.10}$$

where the coefficients a, b, c, d and e are scalar. Using Dirac's equations for the spinors $\bar{u}_f(p_f)$ and $u_i(p_i)$ and the additional condition (22.4′) which may now be written as

$$J_\mu q_\mu = 0, \tag{22.4″}$$

we may simplify (22.10) and reduce it to the form

$$J_\mu = \bar{u}_f(p_f)\,\{B(q^2)\,\gamma_\mu + C(q^2)\,\sigma_{\mu\nu}q_\nu\}\,u_i(p_i) \tag{22.11}$$

(see Appendix V.B). Here $B(q^2)$ and $C(q^2)$ are the only form factors which remain after the simplification of (22.10). For point particles with charge e and anomalous magnetic moment μ we have $B(0)=e$, $C(0)=\mathrm{i}\mu$. Therefore, we use the following normalization for form factors

$$B(q^2) = eF_1(q^2), \tag{22.12}$$
$$C(q^2) = i\mu F_2(q^2), \tag{22.13}$$
$$F_1(0) = F_2(0) = 1. \tag{22.14}$$

Under this normalization, the nucleon current takes on the form

$$J_\mu = \bar{u}_f(p_f)\,\{eF_1(q^2)\,\gamma_\mu + i\mu F_2(q^2)\,\sigma_{\mu\nu}q_\nu\}\,u_i(p_i). \tag{22.15}$$

We shall consider the exchange of momentum in the center-of-mass system of the particle. For the case of elastic scattering of electrons on protons, this will be the system of the center of mass of the electron and proton. The magnitude of the momentum of the particles remains constant in this system, and only the directions change so that $|p_f|=|p_i|$, and correspondingly the fourth component of the vector q is equal to zero [$q=(\mathbf{q},0)$, $\mathbf{q}=\mathbf{q}_f-\mathbf{q}_i$, $q^2=\mathbf{q}^2$]. In this coordinate system we may define the spatial image of the nucleon; that is the spatial form for the distribution of the electric charge density $\rho(\mathbf{X})$ will be

$$\rho(\mathbf{X}) = e\int F_1(\mathbf{q}^2)\exp(\mathrm{i}\mathbf{q}\mathbf{X})\,\frac{\mathrm{d}^3q}{(2\pi)^3} \tag{22.16}$$

and the magnetic moment distribution $\mu(\mathbf{X})$ will be

$$\mu(\mathbf{X}) = \mu\int F_2(\mathbf{q}^2)\exp(\mathrm{i}\mathbf{q}\mathbf{X})\,\frac{\mathrm{d}^3q}{(2\pi)^3} \tag{22.16′}$$

From these equations it follows that

$$\int \rho(\mathbf{X})\, \mathrm{d}^3X = eF_1(0) = e\,, \tag{22.17}$$

$$\int \mu(\mathbf{X})\, \mathrm{d}^3x = \mu F_2(0) = \mu\,. \tag{22.17'}$$

From (22.16) and (22.16′) we may define the rms of the electric radius $\langle r^2\rangle_e$ and the rms of the magnetic radius $\langle r^2\rangle_\mu$ by means of the equations

$$\langle r^2\rangle_e = \frac{1}{e}\int \rho(\mathbf{X})\, r^2\, \mathrm{d}^3X = \frac{1}{6}\left(\frac{\mathrm{d}F_1}{\mathrm{d}q^2}\right)_0, \tag{22.18}$$

$$\langle r^2\rangle_\mu = \frac{1}{\mu}\int \mu(\mathbf{X})\, r^2\, \mathrm{d}^3X = \frac{1}{6}\left(\frac{\mathrm{d}F_1}{\mathrm{d}q^2}\right)_0. \tag{22.18'}$$

The second equation is obtained by using the Fourier expansions (22.17) and (22.17′) differentiated with respect to $\mathbf{q}^2$.

From Equations (22.16) and (22.16′) it follows that the form factor $F_1(\mathbf{q}^2)$ defines the diffusion in space of the electric charge of the nucleon, and the form factor $F_2(\mathbf{q}^2)$ defines the diffusion of its magnetic moment.

Thus the form factors F_1 and F_2 reflect the space-time structure of nucleons. This structure was first discovered experimentally during studies of electron scattering in hydrogen and deuterium by Hofstadter [40-42]. His basic result shows that

$$F_{1\mathrm{p}}(q^2) = F_{2\mathrm{p}}(q^2), \tag{22.19}$$

$$F_{1\mathrm{n}}(q^2) \approx 0\,, \qquad F_{2\mathrm{n}}(q^2) \approx F_{1\mathrm{p}}(q^2), \tag{22.19'}$$

where we denote the form factor for protons by p, and that for neutrons by n. Figure 18 shows the recent results of the electromagnetic form factor for nucleons [43]. The electric radii prove to be the same: for protons $r_e = (\langle r^2\rangle_e)^{1/2} = 0.8\times 10^{-13}$ cm and for neutrons $r_e = 0$. The magnetic radii $r = (\langle r^2\rangle_\mu)^{1/2}$ are equal to r_e.

The spatial structure of nucleons which is defined by Equations (22.16) and (22.16′) has a somewhat arbitrary character. This arbitrariness arises from the fact that, from the point of view of quantum field theory, the electromagnetic structure of the nucleons is described by the vertex function $\Gamma_\mu(\mathbf{X}, x)$ which is a function of two variables $\mathbf{X} = y - \frac{1}{2}(x_1 + x_2)$, $x = x_2 - x_1$, and not by one variable. Furthermore, this structure is described as dynamic, taking into account the nucleon motion in its interaction with the virtual photons.

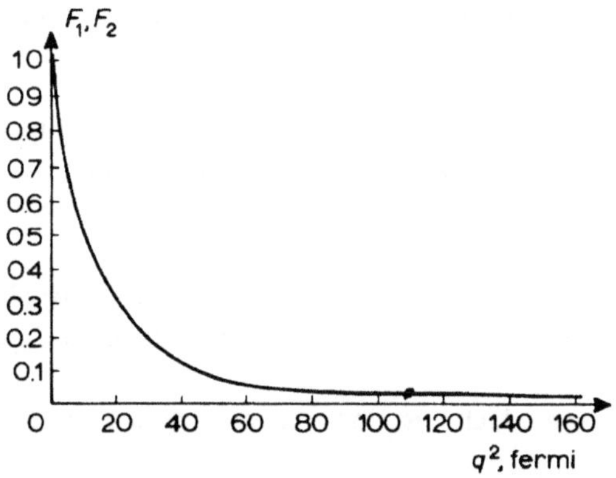

Fig. 18. This graph represents the form factors F_1 and F_2 as functions of the square of the transferred momentum q^2. Within experimental error, F_1 and F_2 are the same.

In Appendix V.A a similar coupling is considered, between the vertex $\Gamma_\mu(\mathbf{X}, x)$ and densities $\rho(\mathbf{X})$, $\mu(\mathbf{X})$, which are defined by Equations (22.16) and (22.16′). We also show there that these equations have a direct physical meaning for heavy nucleons and for small momentum transfers $q^2 \ll 4M^2$ (M is the nucleon mass). In this case the density $\rho(\mathbf{X})$ may be written in the form

$$\rho_\mu(\mathbf{X}) = \int \tilde{\Gamma}_\mu(\mathbf{q}, 2M)\, \mathrm{e}^{i\mathbf{q}\mathbf{X}}\, \mathrm{d}^3 q, \tag{22.20}$$

in which the assumption $q^2 \ll 4M^2$ is accounted for in an obvious way.

We now return to our main theme, to the problem of measuring the location of a nucleon. Because the dimensions of the pion atmosphere of the nucleon are finite even if particles in the beam-probe were localized exactly, one cannot define the location of nucleon by scattering a probe particle off of it, more accurately than to within the limits of its dimensions $\Delta x \approx r_e$. In the case we considered of electromagnetic interactions, the radius of interaction, strictly speaking, is infinitely large. This property of the electromagnetic (Coulomb and magnetic) interaction is in itself a serious obstacle to the defining of the nucleon location in the *gedanken* experiment illustrated in Figure 14. However, this fact is not all important. Instead of the electromagnetic interaction we may consider a weak interaction which is a contact, and point interaction. As we shall now show however, the same form factors $F_1(q^2)$ and $F_2(q^2)$, which characterize the electromagnetic interaction, enter into this interaction.

For definitness, we shall consider the process that is the reverse of the process in which a proton p captures a μ-meson namely,

$$\mathrm{n} + \nu \rightarrow \mathrm{p} + \mu^- . \tag{22.21}$$

where n is a neutron and ν is a neutrino. The initial particles are neutral and therefore the effects which are related to the electromagnetic interaction are absent.

The interaction which describes this process is a contact one and has the form [44]

$$W = \frac{G_F}{\sqrt{2}} J_\mu^s J_\mu^w , \tag{22.22}$$

where G_F is the weak interaction constant. $G_F = g_F/\hbar c(mc/\hbar)^2$, m is the nucleon mass and $g_F = 2 \times 10^{-49}$ erg cm^{-3} is the Fermi constant. This constant may be expressed in terms of the length $\Lambda_F = (g_F/\hbar c)^{1/2} = 0.6 \times$ $\times 10^{-16}$ cm, which is characteristic of weak interactions. J_μ^s is the nucleon "strong" current given by

$$J_\mu^s = \bar{\psi}_N J_\mu \tau^+ \psi_N \qquad (\mu = 1, 2, 3, 4) . \tag{22.23}$$

where $\bar{\psi}_N$, ψ_N are the nucleon wave functions (proton or neutron). The operator $\hat{J}_\mu$ for point nucleons has the form

$$\hat{J}_\mu = e\gamma_\mu + i\mu\sigma_{\mu\nu} q_\nu . \tag{22.24}$$

τ^+ is the operator of the isotopic spin which transforms a neutron into a proton (i.e., it changes the symbol N=n to N=p).

The current J_μ^w is the "weak" current and it has the form

$$J_\mu^w = \bar{\varphi}_\nu \hat{O}_\mu \varphi_\mu , \tag{22.25}$$

where $\bar{\varphi}_\nu$ is the neutrino wave function, φ_μ is the μ-meson wave function, and $\hat{O}_\mu$ is the operator that is composed of the γ_μ and γ_5 matrices (for details see [44]).

The existence of the operator $\tau^+ = \frac{1}{2}(\tau_1 + i\tau_2)$ (for reverse transition the operator $\tau^- = \frac{1}{2}(\tau_1 - i\tau_2)$ would be used) shows that J_μ^s consists of a linear combination of the components of the isotopic current vector $J^s_{\mu 1}, J^s_{\mu 2}, J^s_{\mu 3}$:

$$J^s_{\mu\sigma} = \bar{\psi}_N \hat{J}_\mu \tau_\sigma \psi_N, \qquad (22.23')$$

where $\sigma = 1, 2, 3$. However, the components of the current $J^s_{\mu 3}$ correspond exactly to the electromagnetic current and therefore, the operator $\hat{J}_\mu$ [see (22.24)] has to contain the same form factors $F_1(q^2)$ and $F_2(q^2)$ which are characteristic of the nucleon [see (22.19) and (22.19')]. In other words, the current J^s_μ accounts for the charged pion atmosphere of the nucleon. Because of this, the contact character of the interaction (22.22) is violated and it becomes non-local.

The Feynman diagrams describing this contact interaction (22.22) and the interaction taking into account the finite nucleon dimensions, are shown in Figure 19a and 19b. Due to the formation of a space-time distribution of nucleon charge and currents at the vertex Γ, even in a

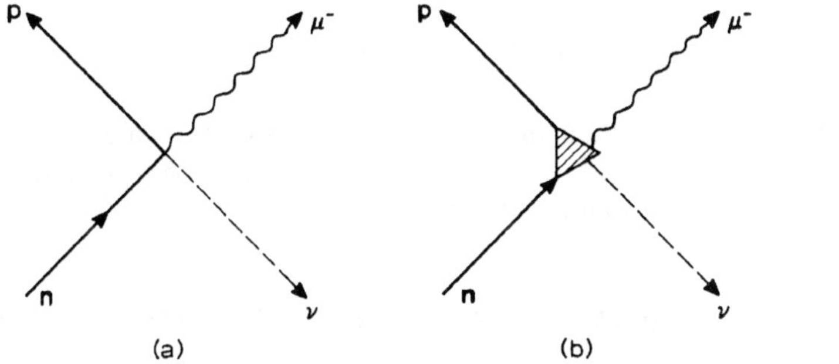

Fig. 19. Diagrams which describe a weak interaction (the process $n + \nu \rightarrow p + \mu^-$). (a) Diagram for a contact interaction. (b) Diagram in which the finite dimensions of the nucleon are taken into account (the form factor).

gedanken experiment which allows the existence of an arbitrarily narrow packet of neutrinos for a probe to define the location of a nucleon, it is impossible to define the location of the scattered neutrino more accurately within the limits of the nucleon dimensions, i.e., in a region $\Delta x \sim r_e$ (or r_μ). Thus we can see that the existence of form factors of elementary particles which relect their space-time structure leads to the fact that it is impossible to localize a particle in space-time more accurately than to within the limits of its dimensions, i.e., than to the accuracy

$$\Delta x \sim r_0, \qquad \Delta t \sim r_0/c, \tag{22.26}$$

where r_0 is the particle radius.

23. The meson structure of nucleons

In this section we turn to the theory of elastic scattering of pions on nucleons in high-energy regions (higher than the threshold for meson production). In this case non-elastic processes play the determining role so that elastic scattering appears as diffraction. Consequently the scattering amplitude $T(E_\pi, \mathbf{q})$ will be imaginary [E_n is the meson energy in the center of mass system, $E=(m^2+\mathbf{k}^2)^{1/2}$ and $\mathbf{q}$ is the momentum transfer, $|\mathbf{q}|=2k \sin \vartheta/2$, ϑ is the scattering angle] and may be expressed directly in terms of differential scattering cross section

$$\frac{\mathrm{d}\sigma}{\mathrm{d}\Omega} = |T(E_\pi, \mathbf{q})|^2 = - T^2(E_\pi, \mathbf{q}). \tag{23.1}$$

Thus we may obtain the amplitude $T(E_\pi, \mathbf{q})$ from experimental data for $\mathrm{d}\sigma/\mathrm{d}\Omega$. On the other hand, this amplitude may be written as a series of Legendre polynomials $P_1(\cos \vartheta)$

$$T(E_\pi, \mathbf{q}) = \frac{1}{2ki} \sum_{l=0}^{\infty} (2l+1)(1 - e^{2i\eta_l})\, P_1(\cos \vartheta), \tag{23.2}$$

where

$$\eta_l(k) = \delta_l(k) + \mathrm{i}\gamma_l(k), \qquad \gamma_l \geqslant 0, \;\ldots \tag{23.2'}$$

is the complex phase. For sufficiently short waves ($\lambda\!\!\!^{-} \ll a$, where a is the dimension of the scattered wave system) the sum over l may be replaced by an integral by assuming $l\Delta l \to b\, \mathrm{d}b/\lambda\!\!\!^{-2} = k^2 b\, \mathrm{d}b$, where b is the impact parameter. Furthermore, for large l and small ϑ the polynomial $P_1(\cos \vartheta)$ may be replaced by the Bessel function

$$J_0(l \sin \vartheta) \approx J_0(bq).$$

Then (23.2) may be rewritten as

$$T(E_\pi, \mathbf{q}) = \frac{k}{2i} \int_0^\infty b \, db \left[1 - e^{-2i\eta(b, k)}\right] J_0(bq). \tag{23.3}$$

Because the Bessel function $J_0(bq)$ is orthogonal, this integral may be inverted to give

$$1 - e^{2i\eta(b, k)} = \frac{2i}{k} \int_0^\infty T(E_\pi, \mathbf{q}) \, J_0(bq) \, q \, dq. \tag{23.4}$$

In this manner, the phase $\eta(b, k)$ may be defined from experimental values as a function of the meson energy $k = (E^2 - m^2)^{1/2}$ and the impact parameter b. On the other hand, from the limits of geometrical optics, the phase changes of the initial waves $2\eta(b, k)$ may be calculated by integrating over the path of the index of refraction n, with the result

$$2\eta(b, k) = \int_A^B n(k, x) \, ds. \tag{23.5}$$

Figure 20 shows that integration path inside a nucleon.

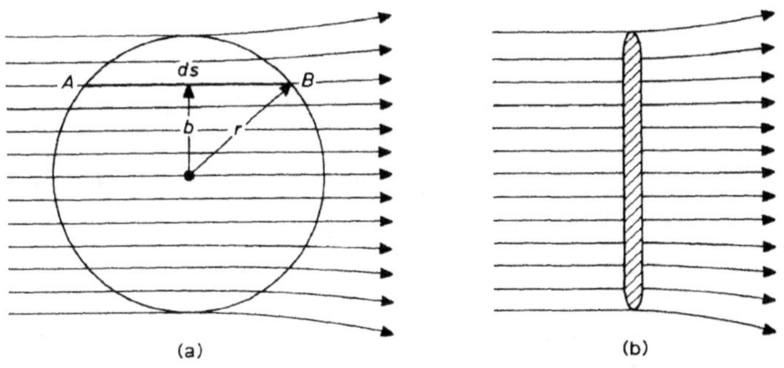

Fig. 20. (a) The sphere represents the nucleon. AB is the path of a beam inside of the nucleon; b is the impact parameter. (b) A partially absorbing screen which is equivalent to the nucleon. This screen gives the same diffraction pattern (for small scattering angles) as the nucleon.

Analyses of the experimental data for pion scattering on nucleons (π^+p) were carried out in [42, 45]. The results of this analysis is shown in Figure 21, where the absorption coefficient $K(r)$ is plotted as the ordinate. The imaginary part of the refraction coefficient n (its real part is extremely small in the case being considered) is plotted on the abscissa as a function of distance from the nucleon center r in the center of mass system. The absorption decreases with distance from the center of the nucleon, and it practically vanishes for values of r greater than

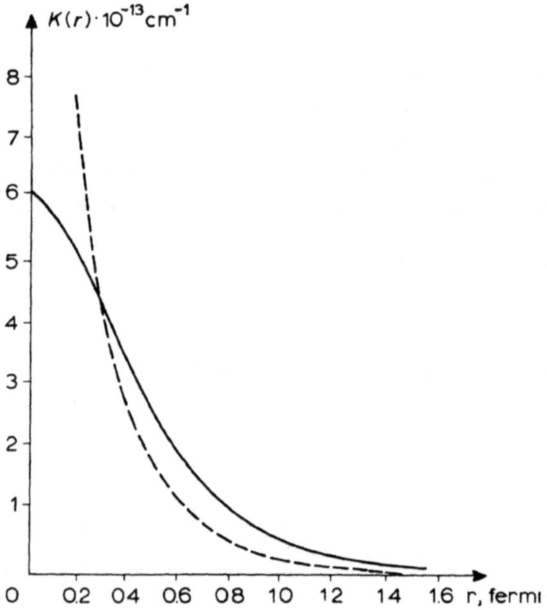

Fig. 21. The continuous curve is the pion absorption coefficient $K(r)$ as a function of the distance from the center of the nucleon r for the pion energy $E = 1.3$ GeV. The dotted curve is the same function for $E = 5$ GeV.

$r \approx 0.7 \times 10^{-13}$ cm. The agreement of this value with data on the electromagnetic radius of a nucleon ($r_e = r_\mu = 0.8 \times 10^{-13}$ cm) shows that the absorption of mesons by nucleons and the distribution of currents and charges in a nucleon are caused by one and the same physical cause, the meson atmosphere of the nucleon.

To what extent can we justify this simple optical picture of the meson-nucleon interaction? Present-day field theory does not give any direct method for calculating strong interactions, to which class belong the interactions of π-mesons with nucleons. Therefore, we can only give the conventional reasons which are characteristic of the acceptance of current theory. These are the assumptions (i) that the interaction constant g is small (this condition is necessary to ensure the convergence of the series), (ii) that it is possible to normalize all of the approximations in the powers of g, (iii) that the final equations obtained may be used without any limitations on the value of the constant g.

Following this line of reasoning, we obtain an equation which has a completely defined structure but which, for practical use, requires further assumptions [46].† As is shown in Appendix VI this equation has the form

$$(E_\pi^2 - \hat{K}^2)\,\psi(\mathbf{x}) = \int U(\mathbf{x}, \mathbf{x}'; E)\,\psi(\mathbf{x}')\,\mathrm{d}^3x', \tag{23.6}$$

where $\hat{K}^2 = -\nabla^2 + m^2$ (m is the meson mass), $E = E_\pi + E_\mathrm{N}$, E_N is the nucleon energy in this system, $U(\mathbf{x}, \mathbf{x}'; E)$ is, generally speaking, the non-localized complex potential which acts from the direction of the nucleon on the meson; $\psi(\mathbf{x})$ is the wave function for relative motion of the nucleon and meson, measured simultaneously so that $\mathbf{x} = \mathbf{x}_2 - \mathbf{x}_1$, $t = t_2 - t_1 = 0$.

Another equivalent equation is

$$(E_\mathrm{N} - \hat{\mathscr{H}}_\mathrm{N}^0)\,\psi(\mathbf{x}) = \int V(\mathbf{x}, \mathbf{x}'; E)\,\psi(\mathbf{x}')\,\mathrm{d}^3x', \tag{23.7}$$

where $\hat{\mathscr{H}}_\mathrm{N}^0$ is the Dirac operator $\hat{\mathscr{H}}_\mathrm{N}^0 = \boldsymbol{\alpha}\mathbf{p} + \beta Mc^2$, and M is the nucleon mass while $V(\mathbf{x}, \mathbf{x}'; E)$ is the potential due to the meson which acts on the nucleon.

The relationship between $U(\mathbf{x}', \mathbf{x}; E)$ and $V(\mathbf{x}, \mathbf{x}'; E)$ is also given in Appendix VI. The incompleteness of the theory, which underlies Equa-

† A similar method is used in the dispersion relation theory. The scattering matrices, which are required for establishing the analytical properties of the scattering amplitude, are constructed using perturbation theory and assuming the interaction to be small. The same dispersion relation is used in the region of strong interactions for all values of the interaction constant.

tions (23.6) and (23.7), is expressed in that these nonlocal and complex potentials can be calculated explicitly only in the form of a series in powers of the interaction constant g. If we remove the assumption that the constant g be small, we must make certain assumptions about the form of U or V. A similar situation exists in the nuclear theory in which the complicated nucleon interaction is replaced by a potential well which describes the interaction in an average manner.

If the wavelength of relative particle motion is significantly greater than the dimensions of the region in which the non-localized potentials U and V differ significantly from zero, then Equations (23.6) and (23.7) reduce to the simplified form

$$(E_\pi^2 - \hat{K}^2)\,\psi(\mathbf{x}) = U(\mathbf{x}, E)\,\psi(\mathbf{x}), \tag{23.6'}$$

$$(E_N - \mathscr{H}_N^0)\,\psi(\mathbf{x}) = V(\mathbf{x}, E)\,\psi(\mathbf{x}), \tag{23.7'}$$

where

$$U(\mathbf{x}, E) = \int U(\mathbf{x}, \mathbf{x}'; E)\,\frac{\psi(\mathbf{x}')}{\psi(\mathbf{x})}\,\mathrm{d}^3x', \tag{23.8}$$

$$V(\mathbf{x}, E) = \int V(\mathbf{x}, \mathbf{x}'; E)\,\frac{\psi(\mathbf{x}')}{\psi(\mathbf{x})}\,\mathrm{d}^3x' \tag{23.8'}$$

and the quantity $\psi(\mathbf{x}')/\psi(\mathbf{x})$ under the integral is replaced by unity. Equations (23.6′) and (23.7′) have the form of Schrödinger's or Dirac's equations with a complex potential $U(\mathbf{x}, E)$ or $V(\mathbf{x}, E)$ that is a function of energy.

Equations (23.6) and (23.7) also simplify in another limiting case, for $\lambda \ll a$. Note that, because of a relationship which is valid for short waves

$$\sigma = \pi a^2 (1 - \beta), \tag{23.9}$$

where σ is the elastic scattering cross section and β $(0 < \beta < 1)$, the transmissivity of the nucleon may be written in the form

$$\frac{a}{\lambda\!\!\!^{-}} = \left[\frac{\sigma}{\pi(1-\beta)}\right]^{1/2} \frac{1}{\lambda\!\!\!^{-}} \gg 1\,. \tag{23.9'}$$

In order to satisfy this condition, we write the wave function in a form characteristic of geometrical optics

$$\psi(\mathbf{x}) = \exp\{ikS(\mathbf{x})\}, \tag{23.10}$$

where $S(\mathbf{x})$ is the action function and $k=1/\lambda$. Substituting (23.10) into (23.6′), we collect together terms according to the order of k^2 (we assume that $k^2\to\infty$). Equating the k^2-order terms gives the eikonal equation

$$(\nabla S)^2 = n^2\,, \tag{23.11}$$

where n is the complex index of refraction that satisfies the equation

$$n^2 - 1 = \frac{1}{k^2}\int U(\mathbf{x}, \mathbf{x}'; E)\exp ik\{S(\mathbf{x}') - S(\mathbf{x})\}\, \mathrm{d}^3x'\,. \tag{23.12}$$

Expanding the difference $S(\mathbf{x}')-S(\mathbf{x})$ and the index of the exponential under the integral in powers of $(\mathbf{x}'-\mathbf{x})$ gives

$$S(\mathbf{x}') - S(\mathbf{x}) = \nabla S(\mathbf{x})(\mathbf{x} - \mathbf{x}') + \cdots = n\boldsymbol{\rho}\cos\theta + \cdots, \tag{23.13}$$

where $\rho=\mathbf{x}-\mathbf{x}'$. Then Equation (23.12) becomes

$$n^2 - 1 = \frac{1}{k^2}\int U(x, x+\rho, E)\, \mathrm{e}^{ikn\rho\cos\theta}\, \mathrm{d}^3\boldsymbol{\rho}\,. \tag{23.14}$$

If we replace the index of refraction n on the right-hand side by $n=1$, we obtain the first approximation for n^2-1. This result may be refined by iteration. By assuming $n=\alpha+i\beta$, it is easy to show that a necessary condition for convergence of the iteration is that $k\beta\to$constant for $k\to\infty$.

On the basis of the optical theorem, according to which the imaginary part of the amplitude $T(E_\pi, q)$ of forward elastic scattering (i.e., for $q=0$) is related to the total cross section σ_T by the equation†

$$\operatorname{Im} T(E_\pi, 0) = \frac{k}{4\pi}\,\sigma_T \tag{23.15}$$

where $E_\pi=(m^2+k^2)^{1/2}$, and using the experimental fact that the total cross section of πN interaction does not increase with energy E_π but tends to a constant value, it is possible to show that the quantity k_β decreases as $k\to\infty$, and that thus the condition necessary for convergence in (23.14) is ensured [47].

This optical approach to describing the spatial structure of a meson atmosphere of a nucleon is only one of the many ways of studying this structure [48]. The qualitative and quantitative conclusion that we ob-

† This theorem is a result of the unitary character of the scattering matrix; see Appendix XII.4.

tained using an optical model of the nucleon are in agreement with the basic quantum theory of meson fields.

24. The Structure of Particles in Quantized Field Theory

We shall study the structure of elementary particles using elastic scattering of probe-particles. In modern theory, there are three types of interactions to be considered: strong, electromagnetic, and weak. Thus we have to consider three types of elementary particle structure that correspond to the three types of interaction.

The concept of elementary particle structure differs from the classical concept of the structure of an object. The reason for this is based on the following fact. From experimental data on elastic particle scattering we construct the scattering amplitude T_{fi} which corresponds to the transition of the scattered particles from some initial state i to a final state f. Reconstructing this amplitude from experimental data in itself presents a serious problem [49], and is known under the name of complete experiment.

The amplitude T_{fi} is a function of two invariants, $s=P^2$ (P is the total momentum of the scattered particle) and $t=-q^2$, where q is the momentum that is transfered and is equal to the difference between the final and initial momenta of one of the particles; $q=p_f-p_i$, i.e.,

$$T_{fi} = A(t, s). \tag{24.1}$$

The information contained in the amplitude relates to both states of the particle being studied, i and f, which are characterized by the momenta p_i and p_f. Thus this information is presented to us in the form of interference between two structures which are related in the classical sense to two different states of motion of the particle being studied. On the basis of the well-known "optical" interpretation, we may count on obtaining information on the space-time structure of an object in a scale of b only in the case that the event are considered as acts of scattering in which a momentum $q \geqslant \hbar/b$ is transferred. We may consider the quantity b as the impact parameter. The duration of the encounter will be defined by a time of the order of $\tau = b/v$ where v is the velocity of the particle-probe which is close to the velocity of light.

In a time τ the particle being studied travels a distance of the order of $l \approx (b/v)(p/m)$, where p is its acquired momentum and M its mass. In order that the "structure" will be related to the initial state of the particle-object, $l \ll b$. It is easy to see that this is equivalent to the requirement that $(\hbar/M)(p/q) \ll b$, and because the momenta p and q are of the same order of magnitude, we obtain

$$b \gg \hbar/Mv, \tag{24.2}$$

i.e., we may study the *classical structure* of a particle only on scales of b that are significantly greater than the Compton wavelength of the particle being studied. This qualitative estimate is in agreement with the better-grounded estimate given in Appendix V in connection with the electromagnetic description of nucleon structure discussed in Section 22.

Secondly, to the degree to which the energy of the particle-probe increases and correspondingly to the degree to which the upper limit of the momentum transfer $|q|_{max} = 2p$ increases (where p is the relative momentum of the particle-probe and the particle-object), the relative role played by inelastic processes increases and the picture of the object that is observed by scattering particles acquires a diffraction character from the strongly absorbing body. As noted in [45], if in some region $b < R$ the absorption becomes total, so that this region may be considered as a black body, all information about the structure of this region is destroyed.† The quantity R is called the "elementary length" that depends upon the dynamics of inelastic processes.

Therefore the behavior of the ratio between elastic scattering processes σ_{el} and total cross section σ_T for an unlimited increase in energy $(s \to \infty)$

$$\lim \frac{\sigma_{el}}{\sigma_T} = ? \tag{24.3}$$

may have fundamental meaning in future theory. The question of the behavior of cross sections of inelastic processes σ_{in} is also important. If $\sigma_{in} \to 0$ for $s \to \infty$, this would indicate that the elementary particle becomes transparent at extremely high energies, and "blackness" would not exist. If $\lim \sigma_{in} > 0$, then, in elementary particle encounters, a macroscopic

† We note that for various interactions this state can arise (if it can in general) when the energies and scales are significantly different.

quantity of material could arise because of their high energies, with no limitations on the amount. In this case elementary processes would be able to generate macroscopic processes and a general field of activity in the micro- and macroworld would be discovered [51]. However "*revenons à nos moutons*".† In order to investigate elastic scattering at high energies more deeply, let us assume that we are able to sum up all the higher Feynman diagrams which contribute to the elastic scattering amplitude. Figure 22 shows such diagrams for a field φ which describes

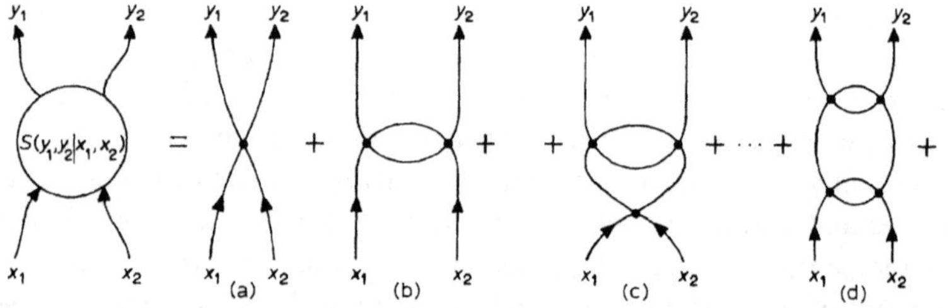

Fig. 22. The left-hand side is the function $s(y_1, y_2 \mid x_1, x_2)$ and the right-hand side is its expansion in terms of Feynman diagrams. (a) is the diagram for the first order in g; (b) is one of the diagrams for the second order (g^2); (c) is one of the third order (g^3); (d) is one of the fourth order (g^4), etc.

uncharged, scalar particles. The energy of interaction in such a field may be written as

$$\hat{W} = \tfrac{1}{3} g \hat{\varphi}^4 , \qquad (24.4)$$

where g is some interaction constant.

Summing such diagrams, which are in the coordinate representation and have two incoming lines (x_1 and x_2) and two outgoing lines (y_1 and y_2), leads to the four-parameter function

$$s = s(x_4, x_3, x_2, x_1), \qquad (24.5)$$

which describes the particle scattering process (two particles initially and two finally).

In Appendix VII we show that the scattering matrix $\hat{S}$ which describes all possible processes with scalar, uncharged particles, may be expressed

† An old french proverb: "Let us return to our sheep".

as a set of scalar functions

$$\hat{S}=\begin{cases} s(x_1), \\ s(x_1, x_2), \\ s(x_1, x_2, x_3), \\ \vdots \\ s(x_1, x_2, \ldots, x_n), \\ \vdots \end{cases} \tag{24.6}$$

Each of these functions is the vacuum average of all possible processes of particle interactions which are accompanied by particle production and annihilation.

Because vacuum is homogeneous, these functions are functions only of the coordinate difference $x_{ik}=x_i-x_k$, in particular $s(x_1)=\text{constant}$. Furthermore, in this case of a scalar field, these functions depend only on the scalar product $x_{ik}x_{mn}$ and hence equally on all x_i.

The matrix elements of the scattering matrix which describes the transition from some initial state $\psi_i(x_1, x_2, \ldots, x_n)$ with n particles to a final state $\psi_f(y_1, y_2, \ldots, y_m)$ with m particles, may be expressed in terms of the functions (24.6) as

$$\begin{aligned} S_{fi} = \mathrm{i}T_{fi} = \int & \bar{\psi}_f(y_1, y_2, \ldots, y_m) \\ & \times s(y_1, y_2, \ldots, y_m \mid x_1, x_2, \ldots, x_n)\, \psi_i(x_1, x_2, \ldots, x_n) \\ & \times \mathrm{d}^4y_1\, \mathrm{d}^4y_2 \cdots \mathrm{d}^4y_m\, \mathrm{d}^4x_1\, \mathrm{d}^4x_2 \cdots \mathrm{d}^4x_n, \end{aligned} \tag{24.7}$$

where the arguments of the function $s(x_1, x_2, \ldots, x_N)$, $N=n+m$, are related to the final state f, denoted now by $y_1, y_2, \ldots, y_m$, and separate characteristics of the arguments $x_1, x_2, \ldots, x_n$, are related to the initial state i. Such a division of the arguments of a function $s(x_1, x_2, \ldots, x_N)$ into parts which belong to the initial and final states allows us to write the arguments in the form of a matrix

$y \backslash x$		x_1	x_1, x_2	x_1, x_2, x_3
	1			
y_1		$s(y_1 \mid x_1)$	$s(y_1 \mid x_1, x_2)$	…
y_1, y_2		$s(y_1, y_2 \mid x_1)$	$s(y_1, y_2 \mid x_1, x_2)$	…
y_1, y_2, y_3		…	…	..

(24.8)

whose columns (y) are related to the final state f, and whose rows (x) to the initial state i. Not all the elements whose sequence is given in the table are different from zero, and not all describe real physical processes with real particles. For example, the interaction (24.4) excludes in general the functions $s(x|y)$ with an odd number of arguments. Elements of the type $s(y_1 \mid x_1, x_2, x_3)$ differ from zero but cannot describe real processes because the rest mass of particles (y_1) has to be greater than $3m$. This contradicts the assumption on particle spectra, in which only particles of rest mass m are allowed. The corresponding element S_{fi} for real particles automatically becomes zero. We shall not consider this characteristic detail.

Let us now return to the most important aspect of the problem. We assume that we can localize a particle in space and time to any degree of accuracy. In other words, we assume that it is possible to take wave functions ψ_f and ψ_i in the form of a product of four-dimensional δ^4-functions

$$\bar{\psi}_f = \prod_{q=1}^{m} \delta^4(y_q - y'_q), \qquad \psi_i = \prod_{q=1}^{n} \delta^4(x_q - x'_q). \tag{24.9}$$

Then the integration in (24.7) may be immediately carried out and we obtain

$$S_{fi} = s(y'_1, y'_2, \ldots, y_m \mid x'_1, x'_2, \ldots, x'_n). \tag{24.10}$$

From this it follows that the function $s(y|x)$ in (24.8) gives the most complete information on the space-time development of the process of particle interaction $N = m + n$. Thus it is this function that must be considered to obtain a complete description of the structure of scalar, elementary particles which interact with each other by emitting and absorbing particles similar to themselves. We may restrict the scope of our description which is related to the concept of particle structure to only those functions $s(y|x)$ from (24.8) that describe elastic scattering, namely, the function $s(y_1, y_2|x_1, x_2)$. In this case we may consider that one of the particles probes the structure of another. This point of view is of course possible, but not obligatory. In one way or another, the description of particle structure based on $\hat{S}$-matrices has a dynamic character that cannot be separated from the processes that cause the structure to become apparent.

This dynamic concept of space-time structure of a particle is shown in Figure 22, using the Feynman diagrams for the special case $\hat{\mathscr{L}} = g\hat{\varphi}^4$.

The function $s(y_1, y_2 \mid x_1, x_2)$ is used as an example. This function is represented by the sum of contributions of the operator (24.4) in powers of g. Each of these contributions is represented as a sum of Feynman diagrams whose number equals the power of g. The first diagram (a) describes the interaction (elastic scattering) of point particles. The other diagrams include virtual particles which form clouds around the original particles $(x_1, x_2 \mid y_1, y_2)$. This cloud has the following dynamic meaning: At each of the vertices particles produce other particles which as a result of successive transformations, begin as the initial particles x_1 and x_2, and result in the final particles y_1 and y_2. Summing all the diagrams and averaging them over vacuum, we account for all possible paths which lead from (x_1, x_2) to (y_1, y_2).

Returning now to real particles, we must consider the three types of interaction. One particle may take part in more than one interaction. For example, a proton interacts strongly with pions, electromagnetically with electrons, and weakly with neutrinos. These interactions correspond to different $\hat{S}$-matrices, and the nature of the particle structure which is reflected in each of these interactions may be unique to one of the interactions. It is highly probable that in the future the inner relationships in these interactions will be found, and then the various structures will be understood as manifestations of the same phenomenon.†

Furthermore, we must remember that the concepts of the space-time structure of particles presented in this section are based on theory which is only well founded for electromagnetic interactions.

Mathematically, this concept is based on the set of functions $s(y|x)$ [see (24.8)], which may be obtained theoretically, but are unobservable quantities. The fact that they are not observable follows from the fact that it is impossible to construct four-dimensional δ-like wave packets in space-time (24.9) and to use them as functions of initial and final state.

In reality the initial state ψ_i and the final state ψ_f are superpositions of plane waves for which $p_0^2 = +(m^2 + p^2)^{1/2}$. Therefore, information about the function $s(y|x)$, which is obtained from knowledge of the matrix elements S_{fi} of the scattering matrix S, is extremely limited. The essence of the matter is contained in the fact that the spectral decomposi-

† In atomic physics the nuclear electron structure is the only basis for understanding of the behavior of atoms in the light of both X-rays and electron beams.

tion of the function s_{fi} in momentum space $\Re(p)$ composes only a small part of the spectral decomposition of the function $s(y|x)$.†

In order to consider this question in greater detail, let us return to the real case of elastic scattering of π-mesons on nucleons. Because of the difference of these particles, we shall write the function $s(y|x)$ in the form

$$s(y\,|\,x) \equiv s(y_2, x_2\,|\,x_1, y_1), \tag{24.11}$$

where x_1 is the coordinate of the nucleon in the initial state

$$\psi_i(x_1) = u_i(p_1)\,\mathrm{e}^{ip_1x_1}, \tag{24.12}$$

and x_2 is its coordinate in the final state

$$\bar{\psi}_f(x^2) = \bar{u}_f(p_2)\,\mathrm{e}^{-ip_2x_2}, \tag{24.13}$$

y_1 and y_2 are the π-meson coordinates in the initial and final states

$$\varphi_i(y_1) = \frac{\mathrm{e}^{ik_1y_1}}{\sqrt{\omega_1}}, \qquad \bar{\varphi}_f(y_2) = \frac{\mathrm{e}^{-ik_2y_2}}{\sqrt{\omega_2}}. \tag{24.14}$$

Let us now introduce the coordinates

$$\begin{aligned} x &= x_1 - x_2, & \xi_1 &= \tfrac{1}{2}X - \tfrac{1}{2}(y_1 + y_2), \\ X &= \tfrac{1}{2}(x_1 + x_2), & \xi_2 &= \tfrac{1}{2}X - (y_1 - y_2), \end{aligned} \tag{24.15}$$

so that

$$s(y_2, x_2\,|\,y_1, x_1) = s(x, \xi_1, \xi_2). \tag{24.16}$$

This last function has the spectral decomposition

$$s(x, \xi_1, \xi_2) = \int \tilde{s}(\alpha, \beta_1, \beta_2)\exp\left[i(\alpha x + \beta_1\xi_1 + \beta_2\xi_2)\right] \times \mathrm{d}^4\alpha\,\mathrm{d}^4\beta_1\,\mathrm{d}^4\beta_2, \tag{24.17}$$

where the integration over the variables α, β_1, β_2 is not restricted to the mass surface.

An evaluation of the matrix element S_{fi} of the function (24.17) for the states

$$\psi_f = \psi_f(x_2)\,\varphi_f(y_2), \qquad \psi_i = \psi_i(x_1)\,\varphi_i(y_1) \tag{24.18}$$

† We already met a similar situation in studying the vertex $\Gamma_\mu(X, x)$ (see Section 22).

leads to the result

$$S_{fi} = \delta^4(p_2 + k_2 - p_1 - k_1)\,\bar{u}_f(p_2) \times \frac{\tilde{s}(-p_1 - k_1, -q, k_1 + \frac{1}{2}q)}{\sqrt{\omega_1\omega_2}}\, u_i(p_1). \tag{24.19}$$

Thus the function $\tilde{s}(\alpha, \beta_1, \beta_2)$ is known only for the values $\alpha = -(p_1 + k_1)$, $\beta_1 = -q$, $\beta_2 = k_1 + \frac{1}{2}q$.

The volume element $d^4\alpha\, d^4\beta_1\, d^4\beta_2$ in (24.17) is

$$d^4\alpha\, d^4\beta_1\, d^4\beta_2 = d^4p_1\, d^4k_1\, d^4q = \frac{\mu\, d\mu\, d^4p_1}{E_N}\frac{m\, dm\, d^4k_1}{E_\pi}\, dq_0\, d^3q. \tag{24.20}$$

where E_N is the energy of the initial nucleon in the center of gravity system, E_π is the energy of the initial meson in the same coordinate system. A substitution of s from (24.19) into (24.17) does not give the complete function $s(x, \xi_1, \xi_2)$, but only its "visible" part Δs as

$$\Delta s(x, \xi_1, \xi_2) = \Delta\Omega \int \tilde{s}\left(-p_1 - k_1, -q, k_1 + \frac{1}{2}q\right) \times \exp\left[-i(p_1 + k_1)x_1 - iq\xi_1 + i\left(k_1 + \frac{q}{2}\right)\xi_2\right] d^3q, \tag{24.21}$$

where $\Delta\Omega$ denotes all the multipliers in (24.20) except for d^3q.

We now consider the special case, assuming $x=0$, $\xi_2=0$ in which the the variable ξ takes on the meaning of the distance between the nucleon and meson, in which

$$\Delta s(0, \xi, 0) = \Delta\Omega \int s(-p_1 - k_1, -q, k_1 + \tfrac{1}{2}q) \times \exp(-iq\xi)\, d^3q. \tag{24.22}$$

If we restrict ourselves to the region in which $k_1 + \frac{1}{2}q \approx k_1$, i.e., the case of a high-energy initial meson and small momentum transfer $\mathbf{q}^2$, then in (24.20) we may neglect q in the expression $\beta_2 = k_1 + \frac{1}{2}q$. Then (24.21) may be considered as a Fourier transform for the variable $\mathbf{q}$. We write

$$\tilde{s}(-p_1 - k_1, -q, k_1)_{q_0=0} = \tilde{s}(E, \mathbf{q}^2), \tag{24.23}$$

where $E = E_N + E_\pi$ is the total energy of the system. The quantity

$$\frac{\Delta s(0, \xi, 0)}{\Delta \Omega} = \int \tilde{s}(E, \mathbf{q}^2) \, e^{iqx} \, d^3 q \tag{24.24}$$

gives us an idea about the structure of nucleon-meson interaction. However, using (24.19), it is more convenient to introduce the quantity

$$V(E, \xi) = \int \tilde{V}(E, \mathbf{q}) \, e^{iq\xi} \, d^3 q \tag{24.25}$$

(where

$$\left.\tilde{V}(E, \mathbf{q}) = \frac{\tilde{s}(E, \mathbf{q}^2)}{E_\pi}\right), \tag{24.26}$$

which is analogous to the interaction potential of the meson and nucleon. In fact, in the limiting case of a small potential $V(E, \xi)$ the scattering amplitude may be written in the form

$$T_{fi}(E, q) = \bar{u}_f(p_f) \, \tilde{V}(E, q) \, u_i(p_i), \tag{24.27}$$

which agrees with (24.19) if the definition (24.26) is used.

We call the quantity $V(E, \xi)$ in (24.25) the *effective potential* of a meson-nucleon interaction [50].

Figure 23 shows experimental data for the differential elastic (π^+p) scattering cross section. If we consider the (π^+p) scattering amplitude as purely imaginary, its real part being small, then we can construct two curves to approximate this experimental data. These curves correspond to the two potentials (24.27):

$$V(E, \xi) = iV_0(E) \exp\left[-\frac{\xi^2}{a^2}\right], \tag{24.28}$$

$$V(E, \xi) = iV_0(E) \frac{\exp(-\xi/a)}{r/a}. \tag{24.28'}$$

The first of these equations agrees better with experimental data than the second (see Figure 23). The length a in both cases is $a = 0.6/0.7 \times 10^{-13}$.

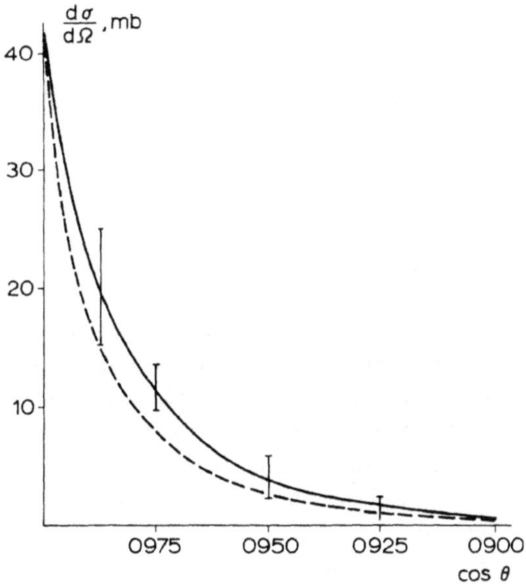

Fig. 23. A comparison of theoretical curves for the differential elastic scattering cross section $d\sigma/d\Omega$ for pions on protons as a function of the scattering angle θ. The solid curve is derived from the effective potential (24.28) and the dotted curve from the potential (24.28′).

Such a value corresponds to the interchange of two π-mesons or a vector ρ-meson.

Thus the "effective potential" $V(E, \xi)$ gives an idea about the structure of a nucleon, which agrees with the concept of a meson atmosphere of a nucleon.

CHAPTER V

CAUSALITY IN QUANTUM THEORY

25. A FEW REMARKS ON CAUSALITY IN THE CLASSICAL THEORY OF FIELDS

The concept of ordering of events contains within it the most important concept of causality which is based on the chronogeometric division of a manifold of events into time-like and space-like parts. Therefore, causality may be considered as a geometric category and the study of causal problems is only one of the possible aspects of geometric analysis. We shall begin our analysis of causality in the microworld with a few remarks on the classical theory of fields.

In the classical theory of linear fields, causality is expressed most directly using a Green's function $\mathfrak{G}$ which represents a field $Q_0(x)$, arising from a point source localized in space and time

$$Q_0(x) = -\,\delta^4(x) \equiv -\,\delta(t)\,\delta^3(\mathbf{x}). \tag{25.1}$$

For a scalar field, the Green's function satisfies the inhomogeneous Klein equation

$$(\Box^2 - m^2)\,\mathfrak{G} = -\,\delta(x). \tag{25.2}$$

Following Appendix I [see (I.12)-(I.14)], we obtain the retarded Green's function $\mathfrak{G}_{\text{ret}}$

$$\mathfrak{G}_{\text{ret}}(x) = \begin{cases} D(x) & \text{for} \quad t > 0, \\ 0 & \text{for} \quad t < 0, \end{cases} \tag{25.3}$$

where the function $D(x)$ coincides with the commutation function defined in (I.3), and the advanced Green's function $\mathfrak{G}_{\text{adv}}$

$$\mathfrak{G}_{\text{adv}}(x) = \begin{cases} 0 & \text{for} \quad t > 0, \\ -\,D(x) & \text{for} \quad t < 0. \end{cases} \tag{25.3'}$$

These equations reduce to a very simple form for the case $m=0$, which corresponds, in quantum theory, to particles with zero rest mass. In this case we obtain from (I.3)

$$\mathfrak{G}_{\text{ret}}(x) = \begin{cases} \dfrac{1}{4\pi}\dfrac{\delta(t-r)}{r} & \text{for} \quad t > 0, \\ 0 & \text{for} \quad t < 0 \end{cases} \tag{25.4}$$

and

$$\mathfrak{G}_{\text{adv}}(x) = \begin{cases} 0 & \text{for} \quad t > 0, \\ -\dfrac{1}{4\pi}\dfrac{\delta(t+r)}{r} & \text{for} \quad t < 0. \end{cases} \tag{25.4'}$$

The solution $\mathfrak{G}_{\text{ret}}(x)$ represents an infinitely narrow impulse which spreads out from the point $r=0$. The solution $\mathfrak{G}_{\text{adv}}(x)$ represents the same impulse that converges into the point $r=0$ (see Figure 24). The first solution describes the impulse which transmits the effect of an event (impulse) from the point $r=0$ to distant points. The second solution

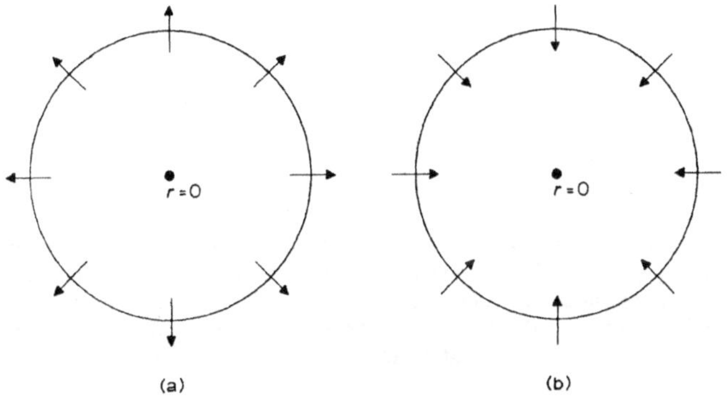

Fig. 24. (a) An infinitely narrow impulse that diverges from the point $r=0$. (b) An infinitely narrow impulse that converges at the point $r=0$

transmits the effect of an event located outside of the point $r=0$ to this point. Where the particle rest mass differs from zero, following this concentrated impulse there extends a "tail" of slower effects which are described by the terms in the function $J_1(-im(x^2)^{1/2})/(im(x^2)^{1/2})$ [see (I.3)]. In corpuscular theory, the existence of this tail might be interpreted as an effect of particles that move with a velocity that is less than the velocity of light c.

Both of the solutions $\mathfrak{G}_{\text{ret}}(x)$ and $\mathfrak{G}_{\text{adv}}(x)$ become zero outside the light cone. This indicates that an interaction at the point $x_1=0$ cannot propagate away from this point with a velocity greater than that of light. Similarly it cannot arrive at the point $x_1=0$ from other points x_2 with a velocity exceeding that of light. Three possible effects are shown in Figure 25. These are (a) the point x_2 lies inside the light cone, in the future relative to the point x_1 (by convention we let $x_2>x_1$); the effect of x_1 on x_2 is described by $\mathfrak{G}_{\text{ret}}(x)$, $x=x_2-x_1>0$; (b) the point x_2 lies inside the cone of absolute future relative to the point x_1 (by convention $x_2<x_1$);

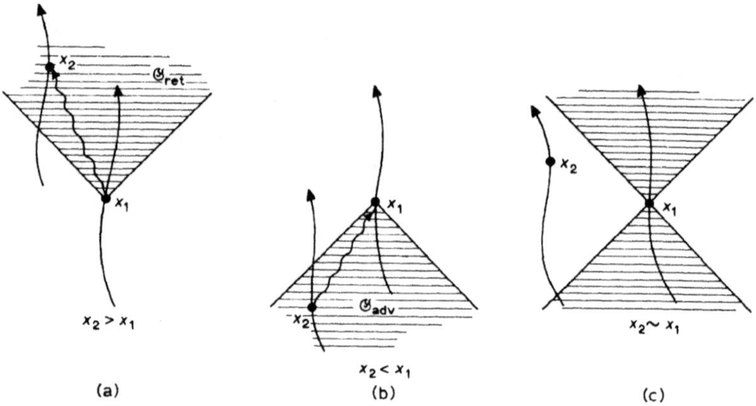

Fig. 25. An illustration of retarded and advanced interactions. (a) A retarded interaction between points x_1 and x_2. The Green's function $\mathfrak{G}_{\text{ret}}$ differs from zero inside the shaded region ($x_2>x_1$). (b) The same situation for an advanced interaction. The Green's function $\mathfrak{G}_{\text{adv}}$ differs from zero inside the shaded region ($x_2<x_1$). (c) The points x_2 and x_1 cannot be related by any interaction.

the effect of x_2 on x_1 in this case is described by $\mathfrak{G}_{\text{adv}}(x)$; (c) when x_2 and x_1 both lie inside the spatial region, i.e. outside both the light cones (by convention $x_2 \sim x_1$), both the Green's functions $\mathfrak{G}_{\text{adv}}(x)$ and $\mathfrak{G}_{\text{ret}}(x)$ become zero so that there is no physical connection between the points x_2 and x_1.

Using the Green's functions we may write the field $\varphi(x)$ which results from any distribution in space and time in terms of a source $Q(x)$. The differential equation for such a field is

$$(\Box^2 - m^2)\,\varphi(x) = -\,Q(x). \tag{25.5}$$

On the other hand, the retarded Green's function $\mathfrak{G}_{\text{ret}}(x_2-x_1)$ satisfies the equation

$$(\Box_2^2 - m^2)\,\mathfrak{G}_{\text{ret}}(x_2 - x_1) = -\,\delta(x_2 - x_1). \tag{25.6}$$

Multiplying this equation by $Q(x_1)$ and integrating over d^4x_1 gives

$$(\Box_2^2 - m^2)\int \mathfrak{G}_{\text{ret}}(x_2 - x_1)\,Q(x_1)\,d^4x_1 = Q(x_2). \tag{25.7}$$

Comparing this with (V.4) shows that the field φ at the point x_2 is equal to

$$\varphi(x_2) = \varphi_0(x_2) + \int \mathfrak{G}_{\text{ret}}(x_2 - x_1)\,Q(x_1)\,d^4x_1, \tag{25.8}$$

where $\varphi_0(x_2)$ is the solution to the homogeneous equation [Eq. (25.5) for $Q=0$].

If the field is formed solely by the source $Q(x_1)$, then $\varphi_0(x_2)\equiv 0$, and in place of (25.8) we have

$$\varphi(x_2) = \int \mathfrak{G}_{\text{ret}}(x_2 - x_1)\,Q(x_1)\,d^4x_1. \tag{25.9}$$

In the simpler case, for $m=0$, on the basis of (25.4) we have

$$\varphi(x_2) = \int \frac{\delta(t_2 - t_1 - r)}{r}\,Q(x_1)\,d^4x_1 \tag{25.10}$$

(here $r=|x_2-x_1|$) or

$$\varphi(x_2) = \int \frac{Q(t - r, \mathbf{x}_1)}{r}\,d^3x_1. \tag{25.11}$$

From this we see that the field $\varphi(x_2)$ at the point $x_2(t_2, \mathbf{x}_2)$ consists of retarded effects that start from the points $\mathbf{x}_1$ at $t_1=t_2-r/c\,(c=1)$. This is the most clear expression of causality in the classical theory of fields.

For $m\neq 0$, according to (I.3) and (25.3) the expression for the field $\varphi(x_2)$ is more complicated, namely,

$$\varphi(x_2) = \int \frac{Q(t_2 - r, \mathbf{x}_1)}{r}\,d^3x_1 + \frac{m}{2} + \int_{\Omega} \frac{J_1\left(m\sqrt{(t_2 - t_1)^2 - r^2}\right)}{\sqrt{(t_2 - t_1)^2 - r^2}}\,Q(t, \mathbf{x})\,dt\,d^3x_1. \tag{25.12}$$

The range of integration in the last integral is limited by conditions $(t_2-t_1)^2>r^2$, $t_2-t_1>0$, in agreement with (25.3). The relationship $v=r/|t_2-t_1|<1\equiv c$ may be considered as the propagation speed of the interactions which occur because the "particles" have a velocity v that is less than the velocity of light.

We now assume that there is a charge distribution $Q_2(x_2)$ in the region around x_2. We defined the quantity

$$W = \int Q_2(x_2)\,\varphi(x_2)\,\mathrm{d}^4x_2 \tag{25.13}$$

as the "action function" (the integral over time of the interaction energy). Denoting the source of the field in the region x_1 by $Q_1(x_1)$, we obtain

$$W = \int Q_2(x_2)\,\mathfrak{G}(x_2-x_1)\,Q_1(x_1)\,\mathrm{d}^4x_2\,\mathrm{d}^4x_1\,, \tag{25.14}$$

where $\mathfrak{G}(x_2-x_1)=\mathfrak{G}_{\text{ret}}(x_2-x_1)-\mathfrak{G}_{\text{adv}}(x_2-x_1)$.

Such a notation accounts symmetrically for the action of one charge on another and, because of (25.3) and (25.3′) it assures a delayed effect of charge located "earlier", on a charge located "later".

We now take the second functional derivative of the action W with respect to the functions of the sources Q_2 and Q_1 at the points x_2 and x_1, respectively. We obtain

$$\frac{\delta^2 W}{\delta Q_2(x_2)\,\delta Q_1(x_1)} = \mathfrak{G}(x_2-x_1)\,. \tag{25.15}$$

From this it follows that

$$\frac{\delta^2 W}{\delta Q_2(x_2)\,\delta Q_1(x_1)} = 0\,, \tag{25.16}$$

if $x_2\sim x_1$, i.e., if the points x_2 and x_1 lie with respect to each other outside the light cone so that the interval $(x_2-x_1)^2$ is spatial. Equation (25.16) expresses the physical fact that the influence of one charge on another cannot propagate with a speed greater than that of light.

In concluding this section we note that the Green's functions, which are convenient for describing causality, are in themselves only mathematical abstractions because the point source

$$Q(x) = g\delta(t)\,\delta^3(\mathbf{x})\,, \tag{25.17}$$

that generates the field described by the Green's function is inconsistent with the concept of charge conservation. The conservation of charge requires that the continuity equation

$$\frac{\partial Q}{\partial t} + \operatorname{div} \mathbf{j} = 0, \tag{25.18}$$

be fulfilled, where $\mathbf{j}$ is the current density corresponding to the charge Q. From this equation it follows that, in the case of a point source $(g\delta(t)\delta^3(\mathbf{x}))$, the current density $\mathbf{j}(x)$ must have the form

$$\mathbf{j}(x) = -g \frac{\mathrm{d}\,\delta(t)}{\mathrm{d}t} \nabla\left(\frac{1}{r}\right), \tag{25.19}$$

because

$$\operatorname{div} \nabla\left(\frac{1}{r}\right) = \nabla^2\left(\frac{1}{r}\right) = \delta^3(\mathbf{x}).$$

Thus the current which corresponds to the point source Q [see (25.17)] appears to be distributed in space, and is highly singular in time (derivative of a δ-function).

A more real case is that of a point source which moves along some trajectory

$$\mathbf{x}(t) = \mathbf{X}(t). \tag{25.20}$$

The charge and current densities in this case have the form

$$Q(x) = g\delta^3[\mathbf{x} - \mathbf{X}(t)], \tag{25.21}$$

$$\mathbf{j}(x) = g \frac{\mathrm{d}\mathbf{X}(t)}{\mathrm{d}t} \delta^3[\mathbf{x} - \mathbf{X}(t)]. \tag{25.21'}$$

It is easy to show that these quantities satisfy the continuity equation (25.18). An insertion of these expressions for $Q(x)$ and $\mathbf{j}(x)$ into (25.9) gives the scalar and vector potentials

$$\varphi(x) = \frac{g}{r[\mathbf{x} - \mathbf{X}(t - r/c)]}, \tag{25.22}$$

$$\mathbf{A}(x) = \frac{g\, \mathrm{d}\mathbf{X}(t - r/c)/\mathrm{d}t}{r[\mathbf{x} - \mathbf{X}(t - r/c)]}, \tag{25.22'}$$

where the points t, $\mathbf{x}$ can be reached by a "signal" from a moving point charge that lies on the surface

$$R^2 = \left[\mathbf{x} - \mathbf{X}\left(t - \frac{r}{c}\right)\right]^2. \tag{25.23}$$

Figure 26 shows the transmission of a signal from one particle (x_1) to another (x_2) by means of a retarded Green's function $\mathfrak{G}_{\text{ret}}(x_2 - x_1)$.

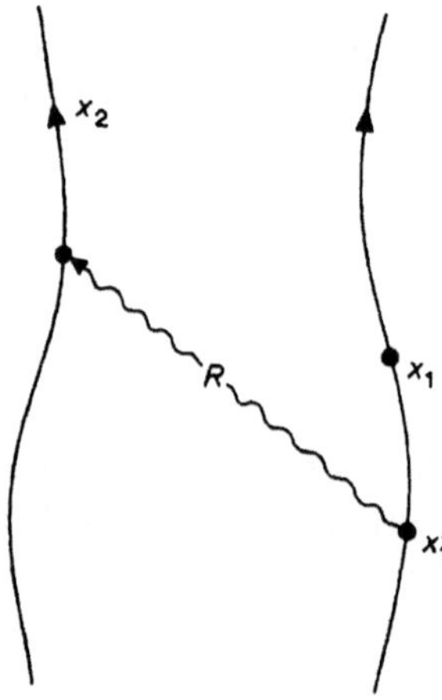

Fig. 26. Retarded interaction between two "point" particles; $x_2 = X_2(t)$ and $x_1 = X_1(t)$ are the locations of these particles at the time $t_2 = t_1 = t$, $x_1' = X_1(t - R/c)$ is the effective location of the particle 1. Here R is the distance between the points x_1' and x_2.

We now consider the potentials $\varphi(x)$ and $\mathbf{A}(x)$ at some distance from the point $\mathbf{X}(t - r/c)$. For definiteness, we assume that the motion of the charge originates in the bounded region $|x| < M$. Then

$$\varphi(x) = \frac{e}{r(\mathbf{x})} - \frac{\mathbf{x}\mathbf{X}(t - r/c)}{r^3} + \cdots, \tag{25.24}$$

$$\mathbf{A}(x) = \frac{e\dot{\mathbf{X}}(t - r/c)}{r(x)}. \tag{25.24'}$$

In this region of space $\mathfrak{R}_4(x)$ the potential $\varphi(x)$ does not have any meaning for causality because the first term $e/r(\mathbf{x})$ is static and, therefore, it is not related to the transmission of any signal. The second term is of the order $O(1/r^3)$. Thus only the vector part of the potential $\mathbf{A}(x)$ can transmit a signal.

We further assume that the current $\mathbf{j} = e\dot{\mathbf{X}}\delta^3(\mathbf{x}-\mathbf{X})$ is generated by an oscillator to which motion is imparted by the force $\mathbf{F}(t)$ of the form

$$\dot{\mathbf{X}} + \gamma\dot{\mathbf{x}} + \omega^2\mathbf{X} = \frac{1}{m}\mathbf{F}(t), \tag{25.25}$$

where ω is the characteristic frequency of the oscillator; m, its mass; and γ the coefficient of friction. The most abrupt motions of the oscillator that we can conceive arises from an instantaneous blow, i.e., in the case when

$$\mathbf{F}(t) = \mathbf{A}\delta(t). \tag{25.26}$$

Noting that

$$\mathbf{X}(t) = \frac{1}{m}\int_{-\infty}^{t} \mathfrak{G}(t-\tau)\,F(\tau)\,\mathrm{d}\tau, \tag{25.27}$$

where $\mathfrak{G}(t)$ is the Green's function of Equation (25.25):

$$\mathfrak{G}(t) = \begin{cases} \dfrac{\exp(-(\frac{1}{2}\gamma)\,t)}{\sqrt{\gamma^2-4\omega^2}}\left(\exp(t\sqrt{\tfrac{1}{4}\gamma^2-\omega^2}) \right. \\ \left. -\exp(-t\sqrt{\tfrac{1}{4}\gamma^2-\omega^2})\right), & t>0, \\ 0, & t<0 \end{cases} \tag{25.28}$$

from (25.26) and (25.27) we find that

$$\mathbf{X}(t) = \frac{\mathbf{A}}{m}\mathfrak{G}(t). \tag{25.29}$$

If $\omega \gg \gamma/2$, is a small friction, then

$$\dot{\mathbf{X}}(t) = \frac{\mathbf{A}}{m}\mathrm{e}^{-\gamma/2t}\cos\omega t; \tag{25.30}$$

and for a large friction

$$\dot{\mathbf{X}}(t) = \frac{\mathbf{A}}{m}\mathrm{e}^{-\gamma t}. \tag{25.31}$$

In the first case, maximum current is obtained for $t \sim 2/\gamma$ and the current flow lasts for about the same length of time. In the second case, maximum current is obtained for $t=0$, and also flows for a time of the order of $t \approx 2/\gamma$. The width of the impulse Δt is defined in both cases as $\gamma/2$.

In this manner the extremely small impulse can be transmitted by an oscillator which has an extremely great friction. This simple estimate illustrates the method by which a source that generates short signals could in practice be obtained.

In classical (macroscopic) physics, no fundamental restrictions are placed on the dimensions of the regions in which sources are localized, or on the frequency of their vibrations, or on their coefficients of friction. In particular the charge may be concentrated in an arbitrarily small region of space [this is expressed in the computation by the δ-function $\delta^3(\mathbf{x}-\mathbf{X}(t))$] and may have an arbitrarily small amplitude of oscillation $\mathbf{A}$ for any frequency ω and any coefficient of friction γ. This situation, which is typical of the consistent theory, does not in itself present any limitations.

The limitations are external ones. Decreasing the size of the sources would in the end lead to sources that have molecular dimensions and thereby would result in a microworld region. The ability to choose sources arbitrarily, independently of their mass, frequency, friction and oscillation amplitude would be significantly decreased. Heisenberg's uncertainty principle, which relates the space-time localization of the source to the uncertainty of its energy ΔE and momentum Δp, is highly significant under these conditions.

26. Causality in Quantum Field Theory

In the quantum theory of fields the propagation of interactions of one region $\mathfrak{R}(x_1)$ to another $\mathfrak{R}(x_2)$ is described by using quantized fields. The operators which represent these fields satisfy the commutation laws [see (14.3) and (16.2)]:
for the scalar complex field $\hat{\varphi}(x)$

$$[\hat{\varphi}(x_2), \hat{\varphi}^+(x_1)] = iD(x_2 - x_1), \tag{26.1}$$

for the scalar real field $\hat{\varphi}(x) = \hat{\varphi}^+(x)$

$$[\hat{\varphi}(x_2), \hat{\varphi}(x_1)] = iD(x_2 - x_1), \tag{26.2}$$

for the vector field $\hat{A}_\mu(x)$

$$[\hat{A}_\mu(x_2), \hat{A}_\nu(x_1)] = iD(x_2 - x_1)\,\delta_{\mu\nu} \tag{26.3}$$

and for the spinor field $\hat{\psi}_\alpha(x)$

$$\{\hat{\psi}_\alpha(x_2), \hat{\bar{\psi}}_\beta(x_1)\} = \mathrm{i}S_{\alpha\beta}(x_2 - x_1). \tag{26.4}$$

The remaining commutators are equal to zero.

The functions $D(x_2-x_1)$ and $S_{\alpha\beta}(x_2-x_1)$ describe the correlation between the fields in various points in space-time. An important property of these functions is the fact that they vanish for space-like intervals, that is for $(t_2-t_1)^2-(\mathbf{x}_2-\mathbf{x}_1)^2<0$. Therefore, the correlation between values of the field that are separated by the spatial interval $x_2 \sim x_1$ is absent.† We may interpret this as an absence of some causal relationship between the values of the field, taken in the regions $\mathfrak{R}(x_2)$ and $\mathfrak{R}(x_1)$ which are located in a space-like manner next to each other so that for all points $x_2 \in \mathfrak{R}(x_2)$ and $x_1 \in \mathfrak{R}(x_1)$, $x_2 \sim x_1$.

In this relationship, the commutation conditions‡ (26.1)–(26.4), which are called the conditions of local commutativity, may be considered as conditions of microcausality in quantum field theory. We note that other formulations of microcausality, which we shall describe later, imply in one way or another that the conditions of local commutativity of the form (26.1)–(26.4) are fulfilled.

The quantum field operators may be divided into two parts; one which contains the operators of particle generation (this is the positive frequency part) and the other which contains the operators of particle annihilation (this is the negative frequency part). In order that we may not make the following discussion cumbersome by using superfluous notation and indices, we turn to the case of a real scalar field $\hat{\varphi}(x)$. The expansion for this field into a positive frequency part $\hat{\varphi}^+(x)$ and a negative frequency part $\hat{\varphi}^-(x)$ has the form

$$\hat{\varphi}(x) = \hat{\varphi}^+(x) + \hat{\varphi}^-(x), \tag{26.5}$$

$$\hat{\varphi}^+(x) = \int \frac{\hat{\mathrm{a}}^+(\mathbf{k})\, \mathrm{e}^{\mathrm{i}(\omega t - \mathbf{kx})}}{\sqrt{2\omega}}\, \mathrm{d}^3k, \tag{26.6}$$

† In the case of the spinor field, the anti-commutator $\{A, B\}$ and not the commutator $[A, B]$ becomes zero. However, using (26.4) it is easy to show that when physically observable bilinear combinations of the form $\psi(x)\, \Gamma \bar{\psi}(x)$ (where Γ is some spinor matrix) are are taken at two points x_1 and x_2, their commutator vanishes.

‡ We shall not discuss here axiomatic studies which allow somewhat less stringent conditions of commutativity with vacuum averages.

$$\hat{\varphi}^-(x) = \int \frac{\hat{a}(\mathbf{k})\, e^{-i(\omega t - \mathbf{kx})}}{\sqrt{2\omega}} d^3k, \tag{26.6'}$$

where $\omega = (\mathbf{k}^2 + m^2)^{1/2}$. From (26.6) and (26.6′) we may see that $\hat{\varphi}^+(x)$ contains only the generation operators $\hat{a}^+(\mathbf{k})$ and $\hat{\varphi}^-(x)$ contains only the annihilation operators $\hat{a}(\mathbf{k})$.† Therefore the operator $\hat{\varphi}^+(x_1)$ is an operator of particle generation at the point x_1, and the operator $\hat{\varphi}^-(x_2)$ is an operator of particle absorption at the point x_2. The operators $\hat{\varphi}^-(x_2)$ $\hat{\varphi}^+(x_1)$ will be the operator of a particle propagating from the point x_1 where it is formed, to the point x_2, where it is absorbed. For this case, $t_2 > t_1$. For $t_1 < t_2$, the operator $\hat{\varphi}^-(x_1)\,\hat{\varphi}^+(x_2)$ describes the same situation for $t_1 > t_2$.

The matrix element of these operators are influence functions in the quantum theory of fields. The only elements that differ from zero are the vacuum averages, i.e.,

$$\langle \Psi_0 |\, \hat{\varphi}^-(x_2)\, \hat{\varphi}^+(x_1)\, | \Psi_0 \rangle \quad \text{for} \quad t_2 > t_1, \tag{26.7}$$

$$\langle \Psi_0 |\, \hat{\varphi}^-(x_1)\, \hat{\varphi}^+(x_2)\, | \Psi_0 \rangle \quad \text{for} \quad t_2 < t_1. \tag{26.7'}$$

Here ψ_0 is a functional of the field which describes the vacuum state. Both expressions may be combined into one if we utilize the concept of chronological ordering of the products, "the T-products," of two operators. By definition, the T-product of a field $\hat{\varphi}(x_1)$ taken at the point x_1 and the field $\hat{\varphi}(x_2)$ taken at the point x_2 is equal to

$$T\hat{\varphi}(x_2)\,\hat{\varphi}(x_1) = \begin{cases} \hat{\varphi}(x_2)\,\hat{\varphi}(x_1) & \text{if} \quad t_2 > t_1, \quad (26.8) \\ \hat{\varphi}(x_1)\,\hat{\varphi}(x_2) & \text{if} \quad t_2 < t_1. \quad (26.8') \end{cases}$$

Note that the T-product is relativistically invariant. In fact, for the time-like interval $x_2 \gtrless x_1$, the inequality $t_1 \gtrless t_2$ is invariant with respect to the Lorentz transformation. For the space-like interval $x_2 \sim x_1$, the operators $\hat{\varphi}(x_1)$ and $\hat{\varphi}(x_2)$, according to (26.2), commute and therefore their order in (26.3) is unimportant. Noting that $\hat{\varphi}^-(x)\,\psi_0(0) = 0$ (the "stability of a vacuum") it is easy to see that both the equations (26.7) and (26.7′) are united in the equation

$$\langle \Psi_0 |\, T\hat{\varphi}(x_2)\,\hat{\varphi}(x_1)\, | \Psi_0 \rangle = iD_c(x_2 - x_1), \tag{26.9}$$

† It is possible to expand any quantized field $\hat{\varphi}$, $\hat{\mathbf{A}}_\mu$, etc. in this manner.

where $D_c(x)$ is the *causal function* that is defined in Appendix I [see (I.5)]. The physical meaning of the equality (26.9) is contained in the fact that the left-hand side is the average of all processes in which a particle is produced at one point and absorbed at another.

The function $D_c(x)$ has the form [see (I.5)]

$$D_c(x) = \frac{1}{4\pi}\,\delta(x^2) - \theta(x^2)\,\frac{m}{8\pi\sqrt{x^2}}\,[I_1(m\sqrt{x^2}) - \mathrm{i}N_1(m\sqrt{x^2})] + \theta(-x^2)\,\frac{im}{4\pi^2\sqrt{-x^2}}\,K_1(m\sqrt{-x^2})\,. \tag{26.10}$$

This function accounts symmetrically for the advanced and retarded interactions, similarly to the classical Green's function $\mathfrak{G}(x)$ in (25.14). It differs significantly from the classical action function in that it is not equal to zero, not even outside the light cone. In fact, for $x^2 = (t_2 - t_1)^2 - (\mathbf{x}_2 - \mathbf{x}_1)^2 < 0$, from (26.10) it follows that

$$D_c(x) = \frac{im}{4\pi\sqrt{-x^2}}\,K_1(m\sqrt{-x^2})\,. \tag{26.10$'$}$$

Note that for $\mathrm{m}(-x^2)^{1/2} \gg 1$, this expression takes on the form

$$D_c(x) = \frac{im}{4\pi\sqrt{-x^2}}\left(\frac{\pi}{2}\,\frac{1}{m\sqrt{-x^2}}\right)^{1/2} \mathrm{e}^{-m\sqrt{-x^2}}, \tag{26.10$''$}$$

that is, outside the light cone the function $D_c(x)$ decreases exponentially over a distance of a Compton wavelength of the particle $\hbar/mc$ [in equations (26.10), (26.10′) and (26.10″) we use $\hbar = 1$, $c = 1$].

For the case $m = 0$, outside the light cone we have

$$D_c(x) = \frac{\mathrm{i}}{4\pi x^2}, \tag{26.11}$$

that is, $D_c(x)$ decreases outside the light cone more slowly than for $m \nless 0$ [see Appendix I, Equations (I.6) and (I.8)].

At first glance, this characteristic of the function $D_c(x)$ may appear to contradict causality, insofar as interactions can propagate outside the light cone. We shall now show that because of Heisenberg's uncertainty principle, such a contradiction does not in fact arise. To prove this important proposition we consider two microsystems (particles or particle systems) A and B that are separated from each other in space.†

Let an event a occur in system A, and an event b in system B. Let event a precede event b so that $x_a < x_b$, where x_a is the coordinate of the event a. We assume that events a and b may be distributed in space time regions $\mathfrak{R}_a$ and $\mathfrak{R}_b$, and so x_a, x_b are points in these regions (all or part).

We shall consider that b is a consequence of event a (cause), if the system A as a result of the event a transmits a "signal" to the system B which initiates the event b in B. Thus we assume that in this causal interaction process the energy E_A of the system A decreases (it looses energy by sending out the signal), and the energy E_B of the system B increases by receiving the signal. Correspondingly, the momenta of the systems must change. In other words, the signal carries a positive energy $\varepsilon = \hbar\omega$ and a corresponding momentum $\mathbf{p} = \hbar\mathbf{k}$. These assertions have meaning only in the case that the energy spread ΔE and momentum spread $\Delta p \approx \Delta E/v$ (v is the velocity of the system) in the systems A and B are small in comparison to the quantum energy and the momentum

$$\Delta E_A \ll \hbar\omega, \qquad \Delta E_B \ll \hbar\omega, \tag{26.12}$$

$$\Delta p_A \approx \frac{\hbar}{L} \ll \frac{\hbar\omega}{c}, \qquad \Delta p_B \approx \frac{\hbar}{L} \ll \frac{\hbar\omega}{c}. \tag{26.12'}$$

where L denotes the dimensions of the region in which the microsystems A and B are located (we consider these dimensions to be the same). If these inequalities are not fulfilled, we cannot specify which of the systems, A or B, sends out the signal and which absorbs it.

We see that a coordination of events in terms of cause and effect requires *simultaneous space-time localization of microsystems and that the energies E and momenta* **p** *be sufficiently well defined.* This fact has a fundamental meaning for causal problems in the microworld. We may study the causal relationship only between two wave packets which are diffuse and are not sharply defined objects in space-time. Therefore, the causal relationship and causal order of events in the microworld may be determined only to within the limits of certain "tolerances", the magnitudes of which are defined by the spatial dimensions of the wave packet $\Delta x = L$. This uncertainty in spatial dimensions corresponds to the un-

† In the opposite case we must consider $A + B$ as one system C that cannot be divided.

certainty in time $\Delta t \gtrsim \Delta x/c$. These considerations have an important meaning for the definition of the concept of macroscopic causality in elementary particle theory (see Chapter 6).

We shall now illustrate these general considerations with a simple example. We shall consider two wave packets (see Figure 27a). The first packet is located inside the space-time region $\mathfrak{R}(x_1)$ and has a spatial extent of $\Delta x \approx L$ and a time duration $\Delta t \approx T$. We shall consider that this wave packet has an energy E to an accuracy of $\Delta E \approx \hbar/T$ and a

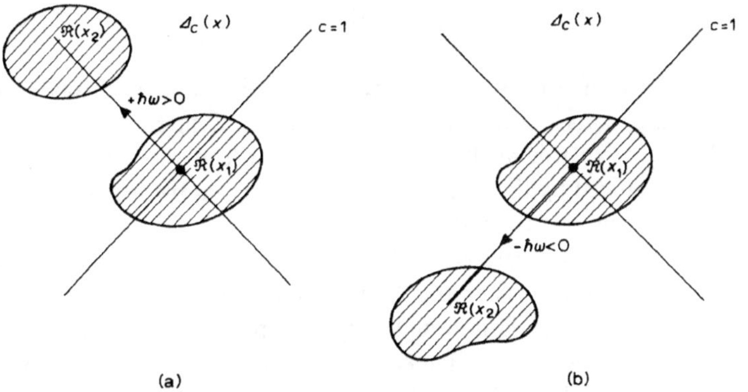

Fig. 27. (a) The packet in the region $\mathfrak{R}(x_2)$ is located later than the packet located in the region $\mathfrak{R}(x_1)$ ($\mathfrak{R}(x_2) > \mathfrak{R}(x_1)$). The interaction of the packets depends on the packet in $\mathfrak{R}(x_1)$ emitting a quantum of energy $\hbar\omega$ (>0) which is then absorbed by the packet in $\mathfrak{R}(x_2)$. (b) The packet in $\mathfrak{R}(x_2)$ is located earlier that the packet in $\mathfrak{R}(x_1)$ ($\mathfrak{R}(x_2) < \mathfrak{R}(x_1)$). In this case the interaction takes place by the packet in $\mathfrak{R}(x_1)$ emitting a quantum of negative energy $\hbar\omega(<0)$ and this being absorbed by the packet in $\mathfrak{R}(x_2)$. This process is equivalent to the packet in $\mathfrak{R}(x_2)$ emitting a quantum with positive energy which is then absorbed by the packet in $\mathfrak{R}(x_1)$.

momentum P_1 to an accuracy $\Delta P_1 \approx \hbar/L$ (we assume that $E_1 \gg \Delta E_1$, $P_1 \gg \Delta P_1$). The second packet is concentrated in the space-time region $\mathfrak{R}(x_2)$. We consider this packet as having the same length L (and duration), an energy $E_2 \gg \Delta E_2 \approx \hbar/T$ and a momentum $P_2 \gg \Delta P_2 \approx \hbar/L$.

The charge of the first packet $Q(x_1)$, which produces the field $\varphi(x_2)$, may be written in the form

$$Q(x_1) = g\left(\frac{t_1}{T}\right)\rho\left(\frac{\mathbf{x}_1}{L}\right)e^{i\omega t_1}, \tag{26.13}$$

where the spatial part – the function $\rho(\mathbf{x}_1)$ – differs significantly from zero in the region $\mathfrak{R}_3(\mathbf{x}_1) \approx L^3$ near the point $\mathbf{x}_1$. ω is the characteristic frequency of this quantum jump which occurs in the wave packet. The function $g(t_1/T)$ differs significantly from zero in the region $\mathfrak{R}_1(t_1) \approx T$, $\mathfrak{R}(x_1) = \mathfrak{R}_3(\mathbf{x}_1) \times \mathfrak{R}_1(t_1)$. We may consider this as the magnitude of the charge of the wave packet, which switches on and off adiabatically (i.e., in a time $T \gg 1/\omega$), about the moment in time $t_1 = 0$. We may, for example, write

$$g\left(\frac{t_1}{T}\right) = \frac{1}{\pi} \frac{T}{t_1^2 + T^2}, \qquad \rho\left(\frac{\mathbf{x}_1}{L}\right) = \frac{1}{\pi} \frac{L}{\mathbf{x}_1^2 + L^2}. \tag{26.14}$$

This process of adiabatic switching on and off of a charge violates the law of charge conservation and, therefore, it has a purely formal meaning. We resort to this artificial process only to separate out the effect which is produced by the field source, not only from the region of space $\mathfrak{R}_3(\mathbf{x}_1) \approx L^3$, but also from the specific time $\approx T$.

A far more realistic case would be that of a wave packet which moves along some trajectory. Then the "switching on" and "switching off" of the region $\mathfrak{R}_3(\mathbf{x}_1)$ would automatically take place in the time $T \approx L/v$, where v is the velocity of the packet. The adiabatic nature of the switching on and off of the region $\mathfrak{R}_3(\mathbf{x}_1)$ will be generated if

$$T \approx \frac{L}{v} \gg \frac{1}{\omega} \tag{26.15}$$

or

$$L \gg \frac{v}{c} \lambda\!\!\!^{-}, \tag{26.15'}$$

where $\lambda\!\!\!^{-}$ is the wavelength of the quantum that is emitted or absorbed by the wave packet.

We shall now consider the quantum analogue of the action function W (25.14). By definition the causal function (26.9), in quantum theory, plays the part of the Green's function which relates events in one region of space-time with events in another. Therefore,

$$W = \mathrm{i} \int Q(x_2)\, D_c(x_2 - x_1)\, Q(x_1)\, \mathrm{d}^4x_2\, \mathrm{d}^4x_1\,. \tag{26.16}$$

From this we obtain

$$\frac{\delta^2 W}{\delta Q(x_2)\,\delta Q(x_1)} = D_c(x_2 - x_1) \tag{26.17}$$

and in contrast to (25.14) this quantity does not vanish outside of the light cone.

In order to illustrate the properties of the quantity W, we compute the field $\varphi(x_2)$ produced by the packet $Q(x_1)$ in the region $R(x_2)$:

$$\varphi(x_2) = \int D_c(x_2 - x_1)\, Q(x_1)\, \mathrm{d}^4 x_1 .^{\dagger} \tag{26.18}$$

The most weakly attenuated function $D_c(x)$ outside of the light cone is one for $m=0$, that is for a field which does not have particles whose rest mass differs from zero. Therefore, we shall use this case in the following computations.

For $m=0$, according to (I.6) and (I.8) we have

$$\begin{aligned} D_c(x_2 - x_1) &= \delta^{+}_{\mathrm{ret}}(x_2 - x_1) + \delta^{-}_{\mathrm{adv}}(x_2 - x_1) \\ &= \frac{1}{2\pi}\int\limits_0^\infty \exp\left[\mathrm{i}\nu(t_2 - t_1 - r)\right] \mathrm{d}\nu \\ &\quad - \frac{1}{2\pi}\int\limits_0^\infty \exp\left[-\,\mathrm{i}\nu(t_2 - t_1 + r)\right] \mathrm{d}\nu , \end{aligned} \tag{26.19}$$

where $r = |\mathbf{x}_2 - \mathbf{x}_1|$. Placing this expression for $D_c(x_2 - x_1)$ in equation (26.18), and using (26.13) we obtain

$$\begin{aligned} \varphi(x_2) = \frac{1}{8\pi}\int \frac{\rho(\mathbf{x}_1/L)\,\mathrm{d}^3 x_1}{r} \int\limits_0^\infty \frac{\mathrm{d}\nu}{2\pi} \int\limits_{-\infty}^{+\infty} \mathrm{d}t_1 \\ \times \left\{\left[\exp(\mathrm{i}\omega t_1 + \mathrm{i}\nu(t_2 - t_1))\exp(-\,\mathrm{i}\omega r)\, g\left(\frac{t_1}{T}\right)\right]\right. \\ \left. + \left(\exp(\mathrm{i}\omega t_1 - \mathrm{i}\nu(t_2 - t_1) - \mathrm{i}\omega r)\, g\left(\frac{t_1}{T}\right)\right)\right\}. \end{aligned} \tag{26.20}$$

† Because of the symmetry properties of the function $D_c(x_2 - x_1)$ we could also have computed the field $\varphi(x_1)$ which is produced by the charge $Q(x_2)$ in $\mathfrak{R}(x_1)$. The result would be symmetric with respect to the regions $\mathfrak{R}(x_1)$ and $\mathfrak{R}(x_2)$.

Carrying out the integration we obtain

$$\varphi(x_2) = \frac{1}{8\pi}\int \frac{\rho(\mathbf{x}_1/L)}{r}\,\mathrm{d}^3x_1\,\exp\left[\mathrm{i}\omega(t_2 - r)\right] g\left(\frac{t_2 - r}{T}\right) + \frac{1}{8\pi}\int \frac{\rho(\mathbf{x}_1/L)}{r}\,\mathrm{d}^3x_1\,\exp\left[-\mathrm{i}\omega(t_2 + r)\right] g\left(\frac{t_2 + r}{T}\right). \tag{26.21}$$

In order to separate the events in the space-time regions $\mathfrak{R}(x_1)$ and $\mathfrak{R}(x_2)$ it is necessary that $t_2 \gg T$ and $r \gg L$. From this it follows that for $t_2 > 0$, only the first term in (26.21) is important for it is the only term that differs noticeably from zero for $t_2 = r$ (to an accuracy $\Delta t \approx T$). Therefore, this term corresponds to the retarded interaction ($t_2 = r/c$, $c = 1$ is the speed of light). The condition $r \gg L$ leads to the inequality $\omega r/c \gg 1$, i.e., $r \gg \lambda\!\!\!^{-}$.

In this manner the necessity for space-time localization to be compatible with the assumption that the energy and momentum of a wave packet may be sufficiently well defined, leads automatically to the fact that causality is distinguished *only in a wave zone.*

We must still consider the case $t_2 < 0$. This does not lead to anything substantially new. In this case the second term in (26.20), which describes advanced effects, is important. However, from (26.21) we see that the sign of the frequency ω of the quantum which transfers the interaction changes to the opposite sign. Therefore, for $t_2 < 0$ in the region $\mathfrak{R}(x_1)$ a quantum is emitted with negative energy from the region $\mathfrak{R}(x_1)$ and goes to the region $\mathfrak{R}(x_2)$ by the advanced field. The packet $Q(x_2)$ "absorbs" the quantum with negative energy $-\hbar\omega$. This description is obviously equivalent to the assertion that for $t_2 < 0$ the wave packet $Q(x_2)$ emits a quantum with positive energy which goes over to the retarded wave packet where it is absorbed.

Thus the function $D_c(x_2 - x_1)$ describes a process of interaction by means of a signal with positive energy which is symmetric with respect to both the space-time regions $\mathfrak{R}(x_1)$ and $\mathfrak{R}(x_2)$. This process is illustrated in Figure 27. As for the "anomalous" behavior of the causal function $D_c(x_2 - x_1)$ outside the light cone, it is obvious from the foregoing calculations that the uncertainty relationship does not allow particles to be found violating any of the laws of causality [52, 53].

27. The Propagation of a Signal "Inside" a Microparticle

In the preceding section we showed that although the causal function $D_c(x)$ defining the propagation of interactions in the quantum field theory does not vanish outside the light cone, it is nevertheless impossible to find signals propagating with velocities greater than the velocity of light, particularly with infinitely large velocities.

This "paradox" of the D_c-function was investigated for the case of the interaction between two point particles.

A similar paradox arises for microparticles that have finite dimensions. The essence of the matter is contained in the fact that the form factor of the particles $F(q)$ (Section 22), where q is the momentum transfer, does not vanish for spacelike values of q and therefore a rigid spatial distribution of charges and currents may be assigned to the particle

$$\rho(\mathbf{x}) = \int F(q)\, e^{i\mathbf{qx}}\, d^3q. \tag{27.1}$$

These are able to transmit the signal with an infinitely high speed. In order to explain this situation, we limit our investigation to the interaction of a virtual photon with a spinor particle.

We shall describe the photon field k by the vector potential $A_\mu(x)$ and the spinor field of the particle, which has a mass, by the operator $\hat{\varphi}(x)$. To the first approximation in the charge e, the interaction operator may be written in the form

$$\hat{W} = \int \hat{\bar{\psi}}(x_3)\, \hat{J}_\mu(x_3, x_2, x_1)\, \hat{\psi}(x_1)\, A_\mu(x_2)\quad d^4x_3\, d^4x_2\, d^4x_1, \tag{27.2}$$

where $\hat{J}_\mu(x_3, x_2, x_1)$ is the density of the current induced by the spinor particle (see Figure 28). The amplitude of the probability of the particle transition from the state i to the state f caused by this interaction is equal to

$$T_{fi} = \frac{i}{\hbar} \langle f|\, \hat{W}\, |i\rangle = \int \langle f|\, J_\mu(x_2)\, |i\rangle\, A_\mu(x_2)\, d^4x_2, \tag{27.3}$$

where $\langle f|J_\mu(x_2)|\, i\rangle$ is the matrix element of the current density $\hat{J}_\mu(x_3,$

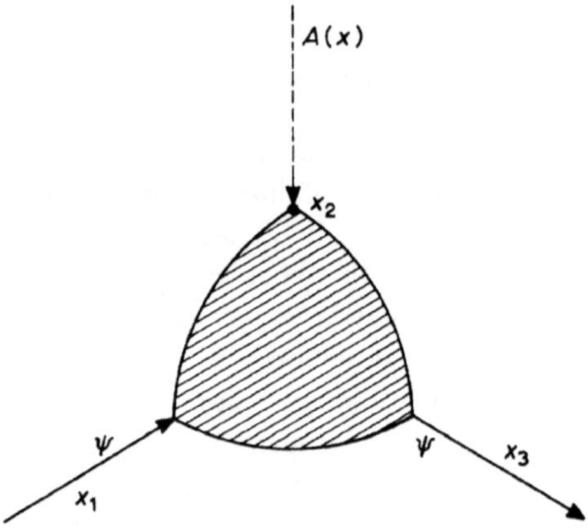

Fig. 28. A graphical representation of the current $J_\mu(x_3, x_2, x_1)$ for a spinor particle which has finite dimensions. x_1 is the particle entry, x_2 its exit, and at x_3, it interacts with the external field $A(x)$.

$x_2, x_1)$ which refers to the transition $i \to f$:

$$\langle f| J_\mu(x_2) |i\rangle = \int \bar{\psi}_f(x_3)\, \hat{J}_\mu(x_3, x_2, x_1)\, \psi_i(x_1)\, d^3x_3\, d^3x_1 . \tag{27.4}$$

We shall consider the wave functions $\bar{\psi}_f$ and ψ_i as arbitrary with the restriction that they are superpositions of states with positive energies:

$$\bar{\psi}_f(x_3) = \int c_f^*(p_3)\, \bar{u}_f(p_3) \exp(-ip_3x_3)\, d^3p_3 , \tag{27.5}$$

$$\psi_i(x_1) = \int c_i(p_1)\, u_i(p_1) \exp(ip_1x_1)\, d^3p_1 ; \tag{27.5'}$$

Here $\bar{u}_f$, u_i are Dirac's bispinors, $\bar{c}_f^*$, c_i are amplitudes of states with momenta p_3, p_1 and p_3^0, $p_1^0 > 0$. Denoting $p_3 - p_1 = q$ and using the relationship

$$\begin{aligned}\langle p_3| \hat{J}_\mu(x_2) |p_1\rangle = & \int \exp(-ip_3x_3)\, \hat{J}_\mu(x_3, x_2, x_1) \\ & \times \exp(-ip_1x_1)\, d^4x_3\, d^4x_2\, d^4x_1 \\ = & \hat{J}_\mu(q) \exp(iqx_2) ,\end{aligned} \tag{27.6}$$

we obtain from (27.4), (27.5) and (27.5′)

$$\langle f|\,\hat{J}_\mu(x_2)\,|i\rangle = \int c_f^*(p_3)\,c_i(p_1)\,\bar{u}_f(p_3)\,\hat{J}_\mu(q)\,u_i(p_1)\,\mathrm{d}^3p_1\,\mathrm{d}^3p_3\,, \tag{27.7}$$

where $\hat{J}_\mu(q)$ is the current operator in the momentum representation. According to (22.15) this operator may be written in the form

$$\hat{J}_\mu(q) = eF_1(q^2)\,\gamma_\mu + \mu\sigma_{\mu\nu}q_\nu F_2(q^2), \tag{27.8}$$

where e is the charge of the particle, μ is its magnetic moment and $F_1(q^2)$ $F_2(q^2)$ are the electric and magnetic form factors of the particle. When defined in this manner, the quantities F_1 and F_2 are dimensionless and must depend only on the ratio q^2/q_m^2 where q_m is some value of q which specifies the rate of decrease of F_1 and F_2 for $|q|\to\infty$. The particle dimension R is related to q_m by the equation $R=\hbar/q_m$.

In order that we may investigate the propagation of a signal inside a particle, we shall consider the states ψ_f and ψ_i as localized wave packets. We shall also consider the photon field A_μ as localized.† For definitness, we write

$$c_f(p) = \frac{b_f^{3/2}}{\pi^{3/2}}\exp\{-\tfrac{1}{2}b_f^2(\mathbf{p}-\mathbf{p}_f)^2 - \mathrm{i}\mathbf{p}\mathbf{x}_f\}, \tag{27.9}$$

$$c_i(p) = \frac{b_i^{3/2}}{\pi^{3/2}}\exp\{-\tfrac{1}{2}b_i^2(\mathbf{p}-\mathbf{p}_i)^2 - \mathrm{i}\mathbf{p}\mathbf{x}_i\}, \tag{27.9′}$$

$$\begin{aligned} A_\mu(x) &= \alpha_\mu\exp\{-\mathrm{i}\,[\mathbf{k}(\mathbf{x}-\mathbf{X}) - k_0(t-T)]\} \\ &\quad\times\exp\left\{-\frac{1}{2a^2}[\mathbf{x}-\mathbf{X}]^2 - \frac{1}{2a^2}[\mathbf{x}-(\mathbf{X}-\mathbf{u}T)]^2\right\}. \end{aligned} \tag{27.9″}$$

The wave functions (27.5) and (27.5′) represent wave packets which have average momenta $\mathbf{p}_f$ and $\mathbf{p}_i$, and are localized at $t=0$ about $\mathbf{x}=\mathbf{x}_f$, $\mathbf{x}_i$ with rms deviations

$$\Delta\overline{(\mathbf{p}-\mathbf{p}_f)^2} = \frac{\hbar^2}{b_f^2}, \qquad \Delta\overline{(\mathbf{x}-\mathbf{x}_f)^2} > \alpha b_f^2 + \beta\frac{\hbar}{mc}b_f\,, \tag{27.10}$$

where the coefficients α and $\beta\approx 1$, and m is the particle mass (see Section 13). The same relationship holds for (27.9′).

† See Appendix VIII, Equation (VIII.4).

The field $A_\mu(x)$ represents a wave packet of virtual photons with the characteristic wave vector $k(\mathbf{k}, k_0)$ and "velocity" of propagation $\mathbf{u}$, which is localized at $T=0$ around $\mathbf{x}=\mathbf{X}$ in a region whose width is of the order of $\Delta x \approx a$, $\Delta t \approx a/c$. The choice of this form for the virtual photon packet corresponds to the assumption that it could be emitted by a real physical particle.

We note that if we could have used infinitely narrow packets in space and time described by the δ-functions $\bar{\psi}_f(x_3)=\delta^4(x_3-x_f)$, $\psi_i(x_1)=\delta^4(x_1-x_i)$, $A\approx\delta^4(x_2-X)$, then from (27.2)–(27.4) the amplitude of the transition would be immediately equal to $J_\mu(x_f, X, x_i)$, and we would have established the fact that signals can propagate with the forbidden velocity.

We may not, however, use wave packets whose dimensions are arbitrarily small. In fact the ability of a particle to receive signals is contained in the *established fact* that its momentum changes from $p_i \to p_f$. This change has to be greater than the momentum dispersion $\Delta p_i \approx \hbar/b_i$, $\Delta p_f \approx \hbar/b_f$, $\Delta k \approx \hbar/a$ in the initial and final packets, i.e., (considering $b_f \approx b_i \approx b$)

$$|\mathbf{p}_f - \mathbf{p}_i| = |q_{fi}| \gg \frac{\hbar}{b}, \qquad |\mathbf{k}| \gg \frac{\hbar}{a}. \tag{27.11}$$

From this we obtain the inequality

$$|\mathbf{p}_f|, |\mathbf{p}_i| \gg \frac{\hbar}{b}. \tag{27.11'}$$

On the other hand, the dimensions of all three of these packets must be smaller than the dimensions of the particles.

$$b_f \ll R, b_i \ll R, a \ll R. \tag{27.12}$$

Because the conditions (27.11)–(27.12) do not contradict quantum mechanics, then at first glance it appears that a causal possibility exists of detecting a signal that travels inside the particle with a velocity that is greater than that of light. However, further calculations show that this is in fact an illusion.

In order to prove this assertion, we calculate the probability amplitudes T_{fi} for the wave packets ψ_f, ψ_i and A that satisfy the conditions (27.11), (27.11′) and (27.12). This amplitude defines the intensity of the signal that is received at the center of the particle and in its peripheral regions.

According to (27.3) we have

$$T_{fi} = \frac{i}{\hbar}\frac{b^3}{\pi^{3/2}} \int d^3x_2\, dt_2 \int d^3p_3\, d^3p_1$$
$$\times \exp\{-\tfrac{1}{2}b^2[(\mathbf{p}_3 - \mathbf{p}_f)^2 + (\mathbf{p}_1 - \mathbf{p}_i)^2] + iqx_2\}$$
$$\times u_f^*(p_3)\, F(q)\, u_i(p_1) \exp[-ik(x_2 - x_1)]$$
$$\times \exp\left[-\frac{1}{2a^2}\{(\mathbf{x}_2 - \mathbf{X})^2 + [(\mathbf{x}_2 - \mathbf{u}t) - (\mathbf{X} - \mathbf{u}T)]^2\}\right]. \qquad (27.13)$$

In order to avoid unnecessary calculations we set $\mathbf{x}_i = \mathbf{x}_f = 0$ and b_f, b_i $=$b. We shall also denote $F(q) = a_\mu \gamma_4 \hat{J}_\mu(q)$. If the dimensions of the packets b and a are much greater than the quantities $\hbar/p_f$, $\hbar/p_i$ and $\hbar/k$, then the integration in (27.13) may be performed immediately and it gives the usual result of scattering theory:

$$T_{fi} = \frac{2\pi i}{\hbar}\, \delta^4(k - q)\, u_f^*(p_f)\, F(q_{fi})\, u_i(p_i)\,. \qquad (27.14)$$

However the coordinates of the place from which the signal is emitted (point x_2) and the places at which it is received (points x_1 and x_3) do not appear in this equation at all and, therefore, they remain unidentified.

We now turn to the case of packets having small dimensions a and b which nevertheless satisfy the conditions (27.11), (27.11′) and (27.12). Because of the assumption $|q_{fi}| \gg \hbar/b \gg q_m$ (the momentum of the particle), the factor $F(q)$ may be taken out from the integral. We also take the slowly varying quantities $u_f^*(p_f)$, $u_i(p_i)$ out from under the integral.

Further, we note (see Appendix VIII) that

$$q_0 = E(\mathbf{p}_3) - E(\mathbf{p}_1) = E_f - E_i + \mathbf{V}_f\boldsymbol{\xi} - \mathbf{V}_i\boldsymbol{\eta} + \cdots, \qquad (27.15)$$

where

$$\boldsymbol{\xi} = \mathbf{p}_3 - \mathbf{p}_f, \quad \boldsymbol{\eta} = \mathbf{p}_1 - \mathbf{p}_i, \quad \mathbf{V}_f = \nabla E_f, \quad \mathbf{V}_i = \nabla E_i\,. \qquad (27.15')$$

Integrating now over ξ and η we obtain

$$T_{fi} = \frac{i}{\hbar}\, u_f^*(p_f)\, F(q_{fi})\, u_i(p_i) \exp[i(\mathbf{k}\mathbf{X} - k_0 T)] \int d^3x\, dt e^{-\Phi(\mathbf{x}, t)}, \qquad (27.16)$$

where

$$\Phi(\mathbf{x}, t) = \mathrm{i}(\mathbf{k} - \mathbf{q})\,\mathbf{x} - \mathrm{i}(k_0 - q_0)\,t + \frac{1}{2b^2}(\mathbf{x} - \mathbf{V}_i t)^2 + \frac{1}{2a^2}(\mathbf{x} - \mathbf{X})^2 + \frac{1}{2a^2}[(\mathbf{x} - \mathbf{u}t) - (\mathbf{X} - \mathbf{u}T)]^2 \tag{27.17}$$

where the notation has been shortened to $\mathbf{q}_{fi} = \mathbf{q}$, $E_f - E_i = q_0$.

In this manner the amplitude T_{fi} is defined by four wave packets which overlap in space and time. This situation is illustrated in Figure 29. It is

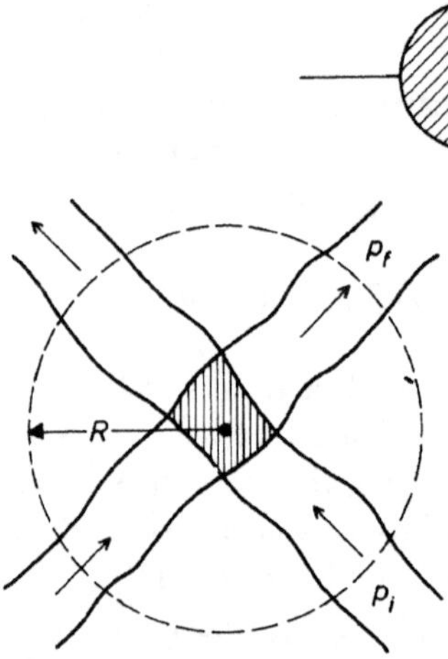

Fig. 29. An illustration of a signal propagation inside a microparticle. Overlapping wave packets in the initial (p_i) and final (p_f) states resemble a δ^4-like source. A represents a packet of photons. The interaction arises when all these packets overlap.

essential that these packets at the points $\mathbf{x}$, $t \approx 0$ and $\mathbf{x} \approx \mathbf{X}$, $t \approx T$ are not only bounded in space but also in time so that they to a certain extent resemble δ like signals of the form $\delta^4(x)$.

Performing the last integral in (27.16) we find that

$$T_{fi} = \frac{\mathrm{i}}{\hbar} u_f^*(\mathbf{p}_f)\,F(q_{fi})\,u_i(\mathbf{p}_i) \exp[\mathrm{i}(\mathbf{kX} - k_0 T)] \times \exp[-\Psi(\mathbf{X}, T)], \tag{27.18}$$

where $\psi(\mathbf{X}, T)$ is equal to (see Appendix VIII)

$$\begin{aligned}\Psi(\mathbf{X}, T) = {} & \frac{1}{2a^2}\mathbf{X}^2 + \frac{1}{2a^2}(\mathbf{X} - \mathbf{u}T)^2 - \frac{C^2}{a^4\Delta^2}[2\mathbf{X} - \mathbf{u}T]^2 \\ & - \frac{A^2}{a^4\Delta^2}(\mathbf{X} - \mathbf{u}T, \mathbf{u})^2 - \frac{B}{a^4\Delta^2}[(2\mathbf{X} - \mathbf{u}T) \\ & \times (\mathbf{X} - \mathbf{u}T, \mathbf{u})] + \frac{C^2}{\Delta^2}(\mathbf{k} - \mathbf{q})^2 \\ & + \frac{A^2}{\Delta^2}(k_0 - q_0)^2 + \operatorname{Im}\Psi. \end{aligned} \tag{27.19}$$

The coefficients A^2, B^2, C^2, Δ^2 are functions of the widths of the packets a and b and of the three velocities $\mathbf{V}_i$, $\mathbf{V}_f$, $\mathbf{u}$. This quadratic form is positive definite and the exponential factor $\exp[-\Psi(\mathbf{X}, t)] \to 0$ for large $|\mathbf{X}|$ as $\exp[-X^2/d^2]$ where $d^2 \approx a^2$ (or $\approx b^2$). This indicates that the transfer of a signal from the periphery to the center of a particle may be observed only if the distance between the periphery and the center $|\mathbf{X}|$ is not greater than the width of the packets a or b.

However, in this case, the wave packets overlap strongly inside the particle and the momentum transfer takes place by direct contact. The fictitious spatial distribution of currents $J_\mu(x_3, x_2, x_1)$ which could transmit a signal with infinitely high speeds does not play any role in this case. It is also impossible to observe a spatially extended particle transmitting a signal which has a velocity greater than that of light [54].

We may consider the results of Section 26 and the present section as showing that in the framework of quantum field theory, it is possible to introduce deviations from the accepted form of causality which, because of the uncertainty relation, will be compatible with the classical concepts of causality.

28. Microcausality in Quantum Field Theory

In quantum theory, the state of a system is defined by the wave function Ψ, which is a vector in Hilbert space and depends both on time and the dynamic variables $Q_1, Q_2, Q_3, \ldots, Q_N$ that characterize the system. In the case of the field theory, the number of such dynamic variables is infinitely large ($N \to \infty^3$) because the instantaneous state of fields in all space $\mathfrak{R}_3(x)$ has to be defined in terms of these variables. Thus in this case the wave function is a functional and depends on some functions $A(x)$, $B(X), \ldots$, which characterize the field. Furthermore, from relativistic considerations it is advisable to consider Ψ, not as function of t, but as a functional of some space-like surface $\sigma \equiv \sigma(x)$, which only in a special case degenerates to the plane $t = \sigma(x) = \text{constant}$.

To stress the functional dependence of the wave functional Ψ on the surface $\sigma(x)$ and fields $A(x)$, $B(x)$, ..., we write the arguments of Ψ in curly brackets.

$$\Psi = \Psi\{\sigma(x), A(x), B(x), \ldots\}. \tag{28.1}$$

The normalization of this functional is defined by the condition

$$(\Psi, \Psi) = \int \bar{\Psi}\Psi \, \mathrm{d}\Omega\{A(x), B(x), \ldots\}, \tag{28.2}$$

where the integral is taken over the space of functions $A(x)$, $B(x)$, ..., must be conserved in time; in other words it must not depend on the choice of surface $\sigma(x)$.

We now consider the value of the field functional Ψ on the surface σ' which differs from the surface σ only in the neighborhood of the point x and it lies after σ (see Figure 30). If the difference between the surfaces is infinitely small

$$\sigma'(x) = \sigma(x) + \delta\sigma(x), \tag{28.3}$$

Fig. 30. The space-like surface $\sigma(x)$ and its variation in the region $(x, x + \mathrm{d}x)$.

then we may write the new functional $\Psi' \equiv \Psi\{\sigma'(x), A(x), B(x), \ldots\}$ in terms of the functional derivative $\delta\Psi/\delta\sigma(x)$ of the functional on the surface $\sigma(x)$ at the point x (see Appendix IX)

$$\Psi' \equiv \Psi\{\sigma'\} = \Psi\{\sigma, A, B, \ldots\} + \int \frac{\delta\Psi\{\sigma, A, B, \ldots\}}{\delta\sigma(x)} \delta\sigma(x) \, \mathrm{d}^3x. \tag{28.4}$$

We note that the quantity $\delta\sigma(x)\, d^3x$ is an infinitely small fourdimensional volume contained between the surfaces σ and σ' (see Figure 30).

We now turn to the principle of causality which we shall formulate classically: *the state Ψ' is uniquely defined in terms of the state Ψ which precedes it.* From this is follows that the increment of the functional defined in (28.4) must be uniquely defined by the same functionals.

Furthermore, because the theory is linear, the relationship between increment of the functional $\Psi' - \Psi$ and the functional Ψ itself must be linear. On the basis of these arguments and using the definition of a functional derivative, we write

$$\frac{\delta \Psi}{\delta \sigma(x)} = i\hat{\eta}\Psi, \tag{28.5}$$

where $\hat{\eta}$ is a linear Hermitian operator. The factor i is dictated by the requirement that the norm of the functional (Ψ, Ψ) be constant. In fact by varying the norm (Ψ, Ψ) it is easy to show that the variation $\delta(\Psi, \Psi)$ is equal to zero only in the case when the operator $i\hat{\eta}$ is anti-hermitian, i.e., if $(i\hat{\eta})^+ = -i\hat{\eta}^+ = -i\hat{\eta}$ so that the operator $\hat{\eta}$ is Hermitian. It may be written in the form

$$\hat{\eta} = -\hat{W}/\hbar, \tag{28.6}$$

where W is by definition the Hermitian operator of the energy of interaction of the field and it is a function of the operators of these fields $\hat{A}(x)$, $\hat{B}(x), \ldots$.

Equation (28.6) may now be written in the form

$$i\hbar \frac{\delta \Psi}{\delta \sigma(x)} = \hat{W}\Psi. \tag{28.7}$$

This is the well-known Tomonaga-Schwinger equation. The functional derivative $\delta\Psi/\delta\sigma(x)$ and the functional Ψ are invariant. From this it follows that the operator $\hat{W}$ has to be relativistically invariant.

For $\hat{W}=0$, the functional Ψ is constant. This means that we have chosen the so-called concept of interaction representation when free, noninteracting fields are described by the constant in the time functional Ψ_0.

We now take the second variation of the functional Ψ and vary the same space-like surface $\sigma(x)$ in the neighborhood of some other point y.

From (28.7) we obtain

$$i\hbar \frac{\delta^2 \Psi}{\delta\sigma(x)\,\delta\sigma(y)} = \hat{W}(x) \frac{\delta\Psi}{\delta\sigma(y)} = \frac{1}{i\hbar} \hat{W}(x)\,\hat{W}(y)\,\Psi. \tag{28.8}$$

Performing the same variation in the inverse order and subtracting the result gives

$$[\hat{W}(x), \hat{W}(y)] = 0 \quad \text{for} \quad x \sim y, \tag{28.9}$$

i.e., the operators of the energy of interaction of the fields, taken at various points x and y, must commute if the points x and y lie on a spatial surface so that they may not be joined by a light signal. Because the operator $\hat{W}$ is a function of the operators of the quantized fields $\hat{A}(x)$, $\hat{B}(x)$ the condition (28.9) will be automatically fulfilled when the conditions of local commutativity of fields are satisfied, i.e., the conditions of microcausality that were described in Section 25. Thus the condition (28.9) may be considered as an important result of the principle of microcausality. It was obtained for the first time in a slightly different form by Bloch [55].

In concluding we remark on one other important result of the causality principle. This is that the change of a wave function Ψ after a finite time, for a transition from spatial surface $\sigma_1(x)$ to $\sigma_2(x)$ may be represented as a unitary transformation with the operator $\hat{U}(\sigma_2, \sigma_1)$ which is a sequence of infinitely small unitary transformations.

In fact, it follows from (28.4) and (28.7) that†

$$\Psi' \equiv \Psi\{\sigma_1 + \delta\sigma(x)\} = \left(1 - \frac{i}{\hbar} \hat{W}(x)\,\delta\sigma(x)\,d^3x\right) \Psi\{\sigma(x)\}. \tag{28.10}$$

By using this operator at all points on the surface $\sigma_1(x)$ we obtain a value for the functional Ψ'' on the surface $\sigma_1''(x)$ which is later, but infinitely close to the surface $\sigma_1(x)$. From (28.8) we see that the order of the operators $(1-(i/\hbar)\,\hat{W}(x)\,\delta_\sigma(x)\,d^3x)$ is unimportant. In this manner we obtain the equation

$$\Psi'' = \Psi\{\sigma_1''(x)\} = \left[1 - \frac{i}{\hbar} \int_{\sigma_1}^{\sigma''_1} \hat{W}(x)\,d^4x\right] \Psi\{\sigma_1(x)\}. \tag{28.10'}$$

† Here we drop the assumption that Ψ depends on the dynamic variables $A(x)$, $B(x)$.

In the same manner we can go from the surface σ_1'' to the surface σ_1''' which lies later than the surface σ_1'' but infinitely close to it. Repeating this operation an infinite number of times gives

$$\Psi\{\sigma_2\} = \hat{U}(\sigma_2, \sigma_1)\,\Psi\{\sigma_1\}, \tag{28.11}$$

where the unitary operator $\hat{U}(\sigma_2, \sigma_1)$ is equal to

$$\hat{U}(\sigma_2, \sigma_1) = T\exp\left(-\frac{i}{\hbar}\int_{\sigma_1}^{\sigma_2}\hat{W}(x)\,d^4x\right). \tag{28.12}$$

where the letter T denotes an ordering in time. The unitarity of the operator $\hat{U}(\sigma_2, \sigma_1)$ follows directly from the fact that the operator $\hat{W}$ is Hermitian. We may consider Equation (28.12) for the operator $\hat{U}(\sigma_2, \sigma_1)$ as the microcausality principle in terms of wave functionals.

From (28.7), (28.11) and (28.12) it follows that the operator $U(\sigma_2, \sigma_1)$ satisfies the Tomonaga-Schwinger equation. We write it in the special case when $\sigma_1 = -\infty$, $\sigma_2 = \sigma$. It is

$$i\hbar\,\frac{\delta\hat{U}(\sigma, -\infty)}{\delta\sigma} = \hat{W}(x)\,\hat{U}(\sigma, -\infty). \tag{28.13}$$

In the special case for a plane spatial surface $\sigma = nx$, where n is a unitary time-like vector ($n^2 = 1, n_0 > 0$), the variation $\delta\sigma$ reduces to a small displacement $d\sigma$ on the whole surface σ in the direction of the vector n, so that $\delta\sigma = d\sigma\, d^3x$. Multiplying Eq. (28.13) by d^3x and integrating over the three-dimensional volume gives

$$i\hbar\,\frac{d\hat{U}(\sigma_1, -\infty)}{d\sigma} = \hat{W}(\sigma)\,\hat{U}(\sigma_1, -\infty), \tag{28.14}$$

where

$$\hat{W}(\sigma) = \int \hat{W}(x)\,d^3x = \int \hat{W}(x)\,\delta(\sigma - nx)\,d^4x. \tag{28.15}$$

In this form the Tomonaga-Schwinger equation resembles the usual Schrödinger equation in the *interaction representation*.†

† The relativistic nature of the interaction representation and the equation in the form (28 14) were first presented in [56].

We now return to the relationship between the operator $\hat{U}$ and the scattering matrix $\hat{S}$. By definition $\hat{S}$ transforms the microsystem state, specified in the infinitely distant past $\Psi(-\infty)$ into the microsystem state in the infinitely distant future $\Psi(+\infty)$:

$$\Psi(+\infty) = \hat{S}\Psi(-\infty). \tag{28.16}$$

From (28.11) it follows that $\Psi(-\infty)$ is a limiting value of $\Psi\{\sigma_1\}$ when the surface σ_1 is moved to $-\infty$, and $\Psi(+\infty)$ is a limiting value of $\Psi\{\sigma_2\}$ when the surface σ_2 is moved to $+\infty$.† Therefore $\hat{S}$ is a unitary operator

$$\hat{S}\hat{S}^+ = \hat{I}, \tag{28.17}$$

and it is the limiting value of the operator $\hat{U}\{\sigma_2, \sigma_1\}$ for $\sigma_2 \to +\infty$, $\sigma_1 \to -\infty$:

$$\hat{S} = \hat{U}\{\sigma_2, \sigma_1\} = T \exp\left(-\frac{i}{\hbar}\int \hat{W}(x)\, d^4x\right), \tag{28.18}$$

where the integration in the exponent is taken over all space-time $\mathfrak{R}_4(x)$.

The matrix $\hat{S}$ is often replaced by the scattering amplitude matrix $\hat{T}$ which is related to $\hat{S}$ by

$$\hat{S} = \hat{I} + i\hat{T}. \tag{28.19}$$

The unitary relationship for the matrix $\hat{T}$, which follows from (28.17), has the form‡

$$i(\hat{T}^+ - \hat{T}) = \hat{T}\hat{T}^+. \tag{28.20}$$

Denoting the particular initial by Ψ_i ($\Psi_i = \Psi(-\infty)$) and the final state by Ψ_f, we obtain the probability amplitude for the transition $i \to f$:

$$S_{fi} = \langle \Psi_f | \hat{S} | \Psi_i \rangle, \tag{28.21}$$

and for $i \neq f$, the probability amplitude for the transition from state i to state f:

$$P_{fi} = |\langle \Psi_f | i\hat{T} | \Psi_i \rangle|^2. \tag{28.22}$$

† In Section 32 we shall similarly consider the physical meaning of the limiting transition $\sigma \to \pm\infty$.

‡ The optical theorem introduced in Section 23 [see (XII.4)] results from this relationship.

29. Microcausality in the Theory of Scattering Matrices

Stuekelberg and his collegues [53] were the first to attempt to develop a theory into which $\hat{S}$-matrices could be introduced directly without using Schrödinger's equation (28.7) which, in contrast to $\hat{S}$-matrices, relates states which are infinitely close in space and time. We noted the principal importance of formulating the causality principle directly for $\hat{S}$-matrices. The main result of this line of investigation was obtained later by Bogolyubov who introduced a somewhat formal, though very effective method for local adiabatic "switching on" and "switching off" interactions [38].

The scattering matrix $\hat{S}$ is a functional of the interaction energy operator. We introduce a second auxiliary operator

$$\hat{W}_g(x) = g(x)\,\hat{W}(x), \tag{29.1}$$

where the function $g(x)$ can have any value in the interval from 0 to 1.

If we construct a scattering matrix using the interaction $W_g(x)$, it will be a functional of $g(x)$ whose magnitude defines the degree to which the interaction is "switched on" in a given region of space-time

$$\hat{S} = \hat{S}\{g(x)\}. \tag{29.2}$$

For $g(0)=0$ the interaction is completely "switched off" and $\hat{S}\{g(x)\}=1$. For $g(x)=1$, the interaction is completely "switched on" and $\hat{S}\{g(x)\} = \hat{S}$, the full scattering matrix (28.18).

We now consider the special case when the switch on function $g(x)$ differs from zero in two regions in space-time $\mathfrak{R}_2(x)$ and $\mathfrak{R}_1(x)$. The region $\mathfrak{R}_2(x)$ lies entirely after some space-like surface $\sigma(x)$, and the region $\mathfrak{R}_1(x)$ lies before it (see Figure 31). It is obvious that

$$g(x) = g_1(x) + g_2(x), \tag{29.3}$$

where $g_1 \neq 0$ in $\mathfrak{R}_1(x)$ and $g_2 \neq 0$ in $\mathfrak{R}_2(x)$. The wave functional $\Psi\{\sigma\}$ on the surface σ may be obtained from the initial state $\Psi(-\infty)$ using the $\hat{S}$-matrix which depends only on the interactions in region $\mathfrak{R}_1(x)$. Interactions in the future, in the region $\mathfrak{R}_2(x)$ do not necessarily influence the state

$$\psi\{\sigma\} = \hat{S}\{g_1(x)\}\,\Psi(-\infty). \tag{29.4}$$

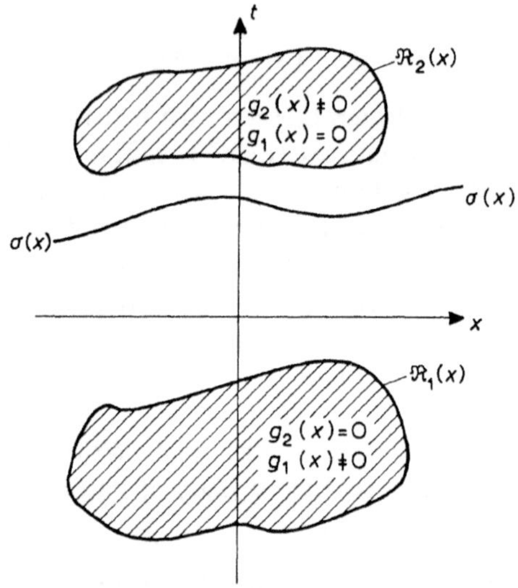

Fig. 31. Two regions $\mathfrak{R}_1(x)$ and $\mathfrak{R}_2(x)$, the second of which lies entirely after (above) the space-like surface $\sigma(x)$. In the region $\mathfrak{R}_1(x)$, the interaction is switched on with a force that is defined by the "charge" $g_1(x)$. In the region $\mathfrak{R}_2(x)$, the same interaction is defined by the charge $g_2(x)$.

The final state $\Psi\{g\}$ is obtained from $\Psi\{\sigma\}$ by using the operator $\hat{S}\{g_2(x)\}$ †

$$\Psi\{g(x)\} = \hat{S}\{g_2(x)\}\,\hat{S}\{g_1(x)\}\,\Psi(-\infty), \tag{29.5}$$

from which it follows that

$$\hat{S}\{g(x)\} = \hat{S}\{g_2(x)\}\,\hat{S}\{g_1(x)\}. \tag{29.6}$$

We now compare two cases that differ from each other only by the interactions in the region $\mathfrak{R}_2(x)$ so that

$$g'(x) = g_2'(x) + g_1'(x) \quad \text{and} \quad g''(x) = g_2''(x) + g_1''(x). \tag{29.7}$$

We form the expression $\hat{S}(g'')\,\hat{S}^+(g')$. According to (29.6) we have

$$\hat{S}^+(g_1 + g_2) = \hat{S}^+(g_1)\,\hat{S}^+(g_2), \tag{29.8}$$

† Note that because the interaction is switched off, on the surface σ the functional $\Psi\{\sigma\} = \Psi\{+\infty, g(x)\}$.

and, therefore,

$$\hat{S}(g'')\,\hat{S}^+(g') = \hat{S}(g_2'')\,\hat{S}^+(g_2'), \tag{29.9}$$

i.e., this expression does not depend on the past history of the system in the region $\mathfrak{R}_1(x)$.

It is convenient to write this result in differential form. We consider that $g_2'(y)$ differ from each other by an infinitely small amount so that $\delta g(y) = g_2''(y) - g_2'(y)$, and $\delta g(y) \neq 0$ in $\mathfrak{R}_2(y)$, i.e., in the region that lies after σ. In this case

$$\hat{S}(g'') = \hat{S}(g') + \int \frac{\delta \hat{S}}{\delta g'}\,\delta g'(y)\,\mathrm{d}y. \tag{29.10}$$

Omitting the primes in (29.9) we may write this equation in the form

$$\hat{S}(g + \delta g)\,\hat{S}^+(g) = 1 + \delta\hat{S}(g)\,\hat{S}^+(g). \tag{29.11}$$

This expression need not depend on the behavior of g in the region $\mathfrak{R}_1(x)$. Therefore, the variational derivative of Equation (29.11) with respect to $g(x)$ [the point $x \in \mathfrak{R}_1(x)$] must be equal to zero. Therefore,

$$\frac{\delta}{\delta g(x)}\left(\frac{\delta \hat{S}(g)}{\delta g(y)}\,\hat{S}^+(g)\right) = 0, \tag{29.12}$$

if the point x precedes the point y in time $(y > x)$. Equation (29.12) is relativistically covariant. In the case of a space-like interval between the points x and y $(x \sim y)$, it is always possible to make y "later" than x by choice of coordinates. Therefore, (29.12) is satisfied even in the case when the points x and y are separated by a space-like interval. Equation (29.12) is thus an expression of the causality principle in terms of the scattering matrix.

As was shown earlier in [38], on the basis of the requirements: (i) of relativistic invariance, (ii) unitarity and (iii) causality [in the form (29.12)], it is possible to construct an $\hat{S}$-matrix without using Schrödinger's equation.† We refer the reader to reference [38] for the necessary proofs and restrict ourselves here to constructing this matrix to an accuracy of the second approximation of the interaction constant. Thus the following computations serve only as an illustration of the ideas presented above.

† There are further mathematical requirements regarding the operators which enter into the $\hat{S}$-matrix structure (see [38]) which must be considered.

We write the matrix $\hat{S}$ in the form of a series of functionals in powers of $g(x)$:

$$\hat{S} = 1 + \frac{1}{1!}\int \hat{S}_1(x_1)\, g(x_1)\, d^4x_1 + \frac{1}{2!}\int \hat{S}_2(x_1, x_2)\, g(x_1)\, g(x_2)\, d^4x_1\, d^4x_2 + \cdots \quad (29.13)$$

Since we are considering that for a completely switched-on interaction $g(x)=1$, we assume that the operators $\hat{S}_1, \hat{S}_2, \ldots, \hat{S}_n$ are proportional to the powers $\varepsilon^1, \varepsilon^2, \ldots, \varepsilon^n$ of some small parameter which has the meaning of a coupling constant ("charge").†

Placing (29.13) for $\hat{S}$ in the unitary condition (28.7) we obtain, to an accuracy of ε^2,

$$\hat{S}_1(x) + \hat{S}_1^+(x) = 0, \quad (29.14)$$

$$\hat{S}_2(x, y) + \hat{S}_2^+(x, y) + \hat{S}_1(x)\, \hat{S}_1^+(y) + \hat{S}_1(y)\, \hat{S}_1^+(x) = 0 \quad (29.15)$$

etc.‡

From the first equation it follows that

$$\hat{S}_1(x) = i\hat{L}(x), \quad (29.16)$$

where $\hat{L}(x)$ is some Hermitian operator

$$\hat{L}(x) = \hat{L}^+(x). \quad (29.17)$$

Condition (29.15) is automatically fulfilled when (29.16) and the causality condition (29.12) are satisfied. Inserting the series (29.13) in (29.12) we compute the causality condition (to an accuracy of ε^2):

$$\hat{S}_2(x, y) + \hat{S}_1(x)\, \hat{S}_1^+(y) = 0 \quad (29.18)$$

for $y > x$.

Taking into account (29.16) we obtain $\hat{S}_2(x, y)$:

$$\hat{S}_2(x, y) = -\hat{L}(x)\, \hat{L}(y) \quad \text{for} \quad y > x. \quad (29.19)$$

† For example, in electrodynamics $\varepsilon = e/(\hbar c)^{1/2}$.

‡ Here we introduced the new notation $x_1 = x$, $x_2 = y$, which is more convenient as we limit ourselves to the second approximation.

From the equality in rights of the points x and y we have

$$\hat{S}_2(y, x) = \hat{S}_2^+(x, y) = -\hat{L}(y)\hat{L}(x) \quad \text{for} \quad y < x. \tag{29.19'}$$

Both of these expressions overlap in the regions $x \sim y$ and are comparable if

$$[\hat{L}(x), \hat{L}(y)] = 0 \quad \text{for} \quad x \sim y. \tag{29.20}$$

This requirement is called the *locality requirement.*

The equations (29.14) and (29.19) may be combined into one if the sign of the T-product is satisfied by

$$\hat{S}_2(x, y) = \mathrm{i}^2 T[\hat{L}(x)\hat{L}(y)]. \tag{29.21}$$

Inserting the expression we obtained for $\hat{S}_1(x)$ and $\hat{S}_2(x, y)$ in (29.13) we find

$$\hat{S}(g) = \hat{I} + \frac{i}{1!}\int \hat{L}(x)\, g(x)\, \mathrm{d}^4x + \frac{i^2}{2!} T\left\{\iint \hat{L}(x)\hat{L}(y)\, g(x)\, g(y)\, \mathrm{d}^4x\, \mathrm{d}^4y\right\} + \cdots. \tag{29.22}$$

Continuing this process, we may obtain the whole series (29.13) which may be summed into the symbolic exponential

$$\hat{S}(g) = T\left\{\exp \mathrm{i} \int \hat{L}(x)\, g(x)\, \mathrm{d}^4x\right\}. \tag{29.23}$$

For $g(x)=1$, this equation agrees with the equation for the $\hat{S}$ matrix (28.18), which was obtained from Schrödinger's equation for a wave functional, if in (29.23) we set

$$\hat{L}(x) = -\frac{1}{\hbar}\hat{W}(x). \tag{29.24}$$

Therefore, the locality requirement (29.20) agrees with the conditions obtained earlier by Bloch (28.9).

From (29.23) it follows that the requirement of relativistic invariance of the $\hat{S}$-matrix will be satisfied if, for the inhomogeneous Lorentz transformation,

$$x'_\mu = \mathbf{L}_{\mu\nu} x_\nu + a_\mu \tag{29.25}$$

(symbolically $x' = \mathbf{L}x + a$), the switch on function $g(x)$ is scalar and the energy of interaction operator $\hat{W}(x)$ is relativistically invariant. This means that, under this transformation, the following equation holds:

$$g(x) \to g'(x') = g(\mathbf{L}^{-1}(x' - a)), \tag{29.26}$$

or in other words, a new function g' has the same value at a given physical point in space-time as the untransformed function g.

Furthermore, the energy of interaction operator $\hat{W}(x)$ is a function of the operators of the fields $\hat{A}(x)$, $\hat{B}(x)$ which transforms under Lorentz transformations according to the equation

$$\hat{A}'(x') = S_L \hat{A}[\mathbf{L}^{-1}(x' - a)], \tag{29.27}$$

The form of the matrix $S_\mathbf{L}$ depends on the covariant nature of the field (for a scalar field, $S_\mathbf{L} = 1$; for a vector field, it coincides with the matrix $\mathbf{L}$ so that $S_\mathbf{L} = \mathbf{L}$ etc.).

Because the operator $\hat{A}'(x')$ gives exactly the same physical information as the untransformed operator $\hat{A}(x)$ according to the general principle of quantum theory, the operators must be related by the unitary transformation

$$\hat{A}'(x') = U_L \hat{A}(x) U_L^{-1}, \tag{29.28}$$

where the unit operator $U_\mathbf{L}$ is completely defined by the Lorentz transformation (29.25).

If the energy operator $\hat{W}$ is invariant, then the transformation will have the form

$$\begin{aligned} \hat{W}' &\equiv W'\{\hat{A}'(x')\} = W\{\hat{A}'(x')\} = W\{U_L \hat{A}(x) U_L^{-1}\} \\ &= U_L W\{\hat{A}(x)\} U_L^{-1} = W\{\hat{A}(x)\} = \hat{W}(x). \end{aligned} \tag{29.29}$$

Because the $\hat{S}$-matrix is a functional (29.23) of the operator $\hat{W}$, it ensures that the matrix $\hat{S}$ is relativistically invariant.

In concluding this section we note that the existence of singular functions like $D(x)$, $S(x)$ which enter into the commutation law as a component of quantized fields, and of the functions $D_c(x)$, $S_c(x)$ which enter into the T-product of quantized fields leads to the fact that the elements of the scattering matrix $\hat{S}$ are divergent. In order to obtain physical sense from the results in the functional series (29.13), we must make certain changes that will exclude contributions from extremely small separations.

These "changes" are called the *laws of renormalization*. These laws are also discussed in existing literature (see [37, 38]) and we shall not repeat them in this monograph. We shall limit ourselves to mentioning that these laws which are brought in from the outside do not violate the basic requirements of relativistic covariance, unitary character and causality when applied to the $\hat{S}$-matrix. However, one cannot get rid of the impression that these renormalization rules support modern theory at the expense of elegance and logical harmony of the theory.

30. Causality and the Analytical Properties of the Scattering Matrix

An important consequence of microcausality is the definition of the analytical properties of the matrix elements of the scattering matrix T_{fi} (the scattering amplitude) which define the probability of quantum transitions from some initial state i to a final state f.

These properties were first discussed in classical electrodynamics by Kronig [57] and Kramers [58] who proceeded from the requirement that the propagation speed of a signal cannot exceed the speed of light in a vacuum and thus obtained the integral relationship between the real part of the refractive index Re $n(\omega)$ (ω is the oscillation frequency) and its imaginary part Im $n(\omega)$:

$$\operatorname{Re} n(\omega) = \operatorname{Re} n(0) - \frac{2\omega^2}{\pi} \int_0^\infty \frac{\operatorname{Im} n(\omega')}{\omega'(\omega'^2 - \omega^2)}. \qquad (30.1)$$

Relationships of this type are called *dispersion relations*. The basis of these dispersion relations is Cauchy's theorem. Their relationship with causality may be illustrated in the following simple example.

Let a microparticle which scatters an incident wave be located at a point $x=0$. We write the incident wave in the form

$$\psi_0\left(t - \frac{x}{c}\right) = \frac{A}{2\pi i} \int_{-\infty}^{+\infty} \frac{\exp[-iz(t - x/z)]}{z - (\omega_0 - i\varepsilon)} dz, \qquad (30.2)$$

where $\varepsilon \to 0$.

We transpose the integration in (30.2) into the complex plane $z = \omega + i\Gamma$

and for $t-x/c>0$ we close the integration path by an infinitely large semicircle $R=|z|\to\infty$ in the lower half-plane $\Gamma<0$ (see Figure 32). There is a "cutting" factor $\exp[\Gamma(t-x/c)]$ in the integral so that for $R\to\infty$, the contribution from the integral taken over the semicircle is equal to zero. For $t-x/c<0$, the contour is closed in the upper half-plane, and the "cutting" factor $\exp[-\Gamma(t-x/c)]$ for $R\to\infty$ makes the

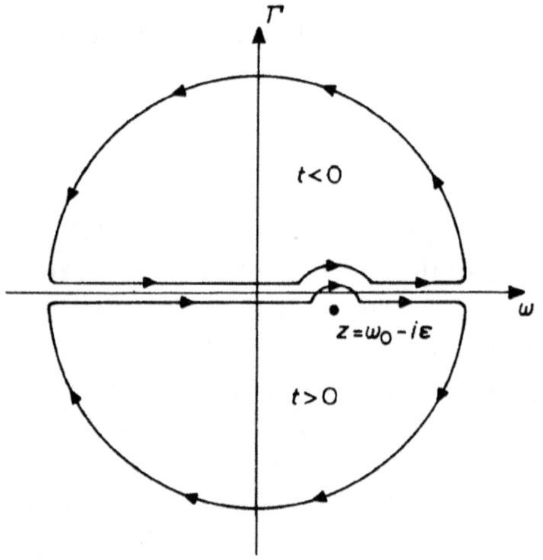

Fig. 32. Integration contours in the complex plane ω for $t>0$ and $t<0$.

contribution from this semicircle also zero. Thus in the first case the integral (30.2) is equal to the contribution from the point $z=\omega_0-\mathrm{i}\varepsilon$, and in the second case, because of the absence of singularities in the upper halfplane, it is equal to zero.

Therefore, for $\varepsilon\to 0$,

$$\psi_0\left(t-\frac{x}{c}\right)=\begin{cases} A\exp\left[-\mathrm{i}\omega_0\left(t-\frac{x}{c}\right)\right], & \text{when} \quad t-\frac{x}{c}>0, \\ 0, & \text{when} \quad t-\frac{x}{c}<0. \end{cases} \tag{30.3}$$

Let us now assume that $T(\omega)$ is an element of the scattering matrix that transforms a plane incident wave $\exp[-\omega(t-x/c)]$ into a scattered wave. We restrict ourselves here to a one-dimensional case. Then the scattered wave has the form

$$\psi(t, x) = \frac{A}{2\pi i} \int_{-\infty}^{+\infty} T(z') \frac{\exp[-iz'(t - x/c)]}{z' - (\omega_0 + i\varepsilon)}, \tag{30.4}$$

where $\varepsilon \to 0$. From the causality principle it follows that the scattered wave $\psi(t, x)$ cannot arise earlier than the time when the incident wave $\psi_0(t - x/c)$ arrives at the scattering particle. In other words,

$$\psi(t, 0) = 0 \quad \text{for} \quad t < 0. \tag{30.5}$$

This requirement will obviously not be fulfilled for all $T(\omega)$.

We shall consider $T(\omega)$ as a function of the complex variable $z = \omega + i\Gamma$. Let $T(z)$ vanish on the circle $R = |z| \to \infty$ and have singularities only on the real axis. Restricting the behavior of the amplitude $T(z)$ in the complex plane $z = \omega + i\Gamma$ allows us to write (30.4) for the point $x = 0$ in the form

$$\psi(t, 0) = \frac{A}{2\pi i} \int T(z) \frac{e^{-izt}}{z - (\omega_0 - i\varepsilon)} dz \tag{30.4'}$$

(here $\varepsilon \to 0$).

Because of our assumptions, the function $T(z)$ is regular on the infinitely large semicircle $R \to \infty$ and it does not have any singularities inside of it. Therefore, we may apply to (30.4′) the same method of integration that we used to compute the integral (30.2). Bypassing any singularities of $T(z)$ on the real axis from above ($z = \omega + i\varepsilon$, $\varepsilon > 0$) and closing the integration contour for $t > 0$ by a semicircle in the lower half-plane, we obtain a value for $\psi(t, 0)$ which is in general different from zero. For $t < 0$, the contour is closed from above and because of the absence of singularities, $\psi(t, 0) = 0$.

Thus if the amplitude $T(\omega)$ has the analytical properties discussed above, then the scattered wave (30.4′) will satisfy the causality condition (30.5). Applying Cauchy's theorem to the amplitude $T(\omega)$ gives

$$T(\omega) = \frac{1}{2\pi i} \oint_C \frac{T(z)\,dz}{z - \omega}. \tag{30.6}$$

We take for the integration contour the line $z = \omega - i\varepsilon$, i.e., a small semicircle with radius ρ, which goes around the point $z=\omega$, and a large semicircle with $R=|z|\to\infty$ which closes from below. Remembering that $T=0$ on the large semicircle, we integrate along the infinitely small semicircle around the point $z=\omega$ and, by separating the real and imaginary parts, we obtain

$$\operatorname{Re} T(\omega) = \frac{1}{\pi} \fint_{-\infty}^{+\infty} \frac{\operatorname{Im} T(\omega')\, d\omega'}{(\omega' - \omega)}, \tag{30.7}$$

where the sign $\fint$ denotes the principal value of the integral around the point $\omega'=\omega\pm\rho$, $\rho\to 0$. Equation (30.7) is the simple dispersion relation for the scattering amplitude $T(\omega)$.

We now assume that the amplitude $T(z)$ has singularities other than those which lie on the real axis. We first consider the case when the amplitude $T(z)$ has an essential singularity at $|z|\to\infty$. We set

$$T(z) = \bar{T}(z)\, e^{-iz\tau}, \tag{30.8}$$

where $\tau=a/c$, a being some length, and the function $T(z)$ has the properties discussed earlier. Then

$$\psi(t, 0) = \frac{A}{2\pi i} \int_{-\infty}^{\infty} \frac{\bar{T}(z)\, e^{-iz(t+\tau)}}{z - (\omega_0 - i\varepsilon)}\, dz. \tag{30.9}$$

Using the method described above to evaluate this integral, we find that $\psi(t, 0)$ is not equal to zero for $t<-\tau$, i.e., the scattered wave arises before the incident wave reaches the microparticle. Such behavior violates causality and it may be interpreted as the presence in the microparticle of some linear dimension a ("an elementary length") along which a signal inside the particle propagates infinitely fast. Thus a violation of the correct behavior of the amplitude $T(z)$ for $|z|\to\infty$ leads to a violation of causality.

Let us now assume that the amplitude $T(z)$ has a pole on the imaginary axis $z=\pm i\Lambda$. It is easy to see that in this case causality will be violated so that now, because of the presence of the pole $z=i\Lambda$ ($\Lambda>0$) in the upper half-plane, the amplitude of the scattered wave $\psi(t, 0)$ for $t<0$ will be equal to

$$\psi(t, 0) = A\, \frac{\operatorname{Res} T(i\Lambda)}{i\Lambda - \omega_0}\, e^{+\Lambda t}, \tag{30.10}$$

This violates causality during the time $|t| \sim 1/\Lambda$. For $t>0$, an additional term with the factor $e^{-\Lambda t}$ appears which is attenuated for $t \gg 1/\Lambda$.

Equation (30.7) shows that the dispersion relation relates the real part of the scattering amplitude to the imaginary part. In some cases the integral in (30.7) may be expressed in terms of observable quantities. Therefore, an experimental examination of the dispersion relation may serve as an investigation of the basic assumptions of the modern local theory of fields and, in particular, the microcausality principle which is important to the entire theory.

However, the properties of the elements of the scattering matrix S_{fi} which result from the general principles of the local theory of fields are difficult to determine, and as yet this problem has not been solved. Significant progress has thus far been made only in the simple case of elastic scattering, by Goldberger [59], Symanzik [60], Bogolyubov [61], *et al.* In particular for the case of forward elastic scattering (for the angle $\theta^0=0$) of π-mesons on nucleons, the dispersion relation does not contain any unobservable quantities.

The most mathematically rigorous proof of these relations was given by Bogolyubov and we refer the reader to reference [61] for details. Here we shall give only an outline of the dispersion relation theory insofar as it is necessary for our subject.

A basic requirement for constructing dispersion relations for πN scattering is the relativistic generalization of the theorems concerning the analyticity of Fourier transformations of functions which vanish for $t<0$.

Let $F(x)$ be a function of $x(\mathbf{x}, t)$ which vanishes outside the absolute future light cone, i.e., for $x \lesssim 0$. Its Fourier transform $\tilde{F}(k)$ is given by

$$\tilde{F}(k) = \int F(x) \exp(ikx)\, d^4x . \tag{30.11}$$

Remembering that $kx=\omega t-\mathbf{kx}$, for the complex value of k, for $\mathbf{k}\to\mathbf{k}+i q$, $\omega\to\omega+i\Gamma$, the exponent in (30.11) takes on the form $\exp[-i(\mathbf{kx}-\omega t)+ +(\mathbf{qx}-\Gamma t)$. In the region where Γ has a significant value, where

$$\Gamma^2 - \mathbf{q}^2 > 0, \qquad \Gamma > 0, \tag{30.12}$$

we may select a coordinate system so that $q=0$. Then the factor e^{ikx} in (30.11) acquires a cut-off factor $e^{-\Gamma t}$. Because of this factor the integral

(30.11) and its derivative with respect to k to any order will converge. This shows the Fourier transform $\tilde{F}(k)$ is an analytical function of the variable k in the region (30.12). We may also show that for $|k|\to\infty$, $\tilde{F}(k)$ increases to a value that is less than some order of $|k|^n$ (see [61]).

The dispersion relations are determined for the matrix elements T_{fi} of the matrix $\hat{T}$ (see Section 24). For elastic scattering of π-mesons on nucleons, wave functions of the initial state (i) and final state (f) may be written in the form described in Section 24 [see (24.14) and (24.18)]:

$$\Psi_i = \frac{e^{iky_1}}{\sqrt{\omega}} u_p \, e^{ipx_1}, \tag{30.13}$$

$$\bar{\Psi}_f^* = \frac{e^{-ik'y_2}}{\sqrt{\omega'}} \bar{u}_{p'} \, e^{-ip'x_2}, \tag{30.14}$$

where $p=(p, p_0=(M^2+\mathbf{p}^2)^{1/2})$ (M is the nucleon mass), $k=(\mathbf{k}, k_0=\omega=(m^2+\mathbf{k}^2)^{1/2})$ (m is the meson mass). In the final state the corresponding quantities are denoted by primes. We did not include the spinor indices of the nucleon nor the isotopic spin indices of the meson so that the equations would not be too cumbersome.

The amplitude T_{fi} is the matrix element of the four branch $s(x_2,y_2 \mid x_1, y_1)$, taken for the state (30.13) and (30.14). By introducing the coordinates $Y=\frac{1}{2}(y_1+y_2)$, $y=y_2-y_1$ and $X=\frac{1}{2}(x_1+x_2)$, $x=x_2-x_1$, it is easy to show that the matrix element of interest, T_{fi}, may be written in the form

$$T_{fi} = (p', k'|\, T \,|p, k) = \frac{2\pi}{\sqrt{4\omega\omega'}} \delta^4 (k' + p' - k - p)\, T_c\left(\frac{k + k'}{2}, p', p\right). \tag{30.15}$$

In the laboratory coordinate system for forward scattering [where $p'=p=(0, 0, 0, M)$ and $k'=k=(\mathbf{k}, \omega)$] we have

$$T_{fi} = \frac{2\pi}{\sqrt{4\omega\omega'}} \delta(k' - k)\, T_c(k). \tag{30.16}$$

The subscript c on the function $T_c(k)$ in (30.15) and (30.16) is used to stress that the function $T_c(k)$ is the Fourier transform of some function

$\tilde{T}_c(k)$ which is analogous in its space-time behavior to the causal function $D_c(x)$†:

$$T_c(k) = \int \tilde{T}_c(x)\, e^{ikx}\, d^4x. \tag{30.17}$$

It is, however, more convenient to investigate the analytical properties of quantities which have retarded or advanced characteristics.

It is possible to show that a retarded function $T_{ret}(k)$ may be obtained which is analogous to the retarded function $D_{ret}(k)$ whose real part, i.e. $p^2 = M^2$ and $k^2 = m^2$, is identical to the function $T_c(k)$ on the energy surface $p + k = p' + k'$ and

$$T_c(k) = T_{ret}(k). \tag{30.18}$$

Together with $T_{ret}(k)$, we also introduce the advanced function $T_{adv}(k)$ which is analogous to $D_{adv}(k)$ and also their difference

$$T(k) = T_{ret}(k) - T_{adv}(k). \tag{30.19}$$

The functions $T_{ret}(k)$ and $T_{adv}(k)$ satisfy some of the symmetry conditions for the inverse vector k, namely,

$$T^*_{ret}(k) = T_{ret}(-k), \tag{30.20}$$

$$T_{ret}(k) = T_{adv}(-k). \tag{30.20'}$$

We will also need to have the amplitude $T(k)$ written as a sum of its dispersion (Hermitian) part D and absorption (anti-Hermitian) part iA:

$$T(k) = D(k) + iA(k). \tag{30.21}$$

In the case of forward scattering in which the nucleon spin and isotopic meson spin are unchanged (i.e., coherent scattering), the quantities $D(\omega)$ and $A(\omega)$ are real because the matrices D and A are Hermitian.

We note one further detail. Because of the conservation laws, the vector k in the exponent of (30.17) contains irrationalities which introduce difficulties. In fact in the scalar product $kx = \omega t - \mathbf{kx}$ the absolute magnitude of the vector $\mathbf{k}$ depends irrationally on ω: $|\mathbf{k}| = (\omega^2 - m^2)^{1/2} = \lambda$. Therefore, by writing the scalar product in the form

$$\omega t - \mathbf{kx} = \omega t - \mathbf{e}\lambda\mathbf{x}, \tag{30.22}$$

† Note that in the second order of perturbation theory $\tilde{T}_c(x) = D_c(x)$.

where $\mathbf{e}=\mathbf{k}/|\mathbf{k}|$, we can remove the irrationality in ω in (30.17) if, instead of (30.17), we consider functions that are symmetric or antisymmetric and that are written in the following form

$$T_s(\mathbf{e}, \lambda) = T(\mathbf{e}, \lambda) + T(-\mathbf{e}, \lambda), \tag{30.23}$$

$$T_a(\mathbf{e}, \lambda) = \frac{1}{\lambda}\int [T(\mathbf{e}, \lambda) - T(-\mathbf{e}, \lambda)], \tag{30.23'}$$

so that they depend only on λ^2. From (30.23) and (30.23′) it follows that

$$T(\mathbf{e}, \lambda) = \tfrac{1}{2}[T_s(\mathbf{e}, \lambda) + \lambda T_a(\mathbf{e}, \lambda)]. \tag{30.24}$$

When the functions are written in this symmetric form, the absorbing part of the amplitude A has a certain symmetry with respect to sign changes, namely,

$$A_s(-\omega) = + A_s(\omega), \tag{30.25}$$

$$A_a(-\omega) = - A_s(\omega). \tag{30.25'}$$

We shall use these equations later.

We shall conclude our studies of analytical properties by considering the functions $T_{\text{ret}}(\omega)$ and $T_{\text{adv}}(\omega)$ written in the symmetric form (30.23)–(30.23′). A very complicated and laborious study leads to the result that the functions $T_{\text{ret}}(\omega)$ and $T_{\text{adv}}(\omega)$ represent one analytic function $T(\omega)$ in the complex plane ω, which is identical to $T_{\text{ret}}(\omega)$ in the region $\operatorname{Im}\omega>0$ and is identical to $T_{\text{adv}}(\omega)$ in the region $\operatorname{Im}\omega<0$. $T(\omega)$ has branches on the real axis in the region $-\infty<\omega<-m$ and $\infty>\omega>-m$ and has poles at

$$\omega = \pm\frac{m^2}{2M} = \pm\,\omega_g. \tag{30.26}$$

It vanishes on the circle $|\omega|=R\to\infty$.

Figure 33 shows the complex plane ω. The heavy line denotes the contour C within which $T(\omega)$ is an analytic function of ω. If the amplitude $T(\omega)$ increases as ω^2, then it can be replaced by the quantity $T'(\omega) = T(\omega)/(\omega-\omega_0)^{n+1}$ which differs from $T(\omega)$ in that it has an additional pole at the point $\omega=\omega_0$ and it tends to zero in the region $|\omega|\to\infty$.

Cauchy's theorem may be applied to $T'(\omega)$ by choosing the integration contour as shown in Figure 33. Taking the limit $R\to\infty$, we obtain the dispersion relation for $T'(\omega)$ in the form

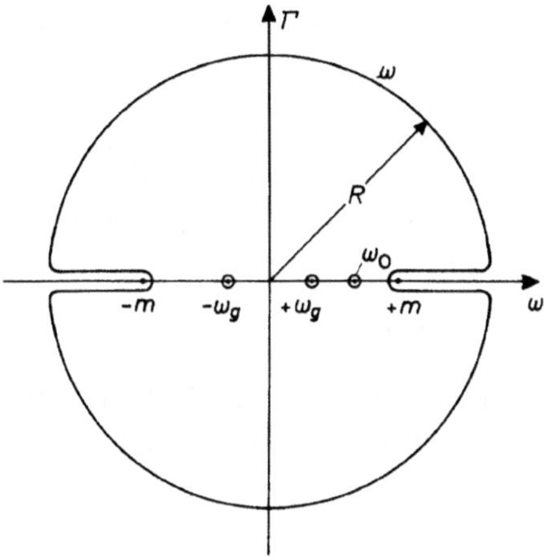

Fig. 33. Integration contour for the amplitude $T(\omega)$ in the complex plane ω. This figure shows the branch cuts from $-R$ to $-m$ and from $+m$ to $+R$ $(R\to\infty)$, and also the simple poles at $\omega = \pm\omega g$. The $(n+1)$th order pole at the point ω_0 is produced by dividing the amplitude by $(\omega-\omega_0)^{n+1}$.

$$\frac{T(\omega)}{(\omega-\omega_0)^{n+1}} = \frac{1}{2\pi i}\int_{-\infty+i\varepsilon}^{\infty+i\varepsilon} \frac{T(\omega')\,d\omega'}{(\omega'-\omega_0)^{n+1}(\omega'-\omega)} + \int_{-\infty-i\varepsilon}^{\infty-i\varepsilon} \frac{T(\omega')\,d\omega'}{(\omega'-\omega_0)^{n+1}(\omega'-\omega)}. \tag{30.27}$$

We denote the discontinuity in $T(\omega)$ on the branch cuts $(-\infty, -m)$ and (m, ∞) by $\Delta T(\omega)$ and, multiplying (30.27) by $(\omega-\omega_0)^{n+1}$ we obtain

$$T(\omega) = \frac{(\omega-\omega_0)^{n+1}}{2\pi i}\int_{-\infty}^{-m} \frac{\Delta T(\omega')}{(\omega'-\omega_0)^{n+1}(\omega'-\omega)}\,d\omega' + \frac{(\omega-\omega_0)^{n+1}}{2\pi i}\int_{m}^{\infty} \frac{\Delta T(\omega')\,d\omega'}{(\omega'-\omega_0)^{n+1}(\omega'-\omega)} + \frac{g^2}{\omega+\omega_g} + \frac{g^2}{\omega-\omega_g} + \sum_{r=0}^{n} c_r\omega^r. \tag{30.28}$$

where $g^2/(\omega \pm \omega_g)$ are residues at the points $\omega = \pm\omega_g$ and the constant coefficients C_r are calculated at the point $\omega = \omega_0$.

The jump $\Delta T(\omega)$ at the branch cuts of $T(\omega)$ is determined by considering that the absorbing part of A, which differs from zero for $\omega < -m$ and $\omega > +m$, changes its sign in going from $\operatorname{Im}\omega > 0$ to $\operatorname{Im}\omega < 0$ while the dispersion part of the amplitude $D(\omega)$ changes continuously. Thus the jump $\Delta T(\omega)$ is equal to

$$\Delta T(\omega) = 2\mathrm{i}A(\omega). \tag{30.29}$$

We may now make use of the symmetry in (30.25) and (30.25′) and replace the integration over negative ω with integration over positive ω. Then we obtain

$$T(\omega) = \frac{2(\omega - \omega_0)^{n+1}}{\pi} \int_m^\infty \frac{A(\omega')\,\mathrm{d}\omega'}{(\omega' - \omega)^{n+1}\,(\omega'^2 - \omega^2)} + \frac{g^2}{(\omega + \omega_g)} + \frac{g^2}{(\omega - \omega_g)} + \sum_{r=0}^{n} c_r \omega^r. \tag{30.30}$$

The choice of index n remains arbitrary, but its smallest possible value must be determined indirectly. According to the optical theorem (XII.4)

$$A(\omega) = \frac{k}{4\pi}\,\sigma_{\mathrm{T}}(\omega), \tag{30.31}$$

where k is the meson momentum in the center of mass system and $\sigma_T(\omega)$ is the total cross section for the process $\pi + \mathrm{N} \to \mathrm{N} + \cdots$. Experiments show that $\sigma_T(\omega)$ tends to a constant value or even decreases with increase in ω. We may, therefore, consider that for high energies $k \approx \omega$, the absorbing part of the amplitude $A(\omega)$ does not increase faster than $\omega \to \infty$, that is for $n \leqslant 1$. Since $|D|^2$ enters into $\sigma_T(\omega)$, the dispersion part of the amplitude D must decrease as ω increases. These considerations allow us to limit ourselves to multiplying $T(\omega)$ by $(\omega - \omega_0)^{-2}$ when constructing the dispersion relation for πN-scattering.

In order now to obtain from (30.30) a specific dispersion relation for scattering of neutral π^0- and charged π^+-mesons, we must determine the dependence of the scattering amplitude on spin and isotopic indices. We shall, however, give only the results of these studies here. We denote the dispersing and absorbing parts of the amplitude for neutral (0) and charged ($\pm$) mesons by D_0, $D_\pm$, A_0, $A_\pm$ respectively. Then the dispersion relations may be written as

$$D_0(\omega) - D_0(\omega_0) = \frac{g^2}{M}\,\omega_g^2\,\frac{(\omega^2-\omega_0^2)}{(\omega^2-\omega_g^2)(\omega_0^2-\omega_g^2)} + \frac{2}{\pi}(\omega^2-\omega_0^2)\int\limits_m^\infty \frac{A_0(\omega')\,\mathrm{d}\omega'}{(\omega'^2-\omega^2)(\omega'^2-\omega_0^2)}, \tag{30.32}$$

$$D_1(\omega) - D_1(\omega_0) = \frac{2g^2}{M}\,\frac{\omega_g^2(\omega^2-\omega_0^2)}{(\omega^2-\omega_g^2)(\omega^2-\omega_0^2)} + \frac{2}{\pi}(\omega^2-\omega_0^2)\int\limits_m^\infty \frac{A_1(\omega')}{(\omega'^2-\omega^2)(\omega'^2-\omega_0^2)}\,\mathrm{d}\omega', \tag{30.32'}$$

$$D_2(\omega) - \frac{\omega}{\omega_0}D_2(\omega_0) = \frac{2g^2}{M}\,\frac{\omega\omega_g^2(\omega^2-\omega_0^2)}{(\omega^2-\omega_g^2)(\omega^2-\omega_0^2)} + \frac{2}{\pi}\,\omega(\omega^2-\omega_0^2)\int\limits_m^\infty \frac{A_2(\omega')}{(\omega'^2-\omega^2)(\omega'^2-\omega_0^2)}\,\mathrm{d}\omega', \tag{30.32''}$$

where

$$D_1(\omega) = D_+(\omega) + D_-(\omega), \tag{30.33}$$

$$D_2(\omega) = D_+(\omega) - D_-(\omega), \tag{30.33'}$$

$$A_1(\omega) = A_+(\omega) + A_-(\omega), \tag{30.33''}$$

$$A_2(\omega) = A_+(\omega) - A_-(\omega). \tag{30.33'''}$$

The two arbitrary constants C_0 and C_1 in the term $\sum_0^n C_r\omega^r$ are eliminated by subtracting the amplitude $D(\omega_0)$ and $(\omega/\omega_0)D(\omega_0)$ taken at the point $\omega=\omega_0$ from the dispersion relation.

The absorbing parts $A_\pm(\omega)$, $A_0(\omega)$ may be expressed in terms of experimentally observed total cross-sections using the optical theorem (30.31). Finally the constant g may be defined from low-energy interactions. In this manner the dispersion relations (30.32)–(30.32″) may be checked experimentally and, therefore, the validity of the local theory of fields can be proved or disproved.

Another important result of the local theory which is also based on the analytical properties of the scattering amplitude, relates to the asymptotic behavior of the total cross section for particles and antiparticles. On the basis of the dispersion relations for forward elastic scattering of π-mesons off nucleons and using the optical theorem (see Section 23), Pomeranchuk [62] showed that the total cross sections for the interaction of π^+- and π^--mesons and nucleons for very high energies ($\omega \to \infty$) have to be equal to

$$\sigma_{\pi+N} = \sigma_{\pi-N} \qquad \text{for} \quad \omega \to \infty . \tag{30.34}$$

Recently another theorem has been used to determine the asymptotic relationship between the cross sections for various processes. This is the theorem concerning the behavior of an analytic function $f(x)$ inside an angle that is formed by rays, as a function of the limits C_1 and C_2 to which it tends along those rays [63]. The theorem states that if the function $f(z)$ which is analytical inside the angle C_1OC_2 had different limits, $C_1 \neq C_2$, along the rays OC_1 and OC_2, then it would increase inside the angle faster than an exponential.

Returning to our problem, we shall consider the plane z as a complex plane of the meson energy $\omega : z = \omega$. If we now take the angle π for the angle C_1OC_2 (upper half-plane), and take the elastic scattering amplitude of a particle a being scattered off a particle b, extended in the upper half-plane, for the function $f(z)$, then for $z = \omega \to \infty$ the function $f(z)$ will correspond to the forward elastic scattering amplitude $T(\omega)$ for the process

$$\mathrm{a} + \mathrm{b} \to \mathrm{a} + \mathrm{b} ; \tag{30.35}$$

For $\omega \to \infty$ and $z = \omega \to -\infty$ this function will correspond to the same amplitude $T(\omega)$ for "cross-over" processes†:

$$\tilde{\mathrm{a}} + \mathrm{b} \to \tilde{\mathrm{a}} + \mathrm{b} , \tag{30.35'}$$

where $\tilde{a}$ denotes the antiparticle of particle a.

Because the forward elastic scattering cross section is analytic in the

† For details of "cross-over" reactions, see Section 31.

upper half-plane and its growth is limited to a polynomial ($\sim\omega^n$), by virtue of the theorem stated above, the extreme values c_1 and c_2 of the amplitude of $f(z)$, is its values at $\omega = \pm\infty$ must be equal. Because the imaginary parts of the amplitudes are equal, from the optical theorem it follows that the total cross section has to be equal. In particular, if particle a is a π-meson and particle b is a nucleon, we obtain (30.34).

Using this same theorem, asymptotic relationships were obtained for many different reactions as well as the total πN scattering cross section and for differential as well as total cross-sections [64].

We shall quote some of these relationships.

(a) For total cross sections with $\omega \to \infty$, identical cross sections are obtained for the following reactions

$$\left.\begin{array}{lcl} \pi^+ + p \to \cdots & \text{and} & \pi^- + p \to \cdots, \\ K^+ + p \to \cdots & \text{and} & K^- + p \to \cdots, \\ p + p \to \cdots & \text{and} & \tilde{p} + p \to \cdots, \\ \Sigma^+ + p \to \cdots & \text{and} & \Sigma^- + p \to \cdots, \end{array}\right\} \quad (30.36)$$

where the dots on the right indicate all possible products of the reaction on the left.

(b) For differential cross sections with $\omega \to \infty$ and with a specific momentum transfer, the following reactions have the same cross sections

$$\left.\begin{array}{lcl} \pi^+ + p \to \pi^+ + p & \text{and} & \pi^- + p \to \pi^- + p, \\ K^+ + p \to K^+ + p & \text{and} & K^- + p \to K^- + p, \\ \pi^+ + p \to K^+ + \Sigma^+ & \text{and} & K^- + p \to \pi^- + \Sigma^+, \\ \pi^- + p \to K^0 + \Lambda & \text{and} & K^0 + p \to \pi^+ + \Lambda. \end{array}\right\} \quad (30.37)$$

In concluding this section, we list the basic principles of local field theory which were used in obtaining the dispersion relations for the scattering amplitude T and the asymptotic relationship for the cross section σ.

I. Microcausality must be fulfilled, which indicates the absence of any influence of one region of space-time $\mathfrak{R}_4'(x)$ on another $\mathfrak{R}_4''(x)$ if the points which belong to these regions are separated by a spatial interval

$$x_2 \sim x_1, \qquad x_1 \in \mathfrak{R}_4'(x), \qquad x_2 \in \mathfrak{R}_4''(x)$$

and if the interval is time-like, i.e., $x^2 > 0$ but $x_2 < x_1$ (the cause must precede the effect!).

II. A spectrum of stable particles with positive energy

$$p_0 = \sqrt{\mathbf{p}^2 + m_s^2}, \tag{30.38}$$

must exist, where m is the mass of the particles ($s = 1, 2, \ldots$).

III. The amplitude $T(\omega)$ for $\omega \to \infty$ must not increase faster than a polynomial:

$$|T(\omega)| < \alpha\, |\omega|^n, \tag{30.39}$$

where n is positive integer.

CHAPTER VI

MACROSCOPIC CAUSALITY

31. Formal $\hat{S}$-matrix theory

We begin this chapter by listing some of the more important properties of the $\hat{S}$-matrix. As we showed in Section 28, the $\hat{S}$-matrix is unitary. It transforms an initial free particle state Ψ_i given at a time $t=-\infty$ into a final free particle state Ψ_f, which is considered at the time $t=+\infty$:

$$\Psi_f = \hat{S}\Psi_i. \tag{31.1}$$

This matrix is usually written in the form [see (29.4)]

$$\hat{S} = \hat{I} + \mathrm{i}\hat{T}. \tag{31.2}$$

The unitary condition

$$\hat{S}\hat{S}^+ = \hat{I} \tag{31.3}$$

may be turned into an identity if we set

$$\hat{S} = \mathrm{e}^{\mathrm{i}\eta}, \tag{31.4}$$

where $\hat{\eta}$ is the Hamiltonian operator

$$\hat{\eta} = \hat{\eta}^+, \tag{31.5}$$

or it may also be written as

$$\hat{S} = \frac{\hat{I} - \mathrm{i}\hat{K}}{\hat{I} + \mathrm{i}\hat{K}}, \tag{31.6}$$

where $\hat{K}$ is another Hermitian operator

$$\hat{K} = \hat{K}^+. \tag{31.7}$$

The concept of the asymptotic states Ψ_i and Ψ_f requires a detailed analysis. In particular, we shall now consider the limit $t\to\pm\infty$ as purely formal and we shall not go into its meaning. We shall return to such questions later.

The asymptotic state itself is defined by a complete set of dynamic variables p which characterize a state of free particles. We choose the momentum of these particles for the set of variables so that

$$\Psi_i \equiv \Psi_i(p), \tag{31.8}$$

$$\Psi_f \equiv \Psi_f(p'), \tag{31.8'}$$

where (p) is the set of momenta in the initial state, and (p') is the set of momenta in the final state. $(p)\equiv(p_n, p_{n-1}, \ldots, p_1)$ if the initial state has n particles and $(p')\equiv(p_{m+n}, p_{m+n-1}, \ldots, p_{n+1})$ if the final state has m particles. For free particles, the four momentum components have specific values, namely,

$$p_{0s} = +\sqrt{\mathbf{p}_s^2 + m_s^2}, \qquad s = 1, 2, \ldots, m + n, \tag{31.9}$$

where m_s is the mass of the sth particle. The set of values of the variables $p=(p_0, \mathbf{p})$ form a three-dimensional manifold $\mathfrak{R}_3(p)$ which corresponds to the surface of a hyperboloid of one sheet

$$p_0^2 - \mathbf{p}^2 \equiv m^2, \qquad p_0 > 0. \tag{31.10}$$

This manifold forms a Lobachevski space,† a scalar curvature of R [compare with (5.13)] is defined in terms of the particle mass

$$R = -1/m^2. \tag{31.11}$$

Note that the second sheet of the hyperboloid

$$p_0 = -\sqrt{m^2 + \mathbf{p}^2} \tag{31.10'}$$

may be used to represent the antiparticle state.

Thus a space in which the asymptotic states i and f are defined may be considered as composition of several Lobachevski spaces:

$$\mathfrak{R}_i(p) = \mathfrak{R}_3(p_1) \otimes \mathfrak{R}_3(p_2) \otimes \cdots \otimes \mathfrak{R}_3(p_n). \tag{31.11'}$$

The space $\mathfrak{R}_f(p')$ may be similarly constructed, so that

$$\mathfrak{R}_f(p') = \mathfrak{R}_3(p'_{n+1}) \otimes \mathfrak{R}_3(p'_{n+2}) \otimes \cdots \otimes \mathfrak{R}_3(p'_{n+m}). \tag{31.11''}$$

We shall call the space $\mathfrak{R}_i(p)$ *the space of initial states* and the space $\mathfrak{R}_f(p')$ *the space of final states*. We shall call both of these states *the space of asymptotic states*.

† This fact was successively used in references [65] and [66].

Note that the state $\mathfrak{R}(p)$, by its very meaning, belongs to the set of characteristic values of the momentum operator $\hat{\mathbf{p}}$ of free particles and the energy operator $\hat{p}_0$. For each characteristic value of the operator $\hat{\mathbf{p}}$ and the sign of the operator $\hat{p}_0$, there exists a characteristic function $\psi_p^{\pm}(x)$. Therefore the set of all the characteristic functions $\psi_p^{\pm}(x)$ of the operators $\hat{\mathbf{p}}$ and $\hat{p}_0$ may be associated with each point in space. Each of these sets forms a Hilbert space $\mathfrak{R}\{\psi\}$. In this space, the functions $\psi_p^{\pm}(x)$ and their products are vectors or tensors which define a complete system of orthonormal functions.

In the representation of the asymptotic states ψ_f and ψ_i that we selected, the rows of the $\hat{S}$-matrix are characterized by the momenta (p') in the final state f, and the columns by the momenta (p) in the initial state i. Therefore, we may write

$$\begin{aligned} S_{fi} &= \langle p'|\,\hat{S}\,|p\rangle \\ &\equiv \langle p'_{m+n}, p'_{m+n-1}, \ldots, p_{n+1}|\,\hat{S}\,|p_n, p_{n-1}, \ldots, p_2, p_1\rangle . \end{aligned} \tag{31.12}$$

It will be more convenient to separate the unitary matrix $\hat{I}$ from the $\hat{S}$-matrix in (31.3). Then we may rewrite (31.12) in the form

$$S_{fi} = \langle p'|\,\hat{I}\,|p\rangle + i\langle p'|\,\hat{T}\,|p\rangle, \tag{31.13}$$

where

$$\langle p'\,|\hat{T}\,|p\rangle = \langle p'_{m+n}, p'_{m+n-1}, \ldots, p'_{n+1}|\,\hat{T}\,|p_n, p_{n-1}, \ldots, p_2, p_1\rangle . \tag{31.14}$$

The $\hat{S}$-matrix must be relativistically invariant (compare to Section 29). Let us transform the momenta (p) using the Lorentz transformation matrix†

$$\bar{p} = \bar{\mathbf{L}}p, \tag{31.15}$$

where $(\bar{p})$ denotes the set of new, transformed momenta. The vectors of the state $\Psi(p)$ undergo unitary transformation using the unitary matrix $\bar{U}_{\mathbf{L}}$.

$$\bar{\Psi}(\bar{p}) = \bar{U}_{\mathrm{L}}\Psi(\bar{\mathbf{L}}^{-1}\bar{p}) = \bar{U}_{\mathrm{L}}\Psi(p), \tag{31.16}$$

where Ψ temporarily denotes the state vector in the new coordinate

† The matrix $\bar{\mathbf{L}}$ differ from the matrix $\mathbf{L}$ in (29.25).

system. Applying this transformation to Equation (31.1) gives

$$\bar{U}_{\mathrm{L}}\Psi_f(\mathbf{L}^{-1}\bar{p}) = \bar{U}_{\mathrm{L}}\hat{S}(\mathbf{L}^{-1}\bar{p})\,\bar{U}_{\mathrm{L}}^{-1}\bar{U}_{\mathrm{L}}\Psi_i(\mathbf{L}^{-1}\bar{p}), \tag{31.17}$$

i.e.,

$$\bar{\Psi}_f = \bar{\hat{S}}(\bar{p})\,\bar{\Psi}_i, \tag{31.18}$$

where

$$\bar{\hat{S}}(\bar{p}) = \bar{U}_{\mathrm{L}}\hat{S}(\mathbf{L}^{-1}\bar{p})\,\bar{U}_{\mathrm{L}}^{-1}. \tag{31.19}$$

Relativistic invariance requires that the same transformation law hold for all reference frames and, therefore, the following relationship must hold:

$$\hat{S}(\mathbf{L}^{-1}\bar{p}) = \hat{S}(\bar{p}). \tag{31.20}$$

This is the condition of relativistic invariance for the theory, expressed in terms of momentum variables.

In the absence of external fields, because of the uniformity of space-time, physical phenomena cannot depend on the position of the origin in space or time. Therefore the $\hat{S}$-matrix must be translationally invariant. From this requirement it follows that

$$\hat{D}\hat{S} - \hat{S}\hat{D} = 0, \tag{31.21}$$

where $\hat{D}$ is a translation operator, which has the form

$$\hat{D} = \exp(\mathrm{i}\hat{P}_\mu a_\mu), \tag{31.22}$$

$\hat{P}_\mu$ is the operator of the total momentum of the system being considered, and a_μ is the vector of an arbitrary displacement. From (31.21) and (31.22) it follows that

$$\hat{P}_\mu\hat{S} - \hat{S}\hat{P}_\mu = 0, \tag{31.23}$$

i.e., the $\hat{S}$-matrix commutes with the total momentum operator $\hat{P}$. Therefore the $\hat{S}$-matrix must be diagonal with respect to the total momentum operator $\hat{P}$. Therefore the matrix elements S_{fi} must be proportional to δ-functions of the complete four-dimensional momentum, i.e., proportional to the function

$$\delta^4(P' - P),$$

where

$$P' = \sum p' \equiv \sum_{s=1}^{m} p_{n+s} \quad \text{and} \quad P = \sum p \equiv \sum_{s=1}^{n} p_s .$$

Note that the matrix element of $\hat{I}$ is equal to

$$\langle p' | \hat{I} | p \rangle \equiv \delta^3 (\mathbf{p}' - \mathbf{p}) \equiv \prod_{s=1}^{n} \delta^3 (\mathbf{p}'_s - p_s),$$

i.e., to the product of three-dimensional δ-functions. For $n \geqslant 2$, it is always possible to obtain a four-dimensional δ-function of the total momentum from this product, so that

$$\langle p' | \hat{I} | p \rangle = \delta^3 (\mathbf{p}' - \mathbf{p}) = \delta^4 (\textstyle\sum p' - \sum p)\, I (p' \mid p)\,; \qquad (31.24)$$

where $I(p' \mid p)$ is expressed in terms of relative momenta. The matrix element $\langle p' | \hat{T} | p \rangle$ must also be proportional to $\delta^4(\sum p' - \sum p)$, i.e.,

$$\langle p' | \hat{T} | p \rangle = \delta^4 (\textstyle\sum p' - \sum p)\, T (p' \mid p). \qquad (31.25)$$

Further details of the form of the matrix elements may be obtained by considering the fact that in a relativistically invariant theory the multiplication law for any two matrices A and B, taken in the momentum representation, must be relativistically invariant. We have

$$\langle p' | \hat{A}\hat{B} | p \rangle = \int \langle p' | \hat{A} | p'' \rangle \, \mathrm{d}^3 p'' \, \langle p'' | \hat{B} | p \rangle. \qquad (31.26)$$

However the volume element $\mathrm{d}^3 p''$ is not invariant. We may ensure the relativistic invariant character of the matrix multiplication if we rewrite the matrix elements in the form

$$\langle p' | \hat{A} | p \rangle = \frac{A (p', p)}{\sqrt{2p'_0} \cdots \sqrt{2p_0}}, \qquad (31.27)$$

where p_0 is the fourth component of the momentum and $A(p', p)$ is an invariant function of the momenta (p') and (p). In fact, in this case in place of (31.26) we obtain

$$\langle p' | \hat{A}\hat{B} | p \rangle = \frac{AB}{\sqrt{2p'_0} \cdots \sqrt{2p_0}} = \int \frac{A (p', p'')}{\sqrt{2p'_0}} \frac{\mathrm{d}^3 p''}{2p''_0} \frac{B (p'', p)}{\sqrt{2p_0}}. \qquad (31.26')$$

The volume element $d^3p''/2p_0''$ that appears here is a volume element in Lobachevski space, which is relativistically invariant.

Finally, we write the elements of the $\hat{S}$-matrix in the form

$$S_{fi} \equiv \langle p'|\, \hat{S}\, |p\rangle = \langle p'|\, \hat{I}\, |p\rangle + i\delta^4 \left(\sum p' - \sum p\right) \frac{T(p' \mid p)}{\sqrt{2p_0'} \ldots \sqrt{2p_0}}, \tag{31.28}$$

where $T(p' \mid p)$ is an invariant function of the momenta of the initial and final states.

We shall now consider in detail the structure of the $\hat{S}$-matrix in the simplest case when the particles are of one type, scalar, uncharged and have a rest mass m. We assume that the interaction between them is such that it neither leads to the formation of stable groups with a mass M that is smaller than the sum of the masses of the original particles, nor to unstable groups which have a mass M^* that is larger than the sum of the masses of the original particles. The possible processes in this simple case are limited to scattering, generation and absorption of particles. In this case there is only one type of scalar particles, and these are able to be scattered, produced and absorbed. For brevity, in Table I the matrix rows (f) are listed according to the number of particles m that are found in the final state and the matrix columns are arranged according to the number of particles found in the initial state n. A matrix element $\langle m\, |\hat{S}|\, n\rangle$, which is defined in the asymptotic space $\mathfrak{R}(p, p') = \mathfrak{R}_i(p) \otimes \mathfrak{R}_f(p')$ which has $3(m+n)$ dimensions is written in each location (m, n) of Table I.

TABLE I
The scattering matrix for the simple case of scalar particles

f \ i	1	2	3	4	5	6	7	8
1	1,1							
2		2,2	2,3	2,4	2,5	2,6	2,7	2,8
3		3,2	3,3	3,4	3,5	3,6	3,7	3,8
4		4,2	4,3	4,4	4,5	4,6	4,7	4,8
5		5,2	5,3	5,4	5,5	5,6	5,7	5,8
6		6,2	6,3	6,4	6,5	6,6	6,7	6,8
7		7,2	7,3	7,4	7,5	7,6	7,7	7,8
8		8,2	8,3	8,4	8,5	8,6	8,7	8,8

The element (1, 1) is equal to one, the diagonal elements (2, 2), (3, 3), ..., (m, m) describe elastic scattering. (We define elastic scattering as the scattering in which particles are neither produced nor absorbed nor do they change their inner state.) The elements for which $m \neq n$ describe transition which are accompanied by production or absorption of particles. The empty locations indicate matrix elements which are equal to zero. Note that all the elements on a diagonal for which $m+n=$constant are described by the function $T(p', p)$ (31.28) with a certain fixed number of variables $3(m+n)$. This fact allows us, at least in some theoretical schemes, to consider all the functions

$$\langle m | \hat{T} | n \rangle = T(p'_1, p'_2, \ldots, p'_m; p_1, p_2, \ldots, p_n)$$

as one function

$$T(p) \equiv T(p_1, p_2, \ldots, p_m; p_{m+1}, \ldots, p_{m+n}), \tag{31.29}$$

in which the properties of the initial and final states are the same. This type of symmetry in the initial and final states is called *crossing symmetry*. In the general case for the reaction in which in the initial state (i) there are n particles, $a, b, c, d, \ldots, n$ and in the final state (f) there are m particles $x, y, z, u, \ldots, m$

$$a + b + c + d + \cdots \rightarrow x + y + z + u + \cdots \tag{31.30}$$

the function $T(p)$ (31.29) will be the same as that for the reaction

$$a + b + c + \cdots + \tilde{x} + \tilde{z} + \cdots \rightarrow \tilde{d} + \tilde{e} + \cdots + u + \cdots, \tag{31.30$'$}$$

where an arbitrary number of particles is transfered from the left to the right of the equation, of vice versa, but the particles are replaced by antiparticles (denoted by ~).

A special case of this situation is shown in Figure 34. The properties of cross symmetry allow us to consider the set of spaces $\mathfrak{R}_i(p)$ and $\mathfrak{R}_f(p')$ as one space

$$\mathfrak{R}(p) = \mathfrak{R}_i(p) \otimes \mathfrak{R}_f(p'). \tag{31.31}$$

We now turn to the case in which, besides the "original" particles with mass m, there also exists a stable particle with mass $m < M < 2m$. Such a particle may be considered the result of two particles with mass m coalescing; this being accompanied by a mass defect of $\Delta = 2m - M > 0$.

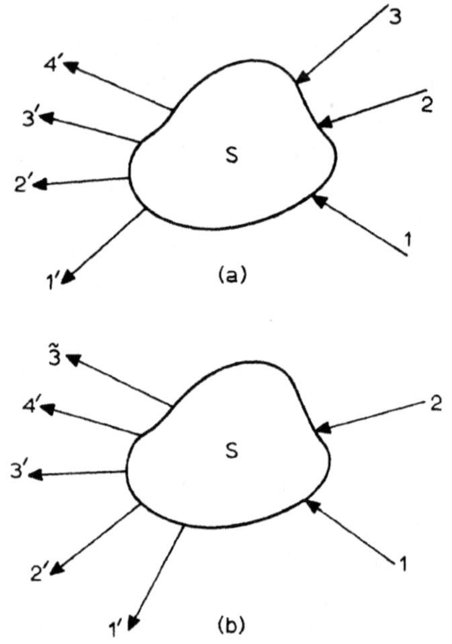

Fig. 34. An illustration of cross symmetry. In (a), 1, 2, 3 are particles in the initial state and 1′, 2′, 3′, 4′ are particles in the final state. In (b), particle 3 is transferred to the final state, but as an antiparticle. The junction s is the same in both diagrams.

The existence of such particles must be taken into account in the asymptotic states by considering that the corresponding Lobochevski space will have the curvature [see (31.11)] $R = -1/M^2$.

The scattering matrix for this case is shown in Table II. The ordering of the rows and columns is defined as before in terms of increasing rest mass of the particles that are found in the initial and final states. In order to simplify the notation the mass of the light particle m is taken to be 1. The existence of mutual transmutation of particles complicates the matrix. Apart from the unitary matrix element (1, 1), there is only one other element (M, M) which describes the motion of the particle M in terms of inertia. The elements $(M, 2)$ and $(2, M)$ are equal to zero because, according to the assumption that $m < M < 2m$, the particle M cannot decompose into two particles m. Similarly, the process in which two particles of mass m coalesce into one M particle is impossible.

TABLE II

The scattering matrix for the case of two types of scalar particles, particles with mass m and with mass M, where $m<M<2m$. The rows and columns are ordered in increasing rest mass.

f \ i	1	M	2	$M+1$	3	$2M$	4	$2M+1$
1	1,1							
M		M, M						
2			2,2	$2, M+1$	2,3	$2,2M$	2,4	$2,2M+1$
$M+1$				$M+1, M+1$	$M+1,3$	$M+1,2M$	$M+1,4$	$M+1, 2M+1$
3			3,2	$3, M+1$	3,3	$3,2M$	3,4	$3,2M+1$
$2M$			$2M, 2$	$2M, M+1$	$2M, 3$	$2M, 2M$	$2M, 4$	$2M, 2M+1$
4			4,2	$4, M+1$	4,3	$4,2M$	4,4	$4,2M+1$
$2M+1$			$2M+1,2$	$2M+1, M+1$	$2M+1,3$	$2M+1, 2M$	$2M+1,4$	$2M+1, 2M+1$

However, such processes are possible if three particles take part, because the third particle can carry off the extra energy which is released in this process. Therefore, the matrix elements $(M+1, 3)$ and $(3, M+1)$ are, in general, not equal to zero. Amongst the matrix elements are those which describe processes that involve only M particles [e.g., the element $(2M, 2M)$ describes elastic scattering of M particles, the element $(3M, 2M)$ describes the formation of M particles during a collision of two M particles, etc.].

We now turn to the case where the M particle has a mass that is greater than $2m$. There are now two possibilities: (i) the particle M is stable and so the processes $M \rightleftarrows 2m$ are forbidden; (ii) the particle M is unstable and in this case the processes $M \rightleftarrows 2m$ are allowed. The scattering matrix for the case of stable M and m particles which can be produced, absorbed or scattered in interactions, is shown in Table III. In contrast to the matrix given in Table II, in this case the rows and colums which contain M particles follow after rows and columns which contain two m particles. For definiteness, we assume that $2m<M<3m$ and, therefore, the third row and third column follow the row M and column M, respectively.

The matrix elements $(M, 2)$ and $(2, M)$ are equal to zero because of the assumption of stability of the M particles. This is a special exclusion.

TABLE III

The structure of the scattering matrix for the case of two scalar particles with mass m and M, where $2m < M < 3m$ and the particle M is assumed to be stable.

f \ i	1	2	M	3	$M+1$	4	$M+2$	$2M$
1	1,1							
2		2,2		2,3	2, $M+1$	2,4	2, $M+2$	2, $2M$
M			MM					
3		3,2		3,3	3, $M+1$	3,4	3, $M+2$	3, $2M$
$M+1$		$M+1$,2		$M+1$,3	$M+1$, $M+1$	$M+1$,4	$M+1$, $M+2$	$M+1$, $2M$
4		4,2		4,3	4, $M+1$	4,4	4, $M+2$	4, $2M$
$M+2$		$M+2$,2		$M+2$,3	$M+2$, $M+1$	$M+2$,4	$M+2$, $M+2$	$M+2$, $2M$
$2M$		$2M$, 2		$2M$, 3	$2M$, $M+1$	$2M$, 4	$2M$, $M+2$	$2M$, $2M$

The processes $M = m \rightleftarrows 2m$ are not considered forbidden; in particular, the process $2m \to M + m$ may be considered as an inelastic impact which is accompanied by the absorption of a light particle m and the formation of a heavy M particle. The processes $3m \rightleftarrows M$ are forbidden because of the condition $M < 3m$ and, therefore, the corresponding matrix elements (3, M) and (M, 3) are equal to zero.

At present the existence of a very large number of unstable particles is known. Therefore, the case in which the M particle is unstable is very important. However, it is impossible to consider this case without a space-time description of the phenomenon. Thus we shall postpone considering case (ii) and turn to some details related to space-time descriptions using the scattering matrix.

32. Space-time descriptions using the $\hat{S}$-matrix

By definition the $\hat{S}$-matrix transforms an initial state Ψ_i which is specified at $t_1 = -\infty$, into a final state Ψ_f which is determined at $t_2 = +\infty$. In Section 28 we showed that [see (28.18)] the $\hat{S}$-matrix is a limit of the unitary operator $U\{\sigma_2, \sigma_1\}$ which transforms the state $\Psi_i\{\sigma_1\}$ which is specified on the spatial surface σ_1 into the state $\Psi_f\{\sigma_2\}$ on the surface σ_2. In the case when the spatial surfaces are planes, these surfaces degenerate into the surfaces $\sigma_2 = t_2$ and $\sigma_1 = t_1$, so that

$$\hat{S} = \lim_{\substack{t_1 \to -\infty \\ t_2 \to -\infty}} \hat{U}(t_2, t_1). \tag{32.1}$$

We must now determine more accurately the meaning of the limits $t_1 \to \infty$ and $t_2 \to \infty$. We shall consider the simple case when there are only two particles A and B in the initial state. We shall consider that the initial state is specified for $t = t_1$ and is described by the wave packets u_A and u_B, respectively. Then

$$\Psi_i(t_1) = u_A(t_1)\, u_B(t_1). \tag{32.2}$$

We may assume that these packets were formed by passing through apertures $\mathscr{A}$ and $\mathscr{B}$ which were opened for some time close to the moment in time $t_1 = -T$. We assume that the initial dimensions of the packets are $a_\perp = a_\parallel = a$ ($a_\perp$ is the dimension perpendicular to the direction of motion and $a_\parallel$ is the parallel dimension). Let the distance between the packets at $t_1 = -T$ be $R \gg a$. We shall consider that the particles have sufficiently well-defined average momenta $P_A \gg \Delta p \approx \hbar/a$, $P_B \gg \Delta p \approx \hbar/a$ and the corresponding average packet velocities are $v_A = \partial E_A/\partial P_A$ and $v_B = \partial E_B/\partial P_B$ where $E_A = (m^2c^4 + c^2p_A^2)^{1/2}$, $E_B = (m^2c^4 + c^2p_B^2)^{1/2}$ are the average energies. In the center-of-mass system the relative velocity of the packets will be $v = v_A - v_B$, and $E_A = E_B$. In the following we shall omit the subscripts A and B because of importance is only the relative motion of packets. If the interaction radius of the particles is b,† then for $b \gg a$ the packets u_A, u_B will interact for a time $T \sim R/v$. In this case the particles may be considered as asymptotically free during the time

$$T = \frac{R}{v} \gg \frac{b}{v}, \tag{32.3}$$

beginning at the time $t = t_1$. In the case when $b < a$, we must remember that the accuracy to within which the particle may be localized and, therefore, for $b < a$, in place of (32.3) we have the condition

$$T = \frac{R}{v} \gg \frac{a}{v}. \tag{32.3'}$$

Because $a \gg \hbar/p = \lambda\!\!\!^{-}$ ($\lambda\!\!\!^{-}$ is the wavelength characteristic of the packets), in general $R \ll \lambda\!\!\!^{-}$ in the asymptotic region.

After the interaction for $t_2 = +T$, the wave function of the final state

† It is well known that the radius of interaction b is not always definable. In particular, in the case of Coulomb interaction, it is infinite. We here exclude this case from our consideration.

$\Psi_f(t_2)$ may be written in the form

$$\Psi_f(t_2) = u_A(t_2)\, u_B(t_2) + u(t_2), \tag{32.4}$$

where the function $u_A(t_2)\, u_B(t_2)$ describes the initial packet in the moment t_2, and the function $u(t_2)$ is the asymptotically scattered wave. The scattered wave u for large separations r of the particles has the form†

$$u(r) = C\frac{e^{ikr}}{r} + D\frac{e^{ikr}}{r^2} + \cdots, \tag{32.5}$$

where $k = 1/\lambda\!\!\!^{-}$ and $C, D, \ldots$ are coefficients that do not depend upon r. The asymptotic expression for this wave for $t_2 = T = R/v$ is

$$u(R) = C\frac{e^{ikr}}{R} \tag{32.6}$$

and we assume the separation $r = R$ is sufficiently large in comparison to $\lambda\!\!\!^{-}$ so that it is possible to neglect the second term (De^{ikR}/R^2) and all higher terms in (32.5). Therefore, the asymptotic region corresponds to the region which is usually called the *wave zone* $(R \gg \lambda\!\!\!^{-})$.
In "the language of time", neglecting terms of the order $1/R^2$ implies that, in taking the limit $t_2 = T \to +\infty$, terms of the order $1/T$ are kept and terms of the order of $1/T^2$ are considered negligibly small.

Note that there also exists an upper limit for T if we want to conserve the space-time description for the collision process. In fact using wave packets in this description assumes that the initial packet (specified for $t_1 = -T$) does not diffuse significantly before the collision, for otherwise the moment of collision would be quite undefined.

According to the relativistic wave-packet theory the packet, which has has a width Δ at the initial moment in time, diffuses in time so that its longitudinal width $\Delta_{\parallel}(t)$ is equal to [see (18.10) and (18.10′)]

$$\Delta_{\parallel} = \sqrt{a^2 + \frac{\lambda\!\!\!^{-2}}{a^2}\frac{m^4}{E^4}v^2t^2} \tag{32.7}$$

and the transverse width $\Delta_{\perp}(t)$

$$\Delta_{\perp} = \sqrt{a^2 + \frac{\lambda\!\!\!^{-2}}{a^2}v^2t^2}\,. \tag{32.7′}$$

† In the general case the scattered wave also describes new particle production in collisions. For simplicity, we shall limit ourselves here to elastic collisions.

Because of the properties of the factor $(m/E)^4$ in (32.7), Eq. (32.7′) is more critical than (32.7). The limits on the diffusion of the packet during the collision process implies that $\Delta_\perp$ at $t=2T$ does not greatly differ from $\Delta_\perp$ at $t=0$. Therefore, $a^4 \gg \lambda\!\!\!^{-2} v^2 T^2 = \lambda\!\!\!^{-2} R^2$. Combining this inequality with (32.3) and (32.3′) gives

$$\frac{a^2}{\lambda\!\!\!^-} \gg R \gg (b \quad \text{or} \quad a) \gg \lambda\!\!\!^-. \tag{32.8}$$

Figure 35a shows the paths of the particles which pass through the apertures $\mathscr{A}$ and $\mathscr{B}$. This figure shows the case when the radius of interaction of the particles is greater than the dimensions of the packet a. Figure 35b shows the case when $b \ll a$ so that the region of possible interaction is defined by the packet dimension a. In both cases the packet diffusion in a time T is assumed to be small. Figure 35c shows the case when the packets diffuse so rapidly that all information on the moment of collision is lost.

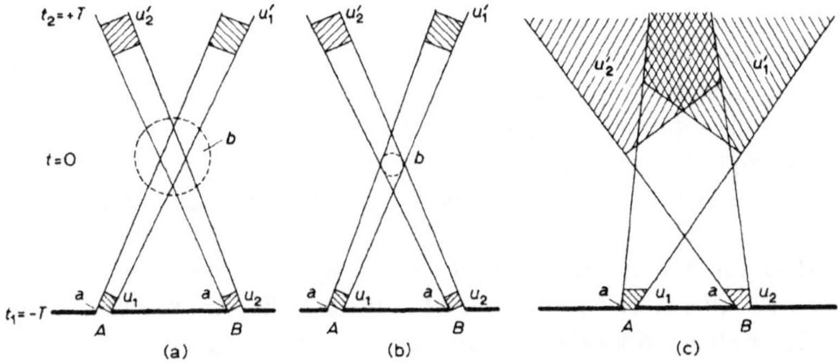

Fig. 35. Wave packets u_1 and u_2 which are emitted from the aperture A and B in an absorbing screen at a time $t_1=-T$ are shown here. Around $t=0$ a collision occurs; u_1' and u_2' are the same packets (partially scattered) at $t_2=+T$. b is the radius of interaction, a is the aperture width, $R=AB=vT$ is the asymptotic separation of the packets. Figure 35a shows the case of $b \gg a$, $R > b$. Figure 35b shows the case $b \ll a$, $R \gg a$ and Figure 35c shows the case of very large diffusion of the packets so that no information on causality exists

These restrictions on the limits $|t_2|$, $|t_1| \to \infty$ apply to all particles. Unstable particles have other restrictions which are significant. Let the initial beam (Ψ_i) include a particle with a decay period of τ_i. For $|t_1| > > \tau_i$, the unstable particle decays and the incident bean changes its character. For $|t_1| \ll \tau_i$ the unstable particle may be considered as stable. Therefore, for the limit $t_1 = -T$ the following restriction holds

$$\tau_i \gg T > \frac{b}{v} \quad \left(\text{or} > \frac{a}{v}\right). \tag{32.9}$$

A similar restriction holds in the case where unstable particles are present in the final state. The instability of the particles may be ignored if

$$\tau_f \gg T > \frac{b}{v} \quad \left(\text{or} > \frac{a}{v}\right). \tag{32.9'}$$

We now assume that in the spectrum of particles there is an unstable particle whose mass in M, where $2m < M < 3m$ so that the particle M may decay into two m particles: $M \to 2m$. Let the decay period be τ. Then there are two possible ways of constructing the $\hat{S}$-matrix. If the asymptotic state is defined for $T \ll \tau$, then the particle M may be considered stable and the $\hat{S}$-matrix has the form given in Table II. In this case the space $\mathfrak{R}(p)$ includes two types of Lobachevskii space. The first type has a curvature that is defined by the mass of the particle m, and the second type has the curvature defined by the mass of the heavy particle M.

We may also consider the case when the asymptotic time $T \gg \tau$. In general unstable M particles will not appear in the asymptotic case because they disintegrate into m particles in the time $T \gg \tau$. In this case the asymptotic space consists entirely of Lobachevski spaces that have curvatures characteristic of particles with mass m. The scattering matrix then has the form given in Table I.

However the "memory" of the possibility of forming unstable M-particles remains and it is expressed in the fact that the matrix element (2, 2) which describes elastic scattering of two m-particles $m + m \to m + m$ has a resonance when the relative energies of the two m-particles is equal to the mass of the M particle. By convention, this resonance could be re-written as a scattering process, which takes place by an instantaneous formation of an M-particle, in the form

$$m + m \to M \to m + m\,. \tag{32.10}$$

The width $\Delta E \sim \hbar/\tau$ of this resonance is defined in terms of the lifetime of the M-particle.†

From these consideration it follows that the character of asymptotic

† For further details on describing unstable particles using an $\hat{S}$-matrix, see Section 34.

space $\mathfrak{R}(p)$ and the character of the asymptotic states Ψ_i and Ψ_f depend essentially on the meaning which is given to the limits $t_1 \to -\infty$ and $t_2 \to +\infty$ in (32.1).

33. THE SCALE FOR THE ASYMPTOTIC TIME T

The interactions of particles are usually divided into strong, electromagnetic and weak interactions. These interactions define the time scale τ which serves as a criterion for selection of the asymptotic time T. In In the case of strong interactions, such as the interactions of baryons with themselves, the radius of the interaction region b is equal to an order of magnitude, to the Compton length of π-mesons, i.e., $b = \hbar/m_\pi c = 1.4 \times \times 10^{-13}$ cm.

The time τ_s during which a group of strongly interaction particles disintegrates, is $\tau_s = b/v$ where v is the velocity of the particles. For $v \approx c$ this gives $\tau_s \approx 5 \times 10^{-24}$ sec. In these interactions, excited states of baryons, baryon resonances and mesons, are produced. These and other particles are unstable. Baryon resonances are described by the equation

$$\mathrm{m} + \mathrm{N} \to \mathrm{N}^* \to \mathrm{N} + \mathrm{m}, \tag{33.1}$$

where m is a meson, N a baryon, and N* is an excited baryon. The corresponding Feynman diagram is shown in Figure 36. Amongst similar reactions is, for example, the reaction

$$\pi + \mathrm{N} \to \mathrm{N}^*_{1/2} \to \pi + \mathrm{N}, \tag{33.2}$$

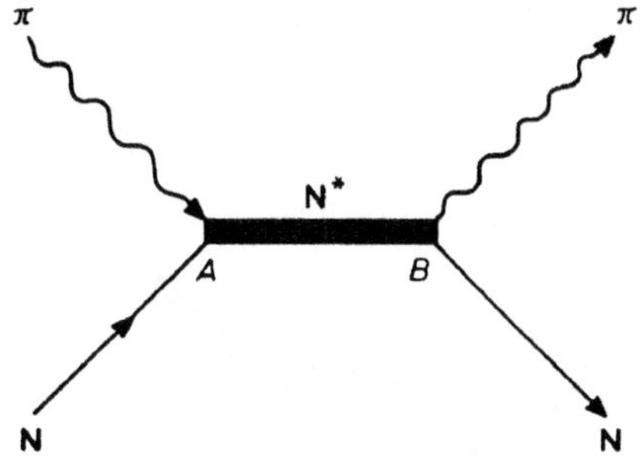

Fig. 36. Diagram of baryon resonance. The wavy line denotes a pion (π), the straight line a baryon (N). The heavy line denotes the excited nucleon (N*). Both vertices A and B are "strong".

where N is a nucleon and $N^*_{1/2}$ is the resonant state which has the quantum numbers $T_3=\frac{1}{2}$ (T_3 is the third component of the isotopic spin) $J=\frac{1}{2}$, is the spin, P is the parity +, and strangeness $s=0$. The energy of this resonance is around 1400 MeV.

At the present time a large number of baryon resonances are known [67]. The width of these resonances Γ is equal to an order of magnitude to ~ 100 MeV so that the lifetime of these particles is equal to

$$\tau^* = \frac{\hbar}{\Gamma} = 6 \times 10^{-24} \quad \text{sec},$$

i.e., it is of the same order of magnitude as the time for strong interactions of the original particles τ_s. The lifetime of mesons which decay in strong interactions, e.g.,

$$\rho \rightarrow \pi + \pi, \qquad \rho^0 \rightarrow \eta + \pi, \qquad K^* \rightarrow K + \pi. \tag{33.3}$$

is of the same order of magnitude. The corresponding diagrams are shown in Figure 37.

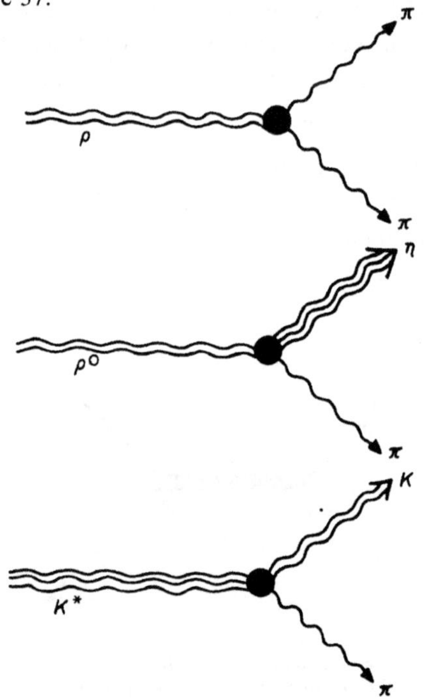

Fig. 37. "Strong" vertices for the disintegration of ρ-, ρ^0- and K-mesons.

Therefore, when we select $T \gg \tau_s$, we must consider baryon resonances and mesons, which decay during strong interactions as unstable particles. They do not enter into the asymptotic space $\mathfrak{R}(p)$.

Note that the lifetime of baryon resonances and mesons that is defined above refers to a system at rest. If the disintegrating particle is moving, then its lifetime τ' increases by a factor of E/m, where E is the energy of the unstable particle and m is its rest mass:

$$\tau' = \frac{E}{m} \tau . \tag{33.4}$$

It is for this reason that ultrarelativistic particles may be observed at very great distances R from the place where they are produced, in a way indicated by

$$R \approx c\tau' = c\tau \frac{E}{m} . \tag{33.5}$$

Therefore, an unstable particle with $E \gg m$ may change to the class of conditionally stable particles (if $E\tau_s/m \gg T \gg \tau_s$) and its existence must then be reflected in the space $\mathfrak{R}(p)$.

Amongst the unstable particles that are formed in strong interactions are π^0-mesons. These mesons decay into two γ-quanta so that electromagnetic interactions take part in the decay mechanism

$$\pi^0 \to \gamma + \gamma \tag{33.6}$$

The diagram for this decay is shown in Figure 38. In addition to the strong vertex A there are two electromagnetic vertices B and B'. These vertices are proportional to the electromagnetic interaction constant $e/(\hbar c)^{1/2} = \alpha^{1/2}$ and, therefore, the factor $\alpha^2 = (e^2/\hbar c) \approx 10^{-4}$ appears in the expression for the probability of such a decay process. Because of this constant, the decay time of $\pi^0 \to \gamma + \gamma'$ is far greater than τ_s. Estimating the lifetime of a π^0-meson from this diagram taking into account the phase space volume of the vertices B and B' gives a value of $\tau_{\gamma\gamma} \approx 10^{-16}$ sec.

For a value of T in the range

$$\tau_{\gamma\gamma} \gg T \gg \tau_s , \tag{33.6'}$$

a π^0-meson may be considered stable. For $T \gg \tau_{\gamma\gamma}$ in the asymptotic space $\mathfrak{R}(p)$ a photon must be included because at some distance from the point of interaction a γ-quantum will be observed instead of the π^0-meson.

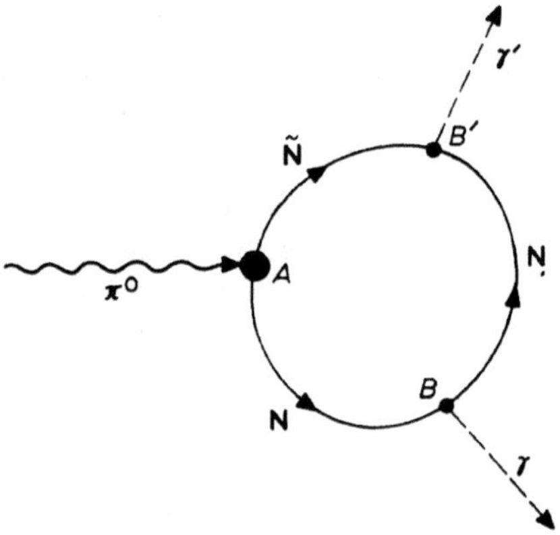

Fig. 38. Diagram for the decay process $\pi^0 \rightarrow \gamma + \gamma'$. The vertex A denotes a strong interaction. The π^0-meson dissociates here into a pair of nucleons (N) and an antinucleon (Ñ). At the point B the nucleon emits a photon γ. At the point B' the nucleon N is anihilated by the antinucleon Ñ and a photon γ' is emitted.

A significant number of mesons and excited baryons whose decay is accompanied by a charge in strangeness ($\Delta s \neq 0$) have a still longer lifetime because weak interactions take part in their decay mechanisms. The decay of $\pi^{\pm}$-mesons is decribed in Figure 39. This diagram consists of one "strong" vertex A and vertices of "weak" interaction W. Because the

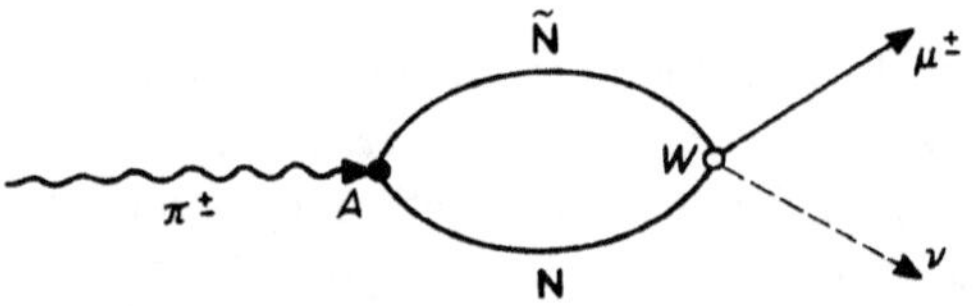

Fig. 39. Diagram for the decay of a $\pi^{\pm}$-meson into a μ-meson and a neutrino ν. A is the "strong" vertex. The vertex W denotes a weak interaction.

constant of the weak interaction is small, the lifetime of the $\pi^{\pm}$-mesons which decay according to

$$\pi^{\pm} \rightarrow \mu^{\pm} + \nu \tag{33.7}$$

(here μ is a mu-meson, ν a neutrino), has a value of $\tau_{\pi^{\pm}} \sim 10^{-8}$ s.

Consequently, if we consider $\pi^{\pm}$-mesons as stable particles, we must take a value of T in the range 10^{-8} sec $\gg T \gg 10^{-24}$ sec.

The decay of baryons (hyperons) which is accompanied by a change in strangeness as, for example, the process

$$\Sigma^- \to n + e^- + \tilde{\nu} \tag{33.8}$$

(Σ^- denotes a hyperon, n a neutron, e^- an electron, $\tilde{\nu}$ an antineutrino) is described by the diagram shown in Figure 40. Here there are two "strong"

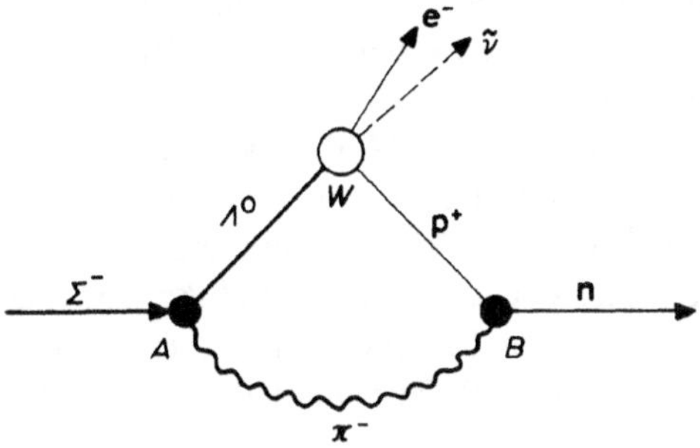

Fig. 40. Diagram describing the decay of a Σ^- hyperon into a neutron n, electron e^- and an antineutrino $\tilde{\nu}$. A and B are the vertices of strong interactions, and W is the vertex of a weak interaction.

vertices A and B and one "weak" vertex W which describes the process

$$\Lambda^0 \to p^+ + e^- + \tilde{\nu}. \tag{33.9}$$

The interaction characteristic of thid process is considered to be an "elementary", weak interaction (see [44]). The lifetimes of particles which decay in such weak processes are $\tau_\Sigma \sim 10^{-10}$ sec.

The foregoing shows that serious limitations must be taken into account when taking the limits $t_2 \to +\infty$ and $t_1 \to -\infty$ in the $\hat{S}$-matrix. These limitations arise from the analysis of the physical processes which are possible and the character of asymptotic space and the asymptotic states Ψ_i and Ψ_f depend on the choice of time T.

34. Unstable particles (resonances)

We now return to the case in which there is an unstable particle with mass $3m > M > 2m$ in the mass spectrum.

As we showed above, if the asymptotic time T is chosen such that $T \ll \tau$, where $\tau = \hbar/G$ is the lifetime of an unstable particle (G is the resonance width) then the M particle may be considered stable and it may be included in the asymptotic space $\mathfrak{R}(p)$. Therefore for $T \ll \tau$, the situation reduces to that discussed earlier in Section 31 where we assumed that two types of stable particles exited.

We shall now consider the opposite case, for $T \gg \tau$. In this case only one type of particle, that with mass $m < M$ will enter into the asymptotic space $\mathfrak{R}(p)$. Particles with mass M will appear in the form of a resonance state which, for $T > t \gg \tau$, has essentially completely disintegrated into the original stable m particles.

We may assign a certain mass M (the resonance width $G \ll M$) to the resonance state and therefore also a total momentum $P = (\mathbf{P}, P_0)$ where $P_0 = +(\mathbf{p}^2 + M^2)^{1/2}$. We may also assign to this state an angular momentum operator $\hat{K}_\mu$ which consists of the spin of the m particles $\hat{S}_{\mu\nu}$ (if they have any) and of the orbital momentum of their relative motions $\hat{L}_{\mu\nu} = \hat{p}_\mu x_\mu - \hat{p}_\nu x_\mu$ (μ, $\nu = 1, 2, 3, 4$). The internal moment of the resonance state $\hat{K}_\mu$ is defined as

$$\hat{K}_\mu = \tfrac{1}{2}\, \varepsilon_{\mu\alpha\beta\lambda} \frac{\hat{M}_{\alpha\beta} \hat{P}_\lambda}{\sqrt{P^2}}, \tag{34.1}$$

where $\varepsilon_{\mu\alpha\beta\lambda}$ is the completely antisymmetric unitary tensor and $\hat{M}_{\alpha\beta}$ is the total angular momentum

$$\hat{M}_{\alpha\beta} = \hat{L}_{\alpha\beta} + \hat{S}_{\alpha\beta}\,. \tag{34.2}$$

In the center of mass system the pseudovector $\hat{K}_\mu$ reduces to the three-dimensional angular momentum operator $\hat{\mathbf{M}}(\hat{M}_{23}, \hat{M}_{31}, \hat{M}_{12})$ [68]. In this coordinate system then, the operator $\hat{K}_\mu$ has a special meaning that is characteristic of non-relativistic theory, namely,

$$K^2 = \hbar J(J+1)\,, \qquad K_z = \hbar m_j\,, \tag{34.3}$$

where

$$-J < m_j < +J\,, \qquad J = 0, 1, 2, \ldots \quad \text{or} \quad J = \tfrac{1}{2}, \tfrac{3}{2}, \ldots. \tag{34.4}$$

Thus the resonant state which for $M \gg G$ represents an unstable particle may be assigned not only a mass M and total momentum P but also a characteristic angular momentum K. In their resonant state mesons are usually called *resonons*. The term *resonance* is reserved for states with baryons and mesons.

Resonons as well as resonances have other properties as well as those already mentioned (isotopic spin, internal symmetry, strangeness). However, we shall not discuss these questions which relate to the classification and spectroscopy of elementary particles, but shall return to considering the simple case of possible resonances in a system of two scalar particles with mass m. We shall use the center of mass system to study resonance, so that $\mathbf{P}=\mathbf{p}_1+\mathbf{p}_2=\mathbf{0}$ ($\mathbf{p}_1$, $\mathbf{p}_2$ are the momenta of the particles). We shall concentrate on the relative motions of these particles which have a specific angular momentum J.

The asymptotic wave function which describes the relative motion of the two m particles may be written as a sum of converging and diverging spherical waves, namely,

$$\psi_J(r, t) = \frac{1}{r}\left[a(s)\,\mathrm{e}^{\mathrm{i}pr} + b(s)\,\mathrm{e}^{-\mathrm{i}pr}\right]\mathrm{e}^{-\mathrm{i}Et}, \tag{34.5}$$

where p is the angular momentum, E the energy of the particle in the center-of-mass system, r the particle separation and t their mutual time (see Appendix X).

In the following it will be convenient to consider the momentum $\mathbf{p}$ and energy E as functions of an invariant energy

$$s = P_0^2 - \mathbf{P}^2. \tag{34.6}$$

In the center of mass we have

$$s = P_0^2 = E^2 = (E_1 + E_2)^2 = 4m^2 + 4\mathbf{p}^2, \tag{34.7}$$

from which

$$\mathbf{p}^2 = \tfrac{1}{4}s - m^2, \tag{34.8}$$

so that

$$\mathrm{i}p(s) = -\sqrt{m^2 - \frac{s}{4}}, \tag{34.9}$$

$$-\mathrm{i}E = -\sqrt{-s} \tag{34.10}$$

and

$$\psi_j(r, t) = \frac{1}{r}\left[a(s)\,\mathrm{e}^{-\sqrt{m^2 - s/4}\,r} + b(s)\,\mathrm{e}^{+\sqrt{m^2 - s/4}\,r}\right]\mathrm{e}^{-\sqrt{-s}\,t}. \tag{34.11}$$

We shall now consider s as a complex variable and in order to remove the ambiguity in the radicals in (34.11) we make a branch cut on the real axis of the complex plane (s) from $s = -\infty$ to $s = 0$ and from $s = 4m^2$ to $s + \infty$ (see Figure 41).

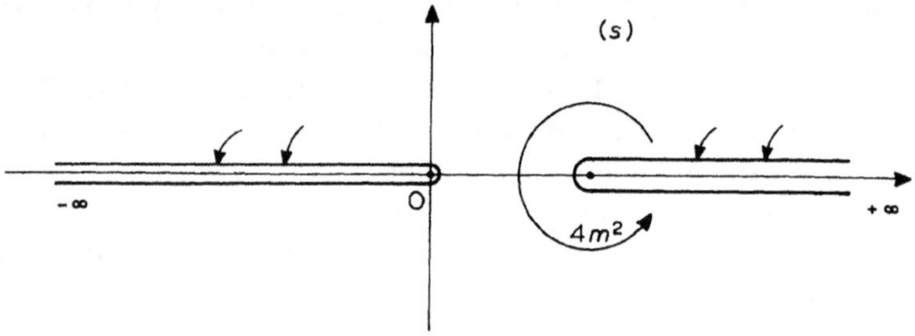

Fig. 41. The complex plane for the variable s. Two branch cuts are shown, from $s = -\infty$ to $s = 0$ and from $s = 4m^2$ to $s = +\infty$.

Using this branch cut we extend the function of the complex variable s onto the second sheet of the Riemann surface. We define the first sheet of this surface (the "physical" one) such that

$$\sqrt{m^2 - \frac{s}{4}} = \varepsilon^{1/2}\, e^{i(\varphi - \pi)/2}, \tag{34.12}$$

where

$$\operatorname{Re}\sqrt{m^2 - \frac{s}{4}} = \varepsilon^{1/2} \sin \tfrac{1}{2}\varphi > 0,$$

$$\varepsilon, \quad \varepsilon^{1/2} > 0 \quad \text{and} \quad m^2 - \frac{s}{4} = \varepsilon\, e^{i\varphi}. \tag{34.13}$$

On the second ("non-physical") sheet we have

$$\sqrt{m^2 - \frac{s}{4}} = \varepsilon^{1/2}\, e^{i(\varphi + \pi)/2}, \tag{34.14}$$

$$\operatorname{Re}\sqrt{m^2 - \frac{s}{4}} = -\,\varepsilon^{1/2} \sin \tfrac{1}{2}\varphi.$$

On the basis of (34.9) and (34.12) the momentum $p(s)$ is defined as

(a) on the first sheet

$$ip(s) = -\sqrt{m^2 - \frac{s}{4}} = i\varepsilon^{1/2}\, e^{i\varphi/2}\,, \tag{34.15}$$

$$p(s) = +\,\varepsilon^{1/2}\, e^{i\varphi/2}\,, \qquad \operatorname{Im} p(s) > 0\,; \tag{34.15'}$$

(b) and on the second, nonphysical sheet

$$ip(s) = -\sqrt{m^2 - \frac{s}{4}} = -\,i\varepsilon^{1/2}\, e^{i\varphi/2}\,, \tag{34.16}$$

$$p(s) = -\,\varepsilon^{1/2}\, e^{i\varphi/2}\,, \qquad \operatorname{Im} p(s) < 0\,. \tag{34.16'}$$

In this manner the physical sheet of the surface (s) maps onto the upper half-plane of the plane (p) and the non-physical sheet maps onto the lower half-plane.

Note also that in the stationary state the component of particle flow J that is directed along the radius vector has to be equal to zero. From this it follows that

$$b(s) = a^*(s^*)\,. \tag{34.17}$$

The condition that defines the discrete levels of our system, stable or resonant, states that

$$b(s) = 0\,, \tag{34.18}$$

and for stable levels

$$-\sqrt{-\,s} = -\,iE \quad \text{and} \quad 0 < E < 2m\,. \tag{34.19}$$

The resonant levels lie in the continuous region of the spectrum. For these, we have

$$-\sqrt{-\,s} = -\,iE - \Gamma\,, \tag{34.20}$$

where $E > 2m$, $G > 0$. The choice of sign in (34.20) guarantees an exponential decay of the resonant level for $t \to \infty$. From (34.20) it follows that

$$s = E^2 - 2iE\Gamma \tag{34.21}$$

(for $G \ll E$) so that $\operatorname{Re} s > 0$, $\operatorname{Im} s < 0$.

For the case of stable levels, from (34.11), (34.18) and (34.19) it follows that

$$\psi_j(r, t) = \frac{1}{r} a(s) \exp(-\varepsilon^{1/2} r - iEt). \tag{34.22}$$

Note that on the second Riemann sheet we would have

$$\psi'_j(r, t) = \frac{1}{r} a(s) \exp(+\varepsilon^{1/2} r - iEt), \tag{34.22'}$$

so that in place of (34.18) we would require that $a(s)=0$. Using (34.17), this condition would give the same level.

We now return to the unstable level. If we remained on the first, physical sheet, then in order to get from the upper half-plane to the lower one to the point $s=E^2-2iEG$ $(G<0)$, which lies below the branch cut, we would have to go around the point $s=4m^2$, along the path indicated by the arrow in Figure 41. The argument of the function $ip(s)$ [see (34.9)] would gain addition in phase of $+2\pi$– i.e., in this case

$$ip(s) = +\varepsilon^{1/2} \exp\left[\frac{i}{2}(\varphi - \pi) + \frac{i2\pi}{2}\right] = -i\varepsilon^{1/2} \exp\left(\frac{i\varphi}{2}\right), \tag{34.23}$$

where φ is defined by the expression

$$\varphi = -\arctan \frac{2\Gamma}{E(1 - 4m^2/E^2)} + 2\pi n + O\left(\frac{\Gamma^2}{E^2}\right), \tag{34.24}$$

in which $n=0$ for the first sheet and $n=1$ for the second. Therefore, on the physical sheet we have

$$p(s) = -p' - i\varkappa', \quad p' = \varepsilon^{1/2} \cos\frac{\varphi}{2}, \quad \varkappa' = -\varepsilon^{1/2} \sin\frac{\varphi}{2} \tag{34.25}$$

so that $p'>0$ and $x'>0$. Consequently, on the first sheet the wave function (34.22) would have the form

$$\psi_j(r, t) = \frac{1}{r} a(s) \exp[-i(p'r + iEt)] \exp[(\varkappa' r - \Gamma t)]. \tag{34.26}$$

It is easy to show that this solution contradicts the continuity equation for the current J because it decreases with increasing $r(x'=0)$ and $t(G>0)$.

The correct wave function $\varphi_j(r, t)$ for unstable levels is found on the second, non-physical sheet. In this case, according to (34.16′) we have

$$p(s) = -\varepsilon^{1/2} \exp\left(i\frac{\varphi + 2\pi}{2}\right) = p'' + i\varkappa'',$$
$$p'' = \varepsilon^{1/2} \cos\frac{\varphi}{2} > 0, \quad \varkappa'' = \varepsilon^{1/2} \sin\frac{\varphi}{2} > 0, \tag{34.27}$$

so that the wave function takes on the form

$$\psi_j(r, t) = \frac{1}{r} a(s) \exp\left[i(p''r - Et)\right] \exp(\varkappa'' r - \Gamma t), \tag{34.28}$$

which behaves correctly on the surface (r, t).

We shall now investigate the space-time properties of the general asymptotic solution (34.5). In order to do this, we shall first change the normalization of (34.5) in such a way that the scattered wave $u(p, r)$ assumes the form

$$u(p, r) = \frac{S_J(p) \exp\left[i(pr - Et)\right]}{2ipr}, \tag{34.29}$$

where $S_j(p)$ is an element of the scattering matrix $S_j(k)$ in a state with total angular moment J for elastic scattering with an energy E and corresponding momentum p. The normalization (34.29) corresponds to the initial asymptotic wave in the form

$$\psi_0(r, t) = \frac{\exp\left[i(pr - Et)\right]}{2ipr} - \frac{\exp\left[-i(pr + Et)\right]}{2ipr}. \tag{34.30}$$

Comparing Equations (34.29) and (34.30) with (34.5) shows that

$$S_J(p) = \frac{a(s)}{b(s)} = \tag{34.31}$$

$$= g(s) \frac{E - (E_R + i\Gamma)}{E - (E_R - i\Gamma)}, \tag{34.32}$$

where we combined the factors that are regular close to the pole $E = E_R - iG$ into the function $g(s)$.

We now assume that the initial wave ψ_0 is specified in the form of a wave packet

$$\psi(r, t) = \int_{-\infty}^{+\infty} c(p)\, \psi_0(r, t)\, \mathrm{d}p. \tag{34.33}$$

Choosing $\psi(r, t)$ to have the form of a wave packet allows the moment of particle impact to be localized about the time $t=0$. By considering that the function $c(p)$ is concentrated about the point $p=p_r$, we may rewrite (34.33) in the form

$$\psi(r, t) = \frac{c}{2\mathrm{i}p_R r} \exp\left[\mathrm{i}(p_R r - E_R t)\right] f(r - v_R t)$$
$$- \frac{c}{2\mathrm{i}p_R r} \exp\left[\mathrm{i}(p_R r - E_R t)\right] f(r + v_R t), \tag{34.34}$$

where $v_R = (\partial E/\partial p)_R$ is the relative velocity of the particles near the resonant energy $E \approx E_R$. The function f decreases outside the interval

$$|r \pm vt| > \hbar/\Delta, \tag{34.35}$$

where Δ is the width of the packet in momentum space

$$c(p) \approx 0 \tag{34.35'}$$

for $|p - p_R| > \Delta$.

From this it follows that the collision takes place in the neighborhood of the time

$$t = \hbar/\Delta v_R$$

(we assume that the width of the packet Δ is greater than the radius of interaction of the particle).

Furthermore, we consider that

$$\frac{\hbar}{\Delta} \ll \frac{\hbar v_R}{\Gamma} = \Lambda, \tag{34.36}$$

where Λ is decay length. The scattered wave which arises in the collision of the packets (34.34) has, in agreement with (34.29) and (34.33), the form

$$u_j(p, r) = \frac{1}{2\mathrm{i}p_R r} \int_{-\infty}^{\infty} c(p) \exp\left[\mathrm{i}(pr - Et)\right] g(s) \frac{E - (E_R + \mathrm{i}\Gamma)}{E - (E_R - \mathrm{i}\Gamma)} \mathrm{d}p. \tag{34.37}$$

We set $p - p_R = \xi$, so then

$$u_j(p, r) \frac{\exp[\mathrm{i}(p_R r - E_R t)]}{2\mathrm{i} p_R r} \int_{-\infty}^{\infty} c(p_R + \xi) \exp[\mathrm{i}\xi(r - v_R t)]$$

$$\times\, g(s) \frac{\xi - \mathrm{i}(\Gamma/v_R)}{\xi + \mathrm{i}(\Gamma/v_R)} \,\mathrm{d}\xi\,. \tag{34.38}$$

From this we see that if $g(s)$ and $c(p)$ are such that the integration contour in (34.38) may be closed with a large semicircle,† then for $(r - v_R t) < 0$ this semicircle will be drawn in the lower half-plane, and for $(r - v_R) > 0$ in the upper half-plane. Evaluating the integral gives

$$u(p, r) = -\frac{ic \exp[\mathrm{i}(p_R r - E_R t)]}{p_R r}$$

$$\times \exp\left[(r - v_R t)\frac{1}{\Lambda}\right] \quad \text{for} \quad r - v_R t < 0 \tag{34.39}$$

and

$$u(p, r) = 0 \qquad \text{for} \qquad r - v_R t > 0\,. \tag{34.39'}$$

In this manner causality is realized in resonant decay. Decay processes do not, in general take place in the region $r > v_R t$ which does not attain resonances that are formed for $r \sim 0$, $t = 0$ (to an accuracy of $\hbar/\Delta$). In the region $r < v_R t$ resonances decay exponentially with a decay constant $G = v_R/\Lambda$ (see Figure 42).

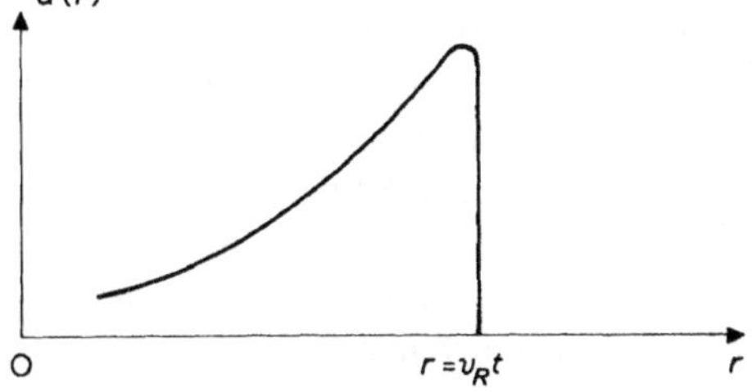

Fig. 42. The decay of resonances is shown near the point $r = 0$. In the region $0 < r < v_R t$, the resonances decay exponentially In the region $r > v_R t$, their wave function $u(r)$ is equal to zero.

† If $c(p)$ decreases sufficiently rapidly with increasing radius of the circle $|p - p_R| \to \infty$, then the behavior of the other functions under the integral sign in (34.38) becomes unimportant.

This analysis of unstable states shows that the $\hat{S}$-matrix can correctly describe the space-time behavior of unstable particles ("resonons" and "resonances"). The computations for non-relativistic resonons which further illustrate this theory are given in Appendix XI.

35. Conditions of Macroscopic Causality for the $\hat{S}$-Matrix

In 1942, Heisenberg put forward his prominent program of further development of quantum theory in which he introduced the concept of describing elementary particle events using the $\hat{S}$-matrix instead of wave functions. The meaning of this was not refuted in the further development of experimental and theoretical physics. *De facto* methods which were based on the wave function concept were replaced by methods based on the analytical properties of the $\hat{S}$-matrix. However, a problem remains in that present-day mathematics of the $\hat{S}$-matrix do not extend to regions in the small, to elementary events.

Frequently, theoretical schemes which work only in $\hat{S}$-matrices are similar to factories with only two departments, on which receives materials and the other which packages the finished products. The department which processes the materials is lacking. The analytic continuation of the $\hat{S}$-matrix across a mass surface and energy surface allows us to some extent to view this "processing department" in which elementary events are "produced". Nevertheless, the methods of analytic continuation are, to all appearances, not able to give a complete picture of physical phenomena in the world of elementary particles.

We must admit that our present-day abilities are very limited. However, there are two facts which support the use of the $\hat{S}$-matrix. These are: (a) there do not appear to exist elementary particles which could not be described in terms of $\hat{S}$-matrices, and (b) the $\hat{S}$-matrix at least in principle is measurable. Thus we have a basis for considering the $\hat{S}$-matrix as a theoretical construction which may be able to preserve its significance in future theory.

Thus we shall proceed in studying the $\hat{S}$-matrix in the hope that it will remain able to account for observational results in the future. We must now formulate the causality principle directly in terms of the $\hat{S}$-matrix without using field theories or microcausality conditions.

At first glance it may appear that there exists an irremovable contradiction in the formulation of the problem. The $\hat{S}$-matrix is given in the momentum-energy space, in the multidimensional Lobachevski space $\mathfrak{R}_3(p)$. However, in defining the causal relationships, by their very meaning, it is necessary to use the space-time variables defined in the space $\mathfrak{R}_4(x)$. Because of this fact, causality can be formulated in terms of the $\hat{S}$-matrix only to the degree of definiteness that is limited by the possibility of using both spaces $\mathfrak{R}_3(p)$ and $\mathfrak{R}_4(x)$. This possibility is characteristic of classical macroscopic physics. Therefore we shall call the causality which we shall study in this section *macroscopic causality* [69, 70].

Simultaneous use of both spaces $\mathfrak{R}_3(p)$ and $\mathfrak{R}_4(x)$ rests upon the principle possibility of constructing a wave packet which in the time $t_1 = -T$ to $t_2 = +T$ (T is the asymptotic time) diffuses so slightly that the whole process of collision of two wave packets in a time $2T$ may be represented by particles that are localized in space and time. The possible existence of such packets was shown in Section 18 (see [30, 74]).

The conditions for describing collision processes using wave packets was also discussed in Section 18. These conditions state that

$$\frac{a^2}{\lambdabar} \gg R \gg a \gg \lambda, \tag{35.1}$$

where $\lambdabar$ is the wavelength characteristic of the packet, a is the packet dimension, R is the distance between the packets $R = Tv$ and v is the relative velocity of the packets. Figure 43 shows two wave packets $u_1(x_1)$ and $u_2(x_2)$ which describe particles in the initial state. In satisfying (35.1) these packets do not change much in the time $t_1 = -T$ to $t_2 = +T$.

We now turn to the mathematical description of the collision of these packets using $\hat{S}$-matrices. We shall write the equation for scalar particles in an explicit form. According to (31.28) the $\hat{S}$-matrix element, written in the momentum representation has the form

$$S_{fi} = \langle p'|\, \hat{I}\, |p\rangle + \mathrm{i}\delta^4(P' - P)\frac{T(p', p)}{\sqrt{2p'_0 \ldots 2p_0}}. \tag{35.2}$$

The functions $\langle p'|I|p\rangle$ and $T(p', p)$ that describe the transition from the state $(p) \equiv (p_n, p_{n-1}, \ldots, p_1)$ to the state $(p') \equiv (p_{m+n}, p_{m+n-1}, \ldots, p_{n+1})$

may be written in a more detailed manner in the form

$$\langle p'|\,\hat{I}\,|p\rangle = \prod_{s=1}^{n} \delta^3(\mathbf{p}_{n+s} - \mathbf{p}_s) \quad \text{for} \quad m = n, \tag{35.3}$$

otherwise $\langle p'|\,\hat{I}\,|p\rangle = 0$ and

$$\frac{T(p', p)}{\sqrt{2p'_0 \ldots 2p_0}} = \frac{T(p_{m+n}, p_{m+n-1}, \ldots, p_2, p_1)}{\sqrt{2p^0_{m+n}}\sqrt{2p^0_{m+n-1}} \cdots \sqrt{2p^0_1}}, \tag{35.3'}$$

where the numerator is an invariant function of the momenta and p^0_1, $p^0_2, \ldots, p^0_{n+n}$ are their fourth components.

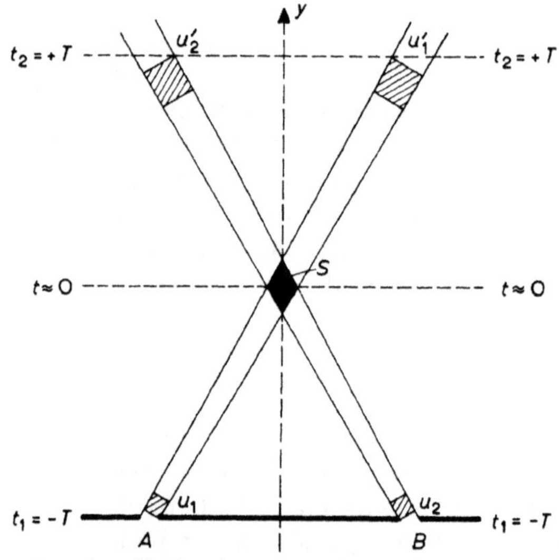

Fig. 43. Description of a collision using wave packets. Packets u_1 and u_2 are shown for $t_1 = -T$ in the apertures A and B. At the time $t \approx 0$, they collide (in the blackened region S). Thereafter the packets separate and gradually diffuse. This increase in packet size is shown for the time $t_2 = +T$. The scattered waves are not shown.

For the sake of definiteness, we consider that there are only two particles in the initial state Ψ_i and that these two particles are described by the wave packets $u_1(x_1)$ and $u_2(x_2)$ which belong to the class of packets which satisfy (35.1). Such packets may be represented by the integrals

$$u(x) = \frac{1}{(2\pi)^{3/2}} \int \tilde{u}(\mathbf{p}) \exp(ipx) \frac{d^3p}{2p^0}, \tag{35.4}$$

where $p^0=(m^2+\mathbf{p}^2)^{1/2}$. The wave function of the initial state Ψ_{in} may be written in the momentum representation in the form

$$\tilde{\Psi}_{\text{in}}(p_2, p_1) = \frac{\tilde{u}_2(\mathbf{p}_2)}{\sqrt{2p_2^0}} \frac{\tilde{u}_1(\mathbf{p}_1)}{\sqrt{2p_1^0}}. \tag{35.5}$$

Then the wave function for the final state $\tilde{\Psi}_{\text{out}}(p_4, p_4)$ for $m=0$ (elastic collision) is obtained from (35.5) using the $\hat{S}$-matrix transformation (35.2):

$$\tilde{\Psi}_{\text{out}}(p_4, p_3) = \tilde{\Psi}_{\text{in}}(p_4, p_3) + i \int \delta^4 (p_4 + p_3 - p_2 - p_1) \times \frac{T(p_4, p_3; p_2, p_1)}{\sqrt{2p_4^0 2p_3^0}} \tilde{u}_2(\mathbf{p}_2)\, \tilde{u}_1(\mathbf{p}_1) \frac{\mathrm{d}^3p_2\, \mathrm{d}^3p_1}{2p_2^0 2p_1^0}, \tag{35.6}$$

and the wave function for $m>4$ (formation of new particles during collision) has the form

$$\Psi_{\text{out}}(p_m, p_{m-1}, \ldots, p_3) = + i \int \delta^4 (p_m + p_{m-1} + \cdots + p_3 - p_2 - p_1) \frac{T(p_m, p_{m-1}, \ldots, p_3; p_2, p_1)}{\sqrt{2p_m^0\, 2p_{m-1}^0 \ldots 2p_3^0}} \times \tilde{u}_2(\mathbf{p}_2)\, \tilde{u}_1(\mathbf{p}_1) \frac{\mathrm{d}^3p_2}{2p_2^0} \frac{\mathrm{d}^3p_1}{2p_1^0}. \tag{35.7}$$

From (35.4) it follows that

$$\frac{\tilde{u}(\mathbf{p})}{2p_0} = \frac{1}{(2\pi)^{3/2}} \int u(x) \exp[-\mathrm{i}px]\, \mathrm{d}^3x. \tag{35.8}$$

This relationship allows us to change over to the coordinate representation in (35.6) and (35.7). In order to do this we express the amplitudes $\tilde{u}_2(\mathbf{p}_2)$ and $\tilde{u}_1(\mathbf{p}_1)$ in (35.6) and (35.7) in terms of $u_1(x_1)$ and $u_2(x_2)$ using (35.8) and then we multiply the left-hand side of (35.6) and (35.7) by the exponential function

$$\frac{1}{(2\pi)^{3/2\,(m-2)}} \frac{\exp[\mathrm{i}(p_m x_m + p_{m-1} x_{m-1} + \cdots + p_3 x_3)]}{\sqrt{2p_m^0 2p_{m-1}^0 \cdots 2p_3^0}} \tag{35.9}$$

and integrate the product with respect to $d^3p_m, d^3p_{m-1}, \ldots, d^3p_3$. The result is

$$\Psi_{out}(x_4, x_3) = \Psi_{in}(x_4, x_3) + i \int \mathfrak{G}(x_4, x_3 \mid x_2, x_1) \times u_2(x_2)\, u_1(x_1)\, d^3x_2\, d^3x_1, \tag{35.10}$$

and in the general case for $m > 4$,

$$\Psi_{out}(x_m, x_{m-1}, \ldots, x_3) = +\, i \int \mathfrak{G}(x_m, x_{m-1}, \ldots, x_3 \mid x_2, x_1) \times u_2(x_2)\, u_1(x_1)\, d^3x_2\, d^3x_1, \tag{35.11}$$

where the function $\mathfrak{G}(x_m, x_{m-1}, \ldots, x_3 | x_2, x_1)$ is defined as

$$\mathfrak{G}(x_m, x_{m-1}, \ldots, x_3 \mid x_2, x_1) = \frac{\partial^2 \mathfrak{G}_0(x_m, x_{m-1}, \ldots, x_3 \mid x_2, x_1)}{\partial t_2\, \partial t_1}. \tag{35.12}$$

where $t_2 = x_2^0$, $t_1 = x_1^0$ are the fourth components of the vectors x_2 and x_1 and the function $\mathfrak{G}_0(x_m, x_{m-1}, \ldots x_3 | x_2, x_1)$ is equal to

$$\begin{aligned}\mathfrak{G}_0(x_m, x_{m-1}, \ldots, x_3 \mid x_2, x_1) &= \int \delta^4(p_m + p_{m-1} + \ldots + p_3 - p_2 - p_1) \\ &\times T(p_m, p_{m-1}, \ldots, p_3; p_2, p_1) \\ &\times \exp\left[i\left(p_m x_m + p_{m-1}x_{m-1} + \cdots + p_3x_3 - p_2x_2 - p_1x_1\right)\right] \\ &\times \frac{d^3p_m\, d^3p_{m-1} \cdots d^3p_1}{2p_m^0 2p_{m-1}^0 \cdots 2p_1^0}.\end{aligned} \tag{35.13}$$

From (35.13) it is obvious that the function $\mathfrak{G}_0$ (3.1), (3.2) is in fact one of the matrix elements T (3.1), (3.2) [see alse (35.2)], written in the coordinate representation. The functions $\mathfrak{G}_1$ and $\mathfrak{G}_0$ satisfy Klein's equation by any of their arguments

$$(\Box_s^2 - m^2)\, \mathfrak{G}_0 = 0, \quad (\Box_s^2 - m^2)\, \mathfrak{G}_1 = 0; \tag{35.14}$$

$s = 1, 2, \ldots, m$. These functions describe the collision process in terms of free particles. They are analogous to the functions $s(x_m, x_{m-1}, \ldots, x_1)$ that we considered in Section 24 [see (24.5) and (24.6)] but they are not identical because the spectral decomposition of the functions $s(x_m, x_{m-1}, \ldots, x_1)$ is not contained within the Lobachevski space $\mathfrak{R}_3(p)$.

The functions $s(x_m, x_{m-1}, x_2, x_1)$ which are based on the concept of scattering matrix $\hat{S}$ in terms of local fields (see Appendix VII) may not exist whereas the function $\mathfrak{G}(x_m, x_{m-1}, \ldots | x_2, x_1)$ is based on the more general assumption of the existence of the scattering matrix $\hat{S}$ taken for real particles and on the energy surface.

We shall now formulate the principle of macroscopic causality. The wave packets $u(x_1)$ and $u(x_2)$ which describe the initial state $\Phi_{\text{in}}(x_2, x_1)$ contribute negligibly to the scattered wave $\Phi_{\text{out}}(x_m, x_{m-1}, \ldots, x_3)$ if in the asymptotic region

$$|\mathbf{x}_i - \mathbf{x}_k| \gtrsim R, \quad |t_i - t_k| \gtrsim T, \tag{35.15}$$

the points $x_1, x_2, \ldots, x_m$ lie outside the region defined by the conditions

$$(x_2 - x_1)^2 = (t_2 - t_1)^2 - (\mathbf{x}_2 - \mathbf{x}_1)^2 > 0 \tag{35.16}$$

and

$$(x_i - x_1)^2 > 0, \quad t_i - t_1 > T, \tag{35.17}$$

$$(x_i - x_2)^2 > 0, \quad t_i - t_2 > T. \tag{35.17'}$$

$(i = 3, \ldots, m-1, m)$. These conditions are illustrated in Figure 44 for elastic collisions $(m=4)$.

We shall denote the region in which the conditions (35.15)-(35.17) hold by $\mathfrak{M}(x)$. Then in terms of the function $\mathfrak{G}(x) = \mathfrak{G}(x_m, x_{m-1} \ldots | x_2, x_1)$, *the principle of macrocausality* may be stated thus: *the functions* $\mathfrak{G}(x)$ *must vanish sufficiently rapidly outside the region* $\mathfrak{M}(x)$

$$\mathfrak{G}(x) \to 0 \qquad \text{inside} \qquad \mathfrak{M}(x). \tag{35.18}$$

In this manner the macroscopic causality condition restricts the behavior of the function $\mathfrak{G}(x)$ only in the asymptotic regions. The words "must vanish sufficiently rapidly" must be understood to mean "significantly faster than the usual causal interactions vanish."

Note that these interactions vanish in a wide zone as $1/R$. Therefore, non-causal interactions must vanish significantly faster than $1/R$. If they vanish in a region that is closer to the wave zone of the signal, i.e., for $R < \lambda\!\!\!^{-}$ where $\lambda\!\!\!^{-}$ is the wavelength, then they will in general be unobservable. Therefore the macrocausal condition given above allows for wide modifications in microcausality and, therefore, for changes in geometry in the small. We may say that the $\hat{S}$-matrix is characterized by its tolerance, its

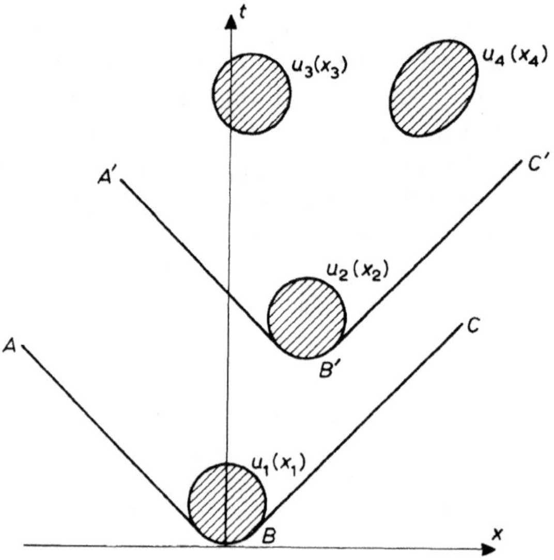

Fig. 44. Macroscopic causality. ABC is the light cone at the wave packet $u_1(x_1)$; $A'B'C'$ is the light cone at the wave packet $u_2(x_2)$. $u_3(x_3)$ and $u_4(x_4)$ are wave packets that describe the scattered waves. The relativistic separation of the four packets shown corresponds to the fulfilling of macroscopic causality.

readiness to assimilate various theoretical schemes for describing elementary particle interactions and their inner structure insofar as such changes take place in small spatial regions with dimensions $\approx b$ and in small times $\approx b/v$. We shall call the theoretical schemes in which microcausality is observed *acausal* (or *non-local*, compare with Section 26). If acausality does not violate macrocausality, we shall call such schemes *locally acausal*

Therefore, to construct such a scheme, it is sufficient that: (a) The $\hat{S}$-matrix remain locally acausal. Furthermore, we must recall the necessary condition that the $\hat{S}$-matrix be unitary (31.3): (b) The $\hat{S}$-matrix must be unitary.† Combining these two conditions in non-trivial and in fact is one of the basic problems of non-local theory. In the next section we shall consider examples of propagation functions that satisfy the condition of local-acausality and in Section 37 we shall give an example of constructing an $\hat{S}$-matrix which satisfies conditions (a) and (b).

† We may consider to what extent the condition of unitarity is really necessary, however omitting this requirement leads to very serious problems (see Appendix XII).

36. Examples of Acausal Functions

We shall first consider the acausal analogue of the causal function $D_c(x)$ which is symmetric with respect to the past and the future. $D_c(x) \neq 0$ for $\varepsilon = \pm 1$. Because of this it is very convenient for constructing an acausal influence function that is symmetric with respect to retardation and advancement.

For simplicity, we shall consider the case in which the rest mass of the field particles that transmit the interaction is equal to zero. In this case, according to Appendix I [Equation (I.6)] we have

$$\left.\begin{aligned} D_c(x) &= \delta(x^2) + \frac{\mathrm{i}}{\pi}\frac{1}{x^2}, \qquad x^2 = t^2 - r^2, \\ t &= t_2 - t_1, \qquad r = |\mathbf{x}_2 - \mathbf{x}_1|. \end{aligned}\right. \tag{36.1}$$

By expanding $\delta(x^2)$ and $1/x^2$ we may rewrite this equation in the form

$$D_c(x) = \frac{1}{2r}\left[\delta^+(t - r) + \delta^-(t + r)\right]. \tag{36.1'}$$

We may now construct the acausal function $D_c(x, a)$ which approximates the causal function $D_c(x)$ for sufficiently small values of the parameter a. In order to do this we first change

$$\delta(x^2) \quad \text{to} \quad \delta(x^2, a) = \frac{a^2}{\pi}\frac{1}{x^4 + a^4} \tag{36.2}$$

and

$$\frac{\mathrm{i}}{\pi}\frac{1}{x^2} \quad \text{to} \quad \mathrm{i}\frac{x^2}{a^2}\,\delta(x^2, a). \tag{36.2'}$$

Then $D_c(x, a)$ may be written as

$$\left.\begin{aligned} D_c(x, a) &= \delta(x^2, a) + \mathrm{i}\frac{x^2}{a^2}\,\delta(x^2, a), \\ \frac{1}{\pi}\frac{1}{x^2 - \mathrm{i}a^2} &= \frac{1}{2\pi\mathrm{i}\sqrt{r^2 + \mathrm{i}a^2}}\left[\frac{1}{t - \tau_1} - \frac{1}{t - \tau_2}\right], \end{aligned}\right\} \tag{36.3}$$

where

$$\tau_1 = +\sqrt{r^2 + \mathrm{i}a^2}, \qquad \tau_2 = -\sqrt{r^2 + \mathrm{i}a^2}. \tag{36.4}$$

We now compute the field $\varphi(x_2)$ which is produced at the point $x_2(\mathbf{x}_2, t_2)$ by the source $Q(x)$ which is localized at the point $x_1(\mathbf{x}_1, t_1)$ to within the space-time region $\Omega(x_1) \approx L^3 T$. Let this source be defined by the equation

$$Q(x_1) = \rho\left(\frac{\mathbf{x}_1}{L}\right) g\left(\frac{t_1}{T}\right) e^{i\omega t_1}, \tag{36.5}$$

where $\omega > 0$, $\omega t \gg 1$. $L \approx cT$. In particular,

$$\rho\left(\frac{\mathbf{x}_1}{L}\right) = \frac{1}{\pi L} \frac{L}{\mathbf{x}_1^2 + L^2}, \quad g\left(\frac{t_1}{T}\right) = \frac{1}{\pi} \frac{T}{t_1^2 + T^2}, \tag{36.6}$$

as in Section 26.

The field $\varphi(x_2)$ which is produced by the source $Q(x_1)$ (36.5) may be computed, using the acausal influence function $D_c(x, a)$ (36.3), from the equation

$$\varphi(x_2) = \int \rho\left(\frac{\mathbf{x}_1}{L}\right) d^3x_1 \int_{-\infty}^{\infty} D_c(x, a)\, g\left(\frac{t_1}{T}\right) e^{i\omega t}\, dt_1 . \tag{36.7}$$

Placing $D_c(x, a)$ from (36.3) into this equation gives

$$\varphi(x_2) = \frac{1}{2\pi i} \int \rho\left(\frac{\mathbf{x}_1}{L}\right) \frac{d^3x_1}{\sqrt{r^2 + ia^2}} \times \int_{-\infty}^{\infty} g\left(\frac{t_1}{T}\right) e^{i\omega t} \left[\frac{1}{t - \tau_1} - \frac{1}{t - \tau_2}\right] dt_1 , \tag{36.8}$$

where $t = t_2 - t_1$. In the complex plane $t_1 = t_1' + it_1''$ there are poles at $t_1 = t_2 - \tau_1$, $t = t - \tau$, and $t = \pm iT$. The integral in (36.8) for $\omega > 0$ may be closed in the upper half-plane with an infinitely large semicircle. The pole $t_1 = iT$ gives a contribution which becomes vanishingly small for $\omega T \to \infty$ as $e^{-\omega T}$ so that only the contribution from the pole $t_1 = t_2 - \tau_1$, $\operatorname{Im} \tau_1 < 0$ remains.

We thus obtain

$$\varphi(x_2) = \int \rho\left(\frac{\mathbf{x}}{L}\right) \frac{d^3x}{\sqrt{r^2 + ia^2}} \exp[i\omega(t_2 - \sqrt{r^2 - ia^2})] \times g\left(\frac{t_2 - \sqrt{r^2 - ia^2}}{T}\right). \tag{36.9}$$

The expression

$$\mathfrak{G}_{\text{ret}}(r, t_2, a) = \frac{\exp\left[\mathrm{i}\omega\left(t_2 - \sqrt{r^2 - \mathrm{i}a^2}\right)\right]}{\sqrt{r^2 + \mathrm{i}a^2}} \tag{36.10}$$

may be considered as the retarded Green's function for an almost periodic source (with frequency ω). This function has no singularity at $r=0$ and for $r \gg a$ it becomes the Green's function $\mathfrak{G}_{\text{ret}}(r, t_2)$ which was computed in Section 26 using $D_c(x)$. The proper character of the causality relation for $r \gg a$ is guaranteed for this case by two facts. These are (a) the proper choice of singularities of the function $D_c(x, a)$ and (b) the choice of the source. The source $Q(x_1)$ is localized in the space-time region $\Omega \approx L^3 T \gg \gg \lambda\!\!\!^{-3}/\omega$ and has a positive frequency $\omega > 0$ which is responsible for emitting a quantum with a positive energy $\hbar\omega > 0$. In the case when $\omega < 0$, $\hbar\omega < 0$ instead of (36.10) we would obtain the expression for $\mathfrak{G}(r, t_2, a)$ which approximates the advanced potential:

$$\mathfrak{G}_{\text{adv}}(r, t_2, a) = \frac{\exp\left[-\mathrm{i}\omega\left(t_2 + \sqrt{r^2 + \mathrm{i}a^2}\right)\right]}{\sqrt{r^2 + \mathrm{i}a^2}} \tag{36.10'}$$

This equation is obtained by closing the integral (36.8) in the lower half-plane where there are poles $t_1 = -T$ and $t_1 = t_2 - \tau_2$, $\operatorname{Im} \tau_2 > 0$. The interpretation of the functions $\mathfrak{G}_{\text{ret}}(r, t_2, a)$ and $\mathfrak{G}_{\text{adv}}(r, t_2, a)$ is the same as that given in Section 26.

Note that if we had taken very small values of L and T, we would have come to a significant violation of macroscopic causality. To explain this, it is sufficient to consider the case of small T ($\omega T \approx 1$). In this case we cannot neglect the contribution to the integral (36.8) of the pole $t_1 = +\mathrm{i}T$. Including this pole gives

$$\begin{aligned} \varphi(x_2) &= \frac{\mathrm{e}^{-\omega T}}{\mathrm{i}\pi} \int \rho\left(\frac{\mathbf{x}_1}{L}\right) \mathrm{d}^3 x_1 \frac{1}{(t_2 - \mathrm{i}T)^2 - (r^2 + \mathrm{i}a^2)} \\ &\approx \frac{\mathrm{e}^{-\omega T}}{\mathrm{i}\pi} \int \rho\left(\frac{\mathbf{x}_1}{L}\right) \frac{\mathrm{d}^3 x_1}{t_2^2 - r^2}. \end{aligned} \tag{36.9'}$$

The nature of the denominator in (36.9′) indicates that $\varphi(x_2)$ contains a symmetric advanced and retarded influence. This violates macroscopic causality and it shows the extent to which it is important for the source

$Q(x_1)$ to be smooth in space-time for formulating causality. It is necessary that the quantity ωT be much greater than one. This condition agrees with the condition that the energy of the source be sufficiently well defined, both in sign and magnitude.

As we shall show in Section 38, one cannot formulate the violation of microcausality in such a way as to ensure *localization* of this violation without using the time-like vector n ($n^2 = 1$). This vector has various meanings. In the case we considered it plays the role of the vector $(\omega, \mathbf{k})$ where ω is the characteristic frequency of the source and $\mathbf{k}$ is the characteristic momentum. In particular we saw that when the sign of ω is changed, the retarded potential is replaced by the advanced one.

We now turn to an example of an acausal influence function in which the vector n is introduced into the influence function itself in an explicit manner. In this example, as in the ordinary causal theory, any influence function may be obtained by superimposing retarded and advanced Green's functions.

In particular

$$D_c(x) = \tfrac{1}{2} D^{+}_{\text{ret}}(x) + \tfrac{1}{2} D^{-}_{\text{adv}}(x), \tag{36.11}$$

and according to (26.16) if we have a field source $Q(y)$ then the field $\varphi(x)$ which arises in the point x is defined by the equation

$$\varphi(x) = \int D_c(x - y)\, Q(y)\, \mathrm{d}^4 y. \tag{36.12}$$

In place of these causal influence functions $D_c(x)$, $D^{+}_{ret}(x)$, $D^{-}_{adv}(x)$, we shall construct acausal functions which are regular at the vertex of the light cone. For this purpose we shall use the form factor $\rho(x, n)$ in which some time-like unit vector n appears as

$$\rho(x, n) = \rho\left(\frac{R}{a}\right), \tag{36.13}$$

where

$$R^2 = 2(nx)^2 - x^2 \geqslant 0 \tag{36.14}$$

is an invariant positive-definite quadratic form of the variable x.† We assume that the form factor (36.13) vanishes rapidly (e.g., exponentially)

† In the frame of reference in which $n = 1, 0, 0, 0$, the invariant $R^2 = t^2 + x^2 \geqslant 0$

for $R/a \to \infty$, so that the quantity a represents the radius of that region within which microcausality is violated. We assume that this form factor does have a singularity at the origin that is stronger than $1/R$.

The acausal influence function is defined by the equations

$$D^{+}_{\text{ret}}(x, a) = \int D^{+}_{\text{ret}}(x - \xi)\, \rho(\xi, n)\, d^4\xi\,, \tag{36.15}$$

$$D^{-}_{\text{adv}}(x, a) = \int D^{-}_{\text{adv}}(x - \xi)\, \rho(\xi, n)\, d^4\xi\,. \tag{36.15'}$$

According to these equations the changed "acausal" field $\varphi(x, a)$ is defined, instead of (36.12), by the equation

$$\varphi(x, a) = \int D_c(x - y, a)\, Q(y)\, d^4y\,, \tag{36.16}$$

where

$$\begin{aligned} D_c(x - y, a) &= \tfrac{1}{2} \int D_c(x - y - \xi)\, \rho(\xi, n)\, d^4\xi \\ &= \tfrac{1}{2} D^{+}_{\text{ret}}(x - y, a) + \tfrac{1}{2} D^{-}_{\text{adv}}(x - y, a)\,. \end{aligned} \tag{36.17}$$

Introducing such a form factor is physically equivalent to assuming that each point in the field source $Q(y)$ is diffused using the form factor (36.13) in a region $\mathfrak{R}(x) \approx a^4$ and, consequently, it produces the same field as that due to a charge that is distributed in a small region with dimensions of the order a^4. Because of this localization of the form factor $\rho(\xi, n)$ the behavior of the acausal function $D_c(x-y, a)$ for large $|x_0 - t_0|$ and $|x-y|$ i.e., for

$$|x_0 - t_0| \gg a, \qquad |\mathbf{x} - \mathbf{y}| \gg a\,, \tag{36.18}$$

agrees with the asymptotic behavior of the acausal function $D_c(x-y)$. Thus it is possible to satisfy the conditions of macroscopic causality (35.15)-(35.17').

37. An Example of Constructing an Acausal Scattering Matrix

We shall proceed from the assumption that there exists a unitary $\hat{S}$-matrix which is constructed on the basis of local theory which assumes that microcausal conditions are satisfied. We shall also assume that all divergences are removed from the matrix by, say, renormalization. In this manner the original $\hat{S}$-matrix will contain only matrix elements which describe real, physical processes. Consequently, the matrix elements will be defined on an energy surface (real particles) and will be written in the form (31.28).

Our problem is to find the acausal scattering matrix $\hat{S}_a$ that satisfies the macrocausality conditions and the unitary condition (compare with Section 35). We shall consider this matrix as "close" to some local matrix $\hat{S}$ (we shall explain the meaning of the word "close" later). We write the initial local scattering matrix $\hat{S}$ in the form

$$\hat{S} = e^{i\hat{\eta}}, \tag{37.1}$$

where $\hat{\eta}$ is the Hermitian phase operator such that

$$\hat{\eta} = \hat{\eta}^+$$

We consider that the interaction is not strong and that it is such that we can write the operators $\hat{S}$ and $\hat{\eta}$ in the form of a series in terms of the interaction constant g

$$\hat{S} = \hat{I} + \sum_{n=1}^{\infty} g^n \hat{S}_n, \tag{37.2}$$

$$\hat{\eta} = \sum_{k=1}^{\infty} g^k \hat{\eta}_k. \tag{37.3}$$

Inserting these series in (37.1) we obtain

$$\hat{S}_n = \sum_{k=1}^{n} \frac{i^k}{k!} \sum_{ab\ldots c} (\hat{\eta}_a \hat{\eta}_b \ldots \hat{\eta}_c)\, \delta_{a+b+\cdots+c,\, n-k-1}, \tag{37.4}$$

where the products $(\eta_a, \eta_b, \ldots, \eta_c)$ contain k factors. From (37.4) in particular, it follows that

$$\hat{S}_1 = i\hat{\eta}_1, \qquad \hat{S}_2 = i\hat{\eta}_2 + \frac{i^2}{2!}\hat{\eta}_1^2 + \cdots \tag{37.4'}$$

In these equations the operator $\hat{S}$ is expressed in terms of the operator $\hat{\eta}$. It is also important to solve the equations to obtain the operator $\hat{\eta}$ in terms of the operator $\hat{S}$. The structure of the matrix elements of the operator $\hat{\eta}$ corresponds to that of the scattering matrix elements [see

(31.28)] and so we may write

$$\langle p'|\,\hat{\eta}\,|p\rangle = \delta^4\left(\sum p' - \sum p\right)\frac{\eta(p', p)}{\sqrt{2p_0'\cdots 2p_0}}, \tag{37.5}$$

where $\eta(p', p)$ is an invariant function of the momenta p and p'. A replacement, of the matrix $\hat{S}$ by the matrix $\hat{\eta}$ is important; because if microcausality is violated in some way, then the unitarity of the matrix $\hat{S}$ may also be violated and one may come to a contradiction. If however we use the operator $\hat{\eta}$, then changing from the microcausal operator $\hat{\eta}$ to the acausal operator $\hat{\eta}_a$ is sufficient to ensure that the new operator is Hermitian. A Hermitian operator satisfies the simple symmetry conditions which can be easily verified. The unitarity of the acausal scattering matrix $\hat{S}_a$ is also ensured by these conditions. It is very difficult and often impossible to prove directly that the matrix $\hat{S}_a$ is unitary.

In order to prove that the matrix $\hat{S}_a$ satisfies the microcausality conditions, we must use the space-time description which we must now formulate in terms of the operator $\hat{\eta}$. For this purpose we recall that, according to Section 35, the macroscopic causality conditions are formulated for the function $\mathfrak{G}(x)$ [see (35.18)], which is defined by Equations (35.12) and (35.13). The matrix element which enters into the second equation is directly related to the $\hat{S}$-matrix by Equations (35.2) and (35.3).

Using (35.2) and (35.4) we may rewrite the function $T(p', p)$ in the form of a series in terms of the interaction constant g and the quantities $\eta_1, \eta_2, \ldots, \eta_k, \ldots$ [see (37.3)]. Using the matrix multiplication law for a Lobachevski space [see (31.26)], it is easy to show that the expansion has the form

$$\begin{aligned} T(p', p) &= g\eta_1(p', p) + g^2\eta_2(p', p) + \frac{\mathrm{i}}{2}g^2\eta_1^2(p', p) + \cdots \\ &= g\eta_1(p', p) + g^2\eta_2(p', p) \\ &\quad + \frac{\mathrm{i}}{2}g^2\int \eta_1(p', p'')\frac{\mathrm{d}^3p''}{2p_0''}\eta_1(p'', p) + \cdots, \end{aligned} \tag{37.6}$$

where p' denotes the set of momenta $p_m, p_{m-1}, \ldots, p_3$ in the final state, and p denotes the set of momenta in the initial state p_2, p_1 [see (35.2) and (35.3)].†

† As in Section 35 we assume that there are only two particles in the initial state. It is easy to generalize this for a greater number of particles.

In this manner Equation (37.6) may also be written as

$$\begin{aligned} T(p_m, p_{m-1}, \ldots, p_3, p_2, p_1) = {} & g\eta_1(p_m, p_{m-1}, \ldots, p_2, p_1) \\ & + g^2\eta_2(p_m, p_{m-1}, \ldots, p_2, p_1) \\ & + \frac{i}{2} g^2 \int \eta_1(p_m, p_{m-1}, \ldots, p''_1, p''_1) \frac{d^3p''_2\, d^3p''_1}{2p''_{20}\, 2p''_{10}} \\ & \times \cdots \eta_1(p''_2, p''_1, \ldots, p_2, p_1) + \cdots . \end{aligned} \tag{37.7}$$

Inserting this expression for the function T in (35.13), we write the function $\mathfrak{G}_0$ in terms of the matrix elements of the phase operator $\hat{\eta}$

$$\begin{aligned} \mathfrak{G}_0(x_m, x_{m-1}, \ldots, x_3, x_2, x_1) = {} & \int \delta^4(p_m + p_{m-1} + \cdots + p_3 - p_2 - p_1) \\ & \times \{g\eta_1(p_m, p_{m-1}, \ldots, p_2, p_1) + g^2\eta_2(p_m, p_{m-1}, \ldots, p_2, p_1) \\ & + \frac{i}{2} g^2 \int \eta_1(p_m, p_{m-1}, \ldots, p''_2, p''_1) \frac{d^3p''_2\, d^3p''_1}{2p''_{20}\, 2p''_{10}} \\ & \times \eta_1(p''_2, p''_1, \ldots, p_2, p_1) + \cdots\} \\ & \times \exp i\,[p_m x_m + p_{m-1}x_{m-1} + \cdots + p_3x_3 - p_2x_2 - p_1x_1] \\ & \times \frac{d^3p^m\, d^3p_{m-1}}{2p^0_m\, 2p^0_{m-1}} \cdots \frac{d^3p^1}{2p^0_1} . \end{aligned} \tag{37.8}$$

The function $\mathfrak{G}(x_m, x_{m-1}, \ldots, x_2, x_1)$ which transforms the incident wave into the scattered wave is obtained from $\mathfrak{G}_0(x_m, x_{m-1}, \ldots, x_2, x_1)$ by differentiating with respect to x^0_2 and x^0_1 [see (35.12)].

Thus the function $\mathfrak{G}(x_m, x_{m-1}, \ldots, x_1)$ is a functional of the matrix elements $\eta_1(p', p)$, $\eta_2(p', p), \ldots$ of the phase operator $\hat{\eta}$. Making use of the properties of these elements and using (37.8) we can verify that the conditions of macroscopic causality (35.15)-(35.17') are fulfilled.

We now turn to a simple example of an acausal matrix $\hat{S}_a$. For definiteness we assume that the original causal function $\hat{S}$ describes a process in the scalar field $\varphi(x)$, in which the interaction energy has the form†

$$\hat{W} = g\!:\!\hat{\varphi}^3(x)\!:, \tag{37.9}$$

where $\hat{\varphi}(x)$ is the scalar field operator, and g is the interaction constant. Using the usual Feynman diagram method, it is easy to show that the

† Compare with Appendix VII. In this appendix we remark that the system with interaction $g\varphi^3$ is not stable. However, if we limit ourselves to perturbation theory, then this fact is unimportant.

matrix elements of the operator $\hat{\eta}$ to the first order are identically equal to zero

$$\eta_1 \equiv 0. \tag{37.10}$$

The second-order matrix elements η_2 which describe the scattering are equal to the matrix elements of the operator $\hat{s}_2$ [compare with (37.4)] and they are described by the sum of the following diagrams

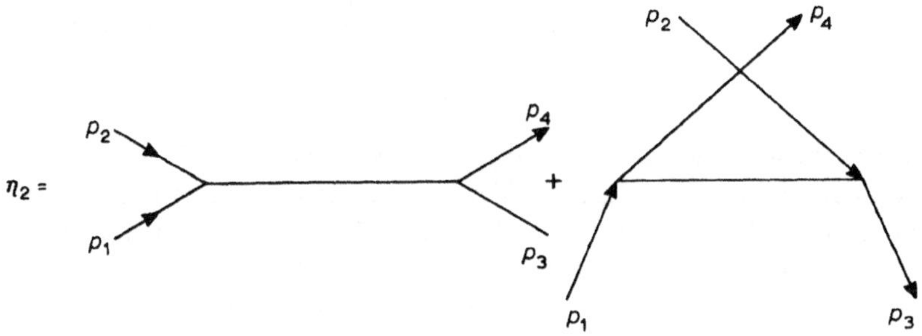

(37.11)

The analytic expression for this element is

$$\eta_2(p_4, p_3, p_2, p_1) = \{\tilde{D}_c(p_3 - p_1) + \tilde{D}_c(p_2 + p_1\}, \tag{37.12}$$

where $\tilde{D}_c(q)$ is the transform of the causal function $D_c(x)$ and q is the momentum that is transferred along the inner line in the diagram (37.11).

According to (37.8) and (37.12) for elastic scattering we have

$$\begin{aligned}\mathfrak{G}_0(x_4, x_3, x_2, x_1) = \int & \delta^4(p_4 + p_3 - p_2 - p_1)\{\tilde{D}_c(p_3 - p_1) \\ & + \tilde{D}_c(p_2 + p_1)\} \exp \mathrm{i}(p_4 x_4 + p_3 x_3 - p_2 x_2 - p_1 x_1) \\ & \times \frac{\mathrm{d}^3 p_4}{2p_4^0} \cdots \frac{\mathrm{d}^3 p_1}{2p_1^0}.\end{aligned} \tag{37.13}$$

We assume that the initial state of the particle $\Psi_{\text{in}}(x_2, x_1)$ may be defined in terms of wave packets $u_2(x_2)$ and $u_1(x_1)$ of the type considered in Section 35 [compare with (35.4)]

$$\Psi_{\text{in}}(x_2, x_1) = u_2(x_2)\, u_1(x_1). \tag{37.14}$$

Using Equation (35.10) we obtain the scattered wave for the elastic collision case

$$\begin{aligned}\Psi_{\text{out}}(x_4, x_3) = u_2(x_4)\, u_1(x_3) &+ \mathrm{i}\int \mathrm{d}^4x\, \mathrm{d}^4y D^+(x_4 - y) \\ &\times D^+(x_3 - x)\, D_c(x - y)\, u_2(y)\, u_1(x) \\ &+ \mathrm{i}\int \mathrm{d}^4x\, \mathrm{d}^4y D^+\left(x_4 - \frac{x+y}{2}\right) \\ &\times D^+\left(x_3 - \frac{x+y}{2}\right) D_c\left(\frac{x-y}{2}\right) u_2(x)\, u_1(x),\end{aligned} \tag{37.15}$$

where the first integral is defined by the first diagram in (37.11) and the second, by the second diagram. The computations leading to this result are given in Appendix XIII. The function $D^+(x)$ is defined in Appendix I, Equation (I.10) and D_c is the causal function, This expression for the scattered wave is based on conventional local theory in which the microcausality conditions are satisfied.

We now replace the function $D_c(x-y)$ which enters into (37.15) by the acausal function $D_c(x-y, a)$

$$D_c(x - y, a) = \int D_c(x - y - \xi)\, \rho(\xi, n)\, \mathrm{d}^4\xi. \tag{37.16}$$

where the form factor $\rho(\xi, n)$ depends on some unit time-like vector n and the elementary length a. This form factor is assumed localized in the region $\mathfrak{R}_4(\xi) \approx a^4$. The vector n will be defined later. As we showed in Section 36, replacing the causal function $D_c(x-y)$ by the acausal function $D_c(x-y, a)$ assures that the macroscopic causality conditions (35.15)-(35.17′) will be satisfied.

From (37.16) we obtain the equation

$$D_c(x, a) = \int \exp \mathrm{i}\, px \tilde{D}_c(p)\, \tilde{\rho}(\rho, n)\, \mathrm{d}^4p, \tag{37.17}$$

where $D_c(p)$ is the Fourier transform of the function $D_c(x)$ and $\rho(p, n)$ is the Fourier transform of the function $\rho(\xi, n)$

$$\rho(\xi, n) = \int \tilde{\rho}(\rho, n) \exp \mathrm{i}p\xi\, \mathrm{d}^4\xi. \tag{37.18}$$

As an example of the form factor $\rho(\xi, n)$ we may take (see [69, 72])

$$\rho(\xi, n) \approx \frac{e^{-R/a}}{R^2}, \tag{37.19}$$

where

$$R^2 = 2(\xi, n)^2 - \xi^2, \tag{37.20}$$

where a is the elementary length. From (37.18)–(37.20) we have

$$\tilde{\rho}(p, n) = \frac{M^2}{2(pn)^2 - p^2 + M^2}, \tag{37.21}$$

where $M = 1/a$ is the large mass corresponding to the small elementary length a. In the coordinate system in which the vector $n = (1, 0, 0, 0)$ we have

$$\tilde{\rho}(p, n) = \frac{M^2}{p_0^2 + \mathbf{p}^2 + M^2}. \tag{37.21'}$$

We now turn to Equation (37.16) for the acausal function $D_c(x, a)$. We assume that the mass of the particles is equal to zero. In this case (see [70])

$$D_c(x) = \frac{1}{r}\,\delta^+(t - r) + \frac{1}{r}\,\delta^-(t + r), \tag{37.22}$$

where

$$\delta^\pm(z) = \frac{1}{2\pi}\int_0^A e^{\pm ivz}\,dv, \qquad A \to +\infty. \tag{37.23}$$

We first compute the part $D_c^+(x, a)$ of the acausal function (37.17) which is derived from the term with $\delta^+(t-r)$ in (37.22). In the integral

$$D_c^+(x, a) = \int D_c^+(x - \xi)\,\rho(\xi, n)\,d^4\xi \tag{37.24}$$

we replace t by $t - \tau$ $(i \equiv \xi_4)$ and $\mathbf{r}(\mathbf{x})$ by $r(\mathbf{x} - \boldsymbol{\xi}) = r(\mathbf{x}) + \boldsymbol{\alpha}\boldsymbol{\xi} + \ldots$, where $\boldsymbol{\alpha}$ is the unit vector in the direction of the vector $\mathbf{x}$. Then using (37.23) we obtain

$$D_c^+(x, a) = \frac{1}{r}\int \rho(\xi, n)\, d^3\xi\, d\tau \int_0^A \exp[iv(t - r - \tau - \boldsymbol{\alpha\xi})]\, dv$$
$$= \frac{1}{r}\int_0^A \exp[iv(t - r)]\, \tilde{\rho}(v, v, \boldsymbol{\alpha} n)\, dv$$
$$= \frac{1}{r}\int_0^A \exp[iv(t - r)]\, \frac{M^2}{2v^2 + M^2}\, dv = \frac{1}{r}\Delta^+ \times (t - r, A, M) \tag{37.25}$$

for $A \to \infty$.

The function $\Delta^+(t-r, \infty, M)$ is computed in Appendix XIV and this leads to the results (XIV.10) and (XIV.11). For $|t-r| \gg a$,

$$\Delta^+\left(t - r, \infty, \frac{1}{a}\right) = \delta^+(t - r) + O\left(\frac{2a^2}{(t - r)^3}\right) + \frac{\pi}{2}\frac{1}{a\sqrt{2}}\exp\left(-\frac{|t - r|}{a\sqrt{2}}\right). \tag{37.26}$$

For $|t-r| \to 0$ this function is regular and has the limits

$$\Delta^+\left(0, \infty, \frac{1}{a}\right) = \frac{\pi}{4}\frac{1}{a\sqrt{2}}. \tag{37.27}$$

The function $\delta^+(t-r)$ has the form [see (I.9)]

$$\delta^+(t - r) = \delta(t - r) + \frac{1}{\pi}\frac{1}{t - r}. \tag{37.28}$$

Thus the second term in (37.26) which decreases as $a^2/(t-r)^3$, and the third exponentially decreasing term may be considered as corrections to the causal function $\delta^+(t-r)$. In this manner one can verify that the conditions of macrocausality are fulfilled.

We must now verify that the condition of unitarity of the scattering matrix is satisfied. According to (37.1) it is sufficient to prove that the phase operator $\hat{\eta}_a$ is Hermitian. Using (37.12) and (37.17), instead of (37.12) we obtain for the acausal operator $\hat{\eta}_a$

$$\eta_2(p_4, p_3, p_2, p_1) = \tilde{D}_c(p_3 - p_1)\,\tilde{\rho}(p_3 - p_1, n) + \tilde{D}_c(p_2 + p_1)\,\tilde{\rho}(p_2 + p_1, n). \qquad (37.12')$$

From this it follows that if the following equations are satisfied

$$\tilde{\rho}(p_3 - p_1, n) = \tilde{\rho}^*(p_1 - p_3, n), \qquad (37.29)$$

$$\tilde{\rho}(p_2 + p_1, n) = \tilde{\rho}^*(p_3 + p_4, n), \qquad (37.30)$$

then the matrix element $\eta(p_4, p_3, p_2, p_1)$ will be Hermitian if the vector n does not change when taking the Hermitian conjugate. For this it is sufficient that this vector be symmetric with respect to interchange of the momenta of the initial and final (p_4, p_3) states. In particular, we may write

$$n = \frac{P}{\sqrt{P^2}}, \qquad (37.31)$$

where $P = p_1 + p_2 = p_3 + p_4$ is the total momentum of the colliding particles. The second possibility is that

$$n = \frac{\Pi}{\sqrt{\Pi^2}}, \qquad (37.32')$$

where $\Pi = p_3 + p_1$. The frame of reference in which the vector $\Pi = 0$ is called the Breit system.

When the vector n is chosen to be symmetric, the acausal matrix $\hat{S}_a$ is both unitary and macroscopically causal. We perform the calculations to an accuracy of g^2. However the calculations can be carried to higher orders in the expansion of the operators $\hat{S}$ and $\hat{\eta}$ in terms of g.

Therefore, we can summarize the results of this section in the following way. If a microcausal unitary scattering matrix exists (at least to an accuracy of $O(g^N)$ where N is some positive integer), then an acausal unitary scattering matrix $\hat{S}_a$ which satisfies the principle of macroscopic causality also exists.

38. The Dispersion Relation for the Acausal $\hat{S}_a$-Matrix

Any violation of microcausality is immediately reflected in the analytic properties of the matrix elements of the $\hat{S}$-matrix. In particular the dispersion relation for forward elastic scattering which we considered in Section 30 would no longer be valid. The nature of the various possible changes in these relations depend on the actual form of the violation of microcausality. We shall consider a form of violating microcausality which can reasonably be called *minimal.*

Let $T(k)$ be the elastic scattering amplitude in the usual local theory and $T_a(k)$ be the acausal amplitude for the same process. We shall consider that the violation of microcausality is minimal if (a) the function $T_a(k)$ has the same spectrum as the function $T(k)$, in other words if the function $T(k)$ differs from zero in some region $\mathfrak{R}_3(k)$, then the spectrum of $T_a(k)$ is also confined to the same region; (b) the symmetry conditions which are characteristic of the amplitudes $T_{\text{ret}}(k)$ and $T_{\text{adv}}(k)$ (30.20), (30.20′) remain in force and, for the corresponding acausal amplitudes $T_a(k)$,

$$T_{a\,\text{ret}}(k) = T_{a\,\text{adv}}(-k), \tag{38.1}$$

$$T_a^*{}_{\text{ret}}(k) = T_{a\,\text{ret}}(-k); \tag{38.1′}$$

(c) the rate of growth of $|T_a(k)|$ for $k \to \infty$ does not exceed that of $|T(k)|$; (d) the conditions of macroscopic causality are fulfilled.

A simple form of acausality (violation of microcausality) that satisfies these conditions is discussed by Blokhintsev and Kolerov [69, 72] and we shall refer to it in the following. Let $\tilde{T}(x)$ and $\tilde{T}_a(x)$ be the amplitudes $T(k)$ and $T_a(k)$ expressed in the coordinate representation

$$\tilde{T}(x) = \int T(k)\, \mathrm{e}^{-ikx}\, \mathrm{d}^4k, \tag{38.2}$$

$$\tilde{T}_a(x) = \int T_a(k)\, \mathrm{e}^{-ikx}\, \mathrm{d}^4k. \tag{38.2′}$$

Following the concepts discussed in Section 37, we write

$$\tilde{T}_a(x) = \int \tilde{T}(x-\xi)\, \rho(\xi, n)\, \mathrm{d}^4\xi, \tag{38.3}$$

where the form factor $\rho(\xi, n)$ depends on some time-like unit vector n. In reference [69] this vector is taken to be parallel to the vector $\Pi = p' + p$, where p is the momentum of the nucleon in the initial state and p' is its momentum in the final state. In Breit's system, when $\mathbf{p}' + \mathbf{p} = 0$, the components of n are $n = 1, 0, 0, 0$. Furthermore, as in Sections 36 and 37, we assume that the form factor $\rho(\xi, n)$ is a function of the ratio R/a where a is the elementary length and R^2 is the positive-definite invariant

$$R^2 = 2(n\xi)^2 - \xi^2 \geqslant 0, \ldots \tag{38.4}$$

which vanishes sufficiently quickly for $R/a \to \infty$ as

$$\rho(\xi, n) \equiv \rho\left(\frac{R}{a}\right) \to 0. \tag{38.5}$$

From conditions (38.5) and (38.3) it follows that if the components of the vector x are the quantities $|x_i| \gg a$ $(i=1, 2, 3, 4)$, then the quantity $\tilde{T}(x-\xi)$ may be taken out from under the integral by setting $\xi=0$ and the integral of $\rho(\xi, n)$ may be normalized to 1 in the space $\mathfrak{R}_4(\xi)$. Thus for $|x_i| \gg a$ we obtain

$$\tilde{T}_a(x) \approx \tilde{T}(x), \tag{38.6}$$

which may be considered as fulfilling macrocausality.

Changing to the momentum representation in (38.3) we obtain

$$T_a(k) = T(k)\,\tilde{\rho}(k, n), \tag{38.7}$$

where $\tilde{\rho}(k, n)$ is the Fourier transform of the form factor $\rho(\xi, n)$. If $\rho(\xi, n)$ is a function of R/a, then it is easy to show that $\tilde{\rho}(k, n)$ is a function of the invariant

$$I = 2(kn)^2 - k^2 \geqslant 0. \tag{38.8}$$

From (38.7) it follows that the spectral character condition (a) is automatically fulfilled. The symmetry condition (b) will also be fulfilled if

$$\tilde{\rho}(k, n) = \tilde{\rho}(-k, n), \tag{38.9}$$

$$\tilde{\rho}^*(k, n) = \tilde{\rho}(-k, n), \tag{38.9$'$}$$

i.e., if $\tilde{\rho}(k, n)$ is even and is a real function of the vector k. Finally, condition (c) will be satisfied if the function $\tilde{\rho}(k, n)$ is bounded for $|k| \to \infty$.

It is sufficient that the conditions (a)–(d) are fulfilled in order to derive the dispersion relation for the acausal amplitude $T_a(k)$ in exactly the same manner as that which is used in deriving the dispersion relation in the usual causal theory (compare with Section 30). Thus in the case of minimal acausality we may write the dispersion relation for the forward elastic scattering amplitude $T_a(\omega)$ using Equation (30.30) directly and differing only in that the singularity which arises from the form factor $\tilde{\rho}(k, n)=\tilde{\rho}(\omega)$ must be taken into account when the integral is taken around the infinitely large semicircle. Then in place of (30.30) we obtain the more general equation:

$$T_a(\omega) = \frac{g^2}{\omega + \omega_g} + \frac{g^2}{\omega - \omega_g} + (\omega - \omega_0)^{n+1} \frac{2}{\pi} \int_m^\infty \frac{A_a(\omega')\, d\omega'}{(\omega' - \omega_0)^{n+1} (\omega'^2 - \omega^2)} + \sum_{r=0}^{n} c_r \omega^r + \psi(\omega), \tag{38.10}$$

where the additional term $\psi(\omega)$ has the form

$$\psi(\omega) = (\omega - \omega_0)^{m+1} \sum_{s=1}^{f} \mathrm{Res} \frac{T_a(\omega_s)}{\omega_s - \omega_0}, \tag{38.11}$$

The sum here is taken over the singularities (poles) of the function $\rho(\omega)$ and $A_a(\omega)$ is the imaginary (absorbing) part of the amplitude $T_a(\omega)$.

Because the function $\rho(\omega)$ is real and even, if it has a pole at the point $\omega=\omega_s$, then it must have a pole at $\omega=-\omega_s$. These poles cannot lie on the real axis because in that case they would describe real states. Therefore, they must have an imaginary part $\mathrm{Im}\, \omega_s \neq 0$ which is sufficiently large. We let $|\mathrm{Im}\, \omega_s| \approx 1/a$ where a is the elementary length.

If the quantity a is small and if the energy ω in the region we are considering is small so that $\omega \ll 1/a$, then we may close the integration contour (see Figure 33) with a circle whose radius R is such that $R<1/a$, yet is also large enough so that it is possible to neglect the contribution to the integral from this circle. Obviously, under these conditions we obtain the usual causal dispersion relation. For $R>1/a$, additional acausal terms appear.

If the elementary length a is sufficiently small, then the acausal matrix $\hat{S}_a$ differs from the causal matrix $\hat{S}$ only in the region where the energy $\omega \gg 1/a=M$. Therefore for $M\to\infty$ the matrix $\hat{S}_a$ is "close" to the original causal matrix $\hat{S}$. We call such a matrix almost local or almost causal.

In concluding this section we shall write out the dispersion relation s for πN forward scattering, taking into account its acausal form. For definiteness we set

$$\rho(\omega) = \frac{M^2}{M^2 + \omega^2}, \qquad M = \frac{1}{a}. \tag{38.12}$$

These equations differ from the causal ones by the additional term $\psi(x)$ which results from the computations of the function $\tilde{\rho}(\omega)$ at points $\omega = \pm \mathrm{i}M$.

Instead of Equations (30.32)–(30.32″) we now have

$$\begin{aligned} D_a(\omega) - D_{a0}(\omega_0) = {} & \frac{g^2}{M^2}\,\omega_g^2\,\frac{(\omega^2 - \omega_0^2)}{(\omega^2 - \omega_g^2)(\omega_0^2 - \omega_g^2)} \\ & + \frac{2}{\pi}(\omega^2 - \omega_0^2)\int_m^\infty \frac{A_{a0}(\omega')\,\mathrm{d}\omega'}{(\omega'^2 - \omega^2)(\omega'^2 - \omega_0^2)} \\ & + \psi_0(\omega), \end{aligned} \tag{38.13}$$

$$\begin{aligned} D_{a1}(\omega) - D_{a1}(\omega_0) = {} & \frac{2g^2}{M}\,\frac{\omega_g(\omega^2 - \omega_0^2)}{(\omega^2 - \omega_g^2)(\omega_0^2 - \omega_g^2)} \\ & + \frac{2}{\pi}\int_m^\infty \frac{A_{a1}(\omega')\,\mathrm{d}\omega'}{(\omega'^2 - \omega^2)(\omega'^2 - \omega_0^2)} + \psi_1(\omega), \end{aligned} \tag{38.13$'$}$$

$$\begin{aligned} \left[D_{a2}(\omega) - D_{a2}(\omega_0)\right]\frac{\omega}{\omega_0} = {} & \frac{g^2}{M}\,\frac{\omega_g(\omega^2 - \omega_0^2)}{(\omega^2 - \omega_g^2)(\omega_0^2 - \omega_g^2)} \\ & + \frac{2}{\pi}\,\omega(\omega^2 - \omega_0^2)\int_m^\infty \frac{A_{a2}(\omega')\,\mathrm{d}\omega'}{(\omega'^2 - \omega^2)(\omega'^2 - \omega_0^2)} \\ & + \psi_2(\omega), \end{aligned} \tag{38.13$''$}$$

where the amplitudes $D_{a0}(\omega)$, $D_{a1}(\omega)$, $D_{a2}(\omega)$, $A_{a0}(\omega)$, $A_{a1}(\omega)$, $A_{a2}(\omega)$ have the same meaning as the amplitudes of the causal theory in equations (30.32)–(30.32″) and the additional terms are equal to

$$\psi_0(\omega) = \frac{\omega^2 - \omega_0^2}{M^2 + \omega_0^2}\,\tilde{\rho}(\omega)\,\mathrm{d}_0(iM), \tag{38.14}$$

$$\psi_1(\omega) = \frac{\omega^2 - \omega_0^2}{M^2 + \omega_0^2}\,\tilde{\rho}(\omega)\,\mathrm{d}_1(iM), \tag{38.14$'$}$$

$$\psi_2(\omega) = \frac{\omega^2 - \omega_0^2}{M^2 + \omega_0^2}\,\tilde{\rho}(\omega)\,\mathrm{d}_2(iM), \tag{38.14''}$$

where d_0, d_1, d_2 are constants whose magnitudes are

$$d_0 = \frac{D_{a0}(\omega)}{\tilde{\rho}(\omega)}, \quad d_1 = \frac{D_{a1}(\omega)}{\tilde{\rho}(\omega)}, \quad d_2 = \frac{D_{a2}(\omega)}{\tilde{\rho}(\omega)}$$

at the point $\omega = iM$. From these equations it is obvious that the additional terms are important only for frequencies $\omega \geqslant M$.

CHAPTER VII

A GENERALIZATION OF CAUSAL RELATIONSHIPS AND GEOMETRY

39. Two possible generalizations

In present-day theory it is assumed that the geometry which we know from macroscopic physics retains its significance in microworld regions. Together with this geometry, the form of causal relationships, which is based on studies of macroscopic phenomena, is carried over into the microworld. Two basic ideas form the basis of modern concepts. These are:

(a) the idea that the Einstein- Minkowski space-time is homogeneous and isotropic, and
(b) the idea about transfer of interaction by physical fields (electromagnetic fields, meson fields, neutron fields, etc.).

The systematic use of these ideas in elementary-particle physics in regions of especially small time and space intervals leads to absurd results. The interaction energy of particles with very small separations, and their characteristic energies are very large. This unwanted result appears in both quantum and classical physics and it indicates that the origin of the difficulties may be the same for both concepts.

We shall discuss this question in greater detail. In classical physics, the propagation of a weak (linear) signal from the world point $P(x_1)$ to the world point $P(x_2)$ is defined by the Green's function $\mathfrak{G}$ (see Section 25). We shall try to construct this function starting from the most general assumptions on geometry and causality.

From the assumption on the homogeneity of space and time, it follows that the Green's function may physically be only the dispersion of the coordinates of the points $Q(x_1)$ and $P(x_2)$, i.e., a function of the separations $\mathbf{x}=\mathbf{x}_2-\mathbf{x}_1$ and $t=t_2-t_1$. The requirement that space-time be isotropic leads to a further restriction. The function $\mathfrak{G}$ must depend on the four-dimensional interval $x^2=t^2-\mathbf{x}^2=(t_2-t_1)^2-(\mathbf{x}_2-\mathbf{x}_1)^2$ as well as on the separations $\mathbf{x}_2-\mathbf{x}_1$, t_2-t_1. Lastly, we may still introduce a direction for the time $\varepsilon = t/|t| = \pm 1$ for $x^2 > 0$ and a direction along the spatial line $\eta = x/|x| = \pm 1$ for $x^2 < 0$. Thus we may write the Green's function in the form

$$\mathfrak{G} = \mathfrak{G}(x^2, \varepsilon, \eta). \tag{39.1}$$

This function is invariant with respect to Lorentz transformation, but may not be conserved for discrete transformation T (inversion in time) and P (inversion in space).

The relationship between the world points $P(x_1)$ and $P(x_2)$ may be expressed, using the Green's function, in the form†

$$\varphi(x_2) = \int \mathfrak{G}(x^2, \varepsilon, \eta)\, Q(x_1)\, \mathrm{d}^4 x_1, \tag{39.2}$$

where $\varphi(x_2)$ is the field at the point $P(x_2)$ and $Q(x_1)$ is the source of the field which is concentrated in the neighborhood of the point $P(x_1)$. We may analogously express the field $\varphi(x_1)$ which arises at the point $P(x_1)$ from the source $Q(x_2)$ which is concentrated in the neighborhood of the point $P(x_2)$. The relationship expressed in (39.2) does not in itself ensure that the relativistic principles of causality will be fulfilled which (a) limit the propagation velocity of all interactions by an upper limit of the velocity of light and (b) require that the cause [source Q located at the point $P(x_1)$] produce a result [field φ at the point $P(x_2)$] at a later moment in time, i.e., $t_2 > t_1$. In order to satisfy these requirements, we obtain the variation in the field $\varphi(x_2)$ at the point $P(x_2)$ for variations in the source $Q(x_1)$ at an arbitrary point $P(x_1)$ or, as it is more convenient, we shall compute the functional derivative of the field $\varphi(x_2)$ with respect to the source $Q(x_1)$ directly (see Appendix IX). From (39.2) we have

$$\frac{\delta\varphi(x_2)}{\delta Q(x_1)} = \mathfrak{G}(x^2, \varepsilon, \eta). \tag{39.3}$$

From this we see that requirements (a) and (b) will be satisfied if the following restrictions are placed on the Green's function:

$$\mathfrak{G} = \mathfrak{G}(x^2, +1), \quad \text{if} \quad x^2 > 0, \varepsilon = +1, \tag{39.4}$$

$$\mathfrak{G} = 0, \quad \begin{cases} \text{if} \quad x^2 > 0, \varepsilon = -1, & (39.4') \\ \text{if} \quad x^2 < 0, \varepsilon = \pm 1 & (39.4'') \end{cases}$$

† Compare with Section 25.

The parameter η does enter into the function $\mathfrak{G}$ here because it equals zero for $x^2 < 0$.

When Equations (39.4)–(39.4″) are satisfied, then so are the relativistic causality requirements, but this does not yet mean that the functions $\mathfrak{G}$ can describe any *physical field.* In order that the function may be considered as a physical field[†] produced by a momentum source, it must satisfy the linear, inhomogeneous differential equation

$$\hat{L}\mathfrak{G}(x^2) = Q(x_1), \tag{39.5}$$

where the source has the form

$$Q(x_1) = \delta^4(x_1) = \delta^3(\mathbf{x}_1)\,\delta(t_1). \tag{39.6}$$

The linear character of the operator $\hat{L}$ follows from the assumption that the signal being considered is *weak*, and that the impulse character of the source Q corresponds to the usual definition of a Green's function as an elementary signal which is produced by a disturbance in the neighborhood of only one point $P(x_1)$.

From the assumption of relativistic invariance, the operator $\hat{L}$ must be a function of the D'Alambertian

$$\Box = -\frac{\partial^2}{\partial t^2} + \frac{\partial^2}{\partial x^2} + \frac{\partial^2}{\partial y^2} + \frac{\partial^2}{\partial z^2}$$

so that

$$\hat{L} \equiv L(\Box). \tag{39.7}$$

The function L may be either a polynomial or an infinite series

$$L(z) = \sum_{n=0}^{N} \frac{C_n z^n}{2n!}; \tag{39.8}$$

Here N is either finite or infinite. In the latter case the operator $L(\Box)$ may be non-local, i.e., by operating on some function $f(x)$, it reproduces the function in some other point $\varphi(x')$, $\mathbf{x}' \neq x$.[‡] The Fourier transforms

$$\mathfrak{G}(x^2, +1) = \int \tilde{\mathfrak{G}}(k^2)\, \mathrm{e}^{ikx}\, \mathrm{d}^4 x, \tag{39.9}$$

[†] Therefore it is possible to have a *generalized interaction* which does not reduce to the concept of a physical field. Such interactions are discussed by the author in [24, 74-77].

[‡] This fact was pointed out in reference [74]. In reference [78] it is shown that this difference depends on the behaviour of the limit $|C_n|^{1/n}$ for $n \to \infty$. The nonlocal operator from the class (39.8) was used to construct the version of the *nonlocal* field theory that is independent of dispersion [79].

of the functions (39.4)–(39.4″) are functions of the invariant $k^2 = \omega^2 - \mathbf{k}^2$: where $kx = \omega t - \mathbf{kx}$. Inserting (39.9) into (39.5) we obtain

$$\hat{L}(-k^2)\,\tilde{\mathfrak{G}}(k^2) = 1. \tag{39.10}$$

From this it follows that the zeros of the function $L(-k^2)$ are poles of the Green's function $\tilde{\mathfrak{G}}(k^2)$.† These poles cause singularities in the function $\mathfrak{G}(x^2)$ on the light cone $x^2 = 0$. In Appendix I, where we derive the Fourier transformations (39.9) for the Fourier transforms $\tilde{\mathfrak{G}}(k^2)$ which have first-order poles, we show that these singularities have the form $\delta(x^2)$, $\ln x^2$, $1/x^2$.

We could have assumed that the situation at the vertex of the light cone changes in the sence that the usual geometric relationships loose their meaning because of changes in the nature of the phenomena themselves which take place within small separations in space and time. In this case there should exist some "elementary length" a which would characterize the scale which restricts the use of the usual causal relationships for large $|\mathbf{x}|$ and $|t|$, in proportion to the approximation of $|\mathbf{x}|$, $|t|$ according to the magnitude of the scale a. The possible significance of such a scale for elementary particle physics was first discussed by Watagin [82] and Markov [83]. Since then, the idea about the existence of a scale which restricts the region of applicability of modern quantum theory and maybe the theory of relativity has been discussed many times and, to this day, theoretical thought returns to it.

A more important fact related to the concept of an elementary length is that the metric of the Einstein-Minkowski space is indefinite. In the region of the manifold $\mathfrak{R}_4(x)$ in which pseudo-Euclidean geometry is applied without question, there does not exist any concept of closeness of two events $P(x_1)$ and $P(x_2)$ except that which is expressed as $x^2 = 0$. This equation only expresses the fact that the points $P(x_1)$ and $P(x_2)$ may be connected by a light signal and that their spatial, $r = |\mathbf{x}_2 - \mathbf{x}_1|$, and temporal, $t = t_2 - t_1$ separations may be of any size. The equality $x^2 = 0$ "does not know" why it is fulfilled; whether it is because, although both the quantities r and t are large, $r = t$, or because $r = 0$ or $t = 0$ [24].

† Equations of the type (39.5) were considered in [80, 81]. For the case $N > 1$, problems due to the fact that the energy-momentum tensor of the particles becomes infinite were discovered.

We now assume that the principle of relativistic causality is satisfied at least approximately. We also assume, first of all, that this applies to the point (a) which restricted the propagation speed of a signal to a value less that the speed of light. With such an assumption, the function $\mathfrak{G}(x^2, \varepsilon, \eta)$ no longer has to vanish outside the light cone and so condition (39.3) is no longer satisfied. If we remove supposition (a), then the function $\mathfrak{G}(x^2, \varepsilon, \eta)$ does not vanish in the spatial region. In this case an interaction in which it is possible to distinguish between directions on the path $\eta = +1$, and therefore not conserve parity, becomes possible.

The departure of $\mathfrak{G}(x^2, \varepsilon, \eta)$ from conditions (39.4)–(39.4″) does not have to be very large. For large time intervals $t \gg a$ where a is the elementary length, Eq. (39.4′) must be satisfies to a high degree of accuracy. We introduce the elementary length into the expression for $\mathfrak{G}(x^2, \varepsilon, \eta)$:

$$\mathfrak{G} = \mathfrak{G}\left(x^2, \varepsilon, \eta, \frac{x^2}{a^2}\right), \tag{39.9′}$$

and we consider that for $|x^2/a^2| \to \infty$

$$\mathfrak{G}\left(x^2, \varepsilon, \eta, \frac{x^2}{a^2}\right) \to \mathfrak{G}(x^2, \varepsilon), \tag{39.9″}$$

where $\mathfrak{G}(x^2, \varepsilon)$ is the Green's function from causality theory (i.e., it vanishes outside the light cone). This correspondance requires that for $|x^2/a^2| \gg 1$ the influence function $\mathfrak{G}$ (39.9) becomes vanishingly small. In fulfilling this requirement, the signal which goes from the point $P(x_1)$ to $P(x_2)$ will propagate with a velocity $\mathbf{v} = (\mathbf{x}_2 - \mathbf{x}_1)/(t_2 - t_1) = \mathbf{x}/t$ which is practically equal to the velocity of light $c = 1$.

In fact, from the condition

$$|x^2| = |t^2 - \mathbf{x}^2| \leqslant a^2 \tag{39.11}$$

it follows that

$$|1 - \mathbf{v}^2| < \frac{a^2}{t^2}, \tag{39.12}$$

so that for $|t| \gg a$, $\mathbf{v}^2 = 1$. This condition is relativistically invariant. In fact, in a reference frame that moves relative to the first along the Ox axis with a velocity u, we have

$$x' = \gamma(x - ut), \quad t' = \gamma(t - ux), \quad \gamma = (1 - u^2)^{-1/2}. \tag{39.13}$$

Correspondingly,

$$|1 - \mathbf{v}'^2| \leqslant \frac{a^2}{t'^2} = \frac{a^2(1 - \mathbf{u}^2)}{t^2(1 - \mathbf{uv})^2} \tag{39.14}$$

where $\mathbf{v}' = \mathbf{x}'/t'$. If $|\mathbf{x}|$ and $|t| \sim a$ and $|u|$ is small, then $|\mathbf{x}'|$ and $|t'|$ are also $\sim a$ for $u \approx 1$. From (39.14) it follows that $|\mathbf{v}'|$ is close to 1 except for the point $\mathbf{1} - \mathbf{uv} = \mathbf{0}$. However, at that point

$$t' = 0, \quad x' = \gamma \frac{tv}{u}(1 - u^2) = \frac{t}{u^2}(1 - u^2)^{1/2},$$

so that $|t'|$ and $|\mathbf{x}'|$ are small (on the order of a). If $|t'|$ and $|\mathbf{x}'| \gg a$, then from (39.14) we have $\mathbf{v}'^2 \approx 1$.

Therefore, if we violate the propagation law for signals in the region $|\mathbf{x}|$, $|t| \sim a$, we may nevertheless conserve the restriction $|\mathbf{v}| \ll c = 1$ for $|\mathbf{x}|$ and $|t| \gg a$, i.e., for macroscopic scales [85].

The second causality condition (b) requires special consideration. According to this condition, $t_2 > t_1$ so that $\varepsilon = +1$ [see (39.4)]. The difficulty lies in the fact that the difference between $\varepsilon = +1$ and $\varepsilon = -1$ has meaning only inside of the light cone $x^2 > 0$ and it cannot be formulated for Green's functions which differ from zero outside of the light cone, i.e., for $x^2 < 0$. Therefore, the limiting transition (39.10) cannot be realized if the retarded nature of the interaction is preserved.

The origin of this difficulty is directly related to the fact that the metric of Einstein-Minkowski space is indefinite and that the concept of closeness of two space-time points is lacking. We can surmount this difficulty only by taking into account the existence of matter as sources of fields.

The essence of the matter is contained in the fact that, in order to define the closeness of two points $P(x_1)$ and $P(x_2)$ in the Einstein-Minkowski space, apart from the radius vector ($\mathbf{x} = \mathbf{x}_2 - \mathbf{x}_1$, $t_2 - t_1 = t$) which connects the two world points $P(x_1)$, $P(x_2)$, we must have some other time-like vector n (this vector can be considered as unitary, $n_2 = 1$, without any restrictions†). Using such a vector, we may form the invariant (xn) in addition to the invariants x^2, $n^2 = 1$, and using these we may construct the positive-definite quantity (also invariant):

$$R^2 = 2(xn)^2 - x^2 \geqslant 0. \tag{39.15}$$

† We already used such a vector in considering causality [see (37.24) and (38.4)].

In the coordinate system in which $n=(1, 0, 0, 0)$, the invariant R is transformed into the quadratic form

$$R^2 = r^2 + t^2, \tag{39.15'}$$

which is the positive-definite Euclidean distance between the points $P(x_1)$ and $P(x_2)$. Using this "distance," the concept of closeness of the two world points $P(x_1)$ and $P(x_2)$ may be defined as follows. The points $P(x_1)$, $P(x_2)$ are considered close if

$$R \lesssim a, \tag{39.16}$$

and distant if

$$R \gg a. \tag{39.17}$$

This is the formal statement of the matter. The physical meaning of the time-like vector n can be quite varied. We shall distinguish between two principally different classes of vectors n. In the first, (a) the vector n is related to the motion of matter located at points $P(x_1)$ and $P(x_2)$. For example, this vector may be identical to the four-dimensional velocity U_Q of the source of the field Q which is located at the point $P(x_1)$, or to the velocity U_A of the field reciever located at the point $P(x_2)$, or to the velocity of their center of mass U_m.

In all these cases, the vector n is related to the motion of matter that is located at the points $P(x_1)$, $P(x_2)$. We call the vector n *internal* in this case. Introduction of the internal vector n into the theory does not violate any of the relativistic relationships for free particles. It does not violate the homogineity or isotropy of space, and it does not result in any one coordinate system being preferred.

The second possibility (b) is that the vector n selects out some prefered coordinate system. In this case we call the vector n *external*. Such a vector may be related to the selection of any coordinate system [105], for example, to the coordinate system that moves with the matter in an expanding Universe,† or to the system that is related to physical vacuum (if this vacuum is considered as "ether"). This coordinate system could be related to discrete quantum space [94, 104], or to the reference frame of the observer [91].

† This classification of possible types of the vector n was introduced in [69, 72].

In concluding, we note that the Einstein interpretation of the variables $\mathbf{x}$ and t looses its meaning in all acausal cases because, in the region $|\mathbf{x}|$, $|t| \approx a$, there is no signal which has a velocity $c=1$. However the finite dimensions of real particles themselves exclude the possibility of measuring the variables $\mathbf{x}$ and t inside of particles and, therefore, place their physical meaning in question. The space-time description of particle structure may be no less of a fictitious construction than the classical electron orbits inside the atoms.

40. EUCLIDEAN GEOMETRY IN THE MICROWORLD

In this section we shall consider a model of causal relations which, for large distances, reduces to the usual causal relations that are based on the principle that the speed of light is constant in space, and for small distances, changes their properties to such an extent that the velocity of light can become imaginary. The first type of causal relation corresponds to the pseudo-Euclidean geometry of the Einstein-Minkowski space, and the second type, to Euclidean geometry of four-dimensional space. Such a model should seem neither too strange nor even too fantastic if we recall, for example, the peculiarities of nonlinear field theory.

In Section 7 we showed that, under certain conditions in the regions of a large field φ and large gradients of this field $p=\partial\varphi/\partial t$, $q=\partial\varphi/\partial x$, the characteristic equations of the field can become imaginary. In this situation there arises what we have what we called the "lump" of events, the set of realities which are related to each other but do not result from each other. In other words, such realities cannot be ordered in time. The concepts "earlier" and "later" do not exist for them. Note that if an elementary particle cannot be decomposed into constituent particles which could be joined by a light signal, then there is no basis for assuming that the causal relation inside of these elementary particles, or inside of close groupings of them formed in collisions, will be the same as that which is characteristic of events which are separated from each other by distances which exceed the elementary particle dimensions.

In the following we shall consider the particle groups which, in experimentalists terminology are called "events" or "stars". Each such "event" or "star" is characterized by its rays which mathematically are described by the momenta p_k of the particles which enter and leave (for a star with N rays, $k=1, 2, \ldots, N$).† These momenta (vectors) may be displaced parallel to themselves within the bounds of the complex, so that they all originate from the same point, for example the center of mass. We call such points the "focus" of the events.

The existence of a focus violates the homogeneity of vacuum, and the vector p_k introduces a specific direction which violates its isotropy. However, we shall assume that the "star" as a whole may be displaced in space and time to any value, and rotated to any direction without changing anything inside of it. This means that we consider vacuum itself as homogeneous and isotropic, and that it is only the existence of the "stars" that *de facto* introduces a special point (the "focus of the star") and direction in it.

We shall now formulate these ideas mathematically. In modern field theory, particles are described by the functions $s(x_1, x_2, \ldots, x_N)$ in the expansion of the matrix $\hat{S}$ in terms of the operators of the field $\varphi(x)$ (see Section 29). The matrix element of such a coefficient function

$$\langle p_N, p_{N-1}, \ldots, p_{N-m} | s(x_1, x_2, \ldots, x_N) | p_{N-m-1}, p_{N-m-2}, \ldots, p_2, p_1 \rangle$$
$$= \int d^4 x_N \cdots d^4 x_1 s(x_1, x_2, \ldots, x_N) \exp i (p_N x_N + \cdots$$
$$+ p_{N-m} x_{N-m} - p_{N-m-1} x_{N-m-1} - \cdots - p_1 x_1) \tag{40.1}$$

is equal to the probability amplitude of particle transitions from an initial state, characterized by the momenta $p_1, p_2, \ldots, p_{N-m-1}$, to a final state, characterized by the momenta $p_{N-m}, \ldots, p_N$.

The functions $s(x_1, x_2, \ldots, x_N)$ are translationally invariant. In order to express this characteristic explicitly, we introduce, in place of the coordinates $x_1, x_2, \ldots, x_N$, the new coordinates which are related to the old by the linear transformations

$$X_\alpha = \sum_k A^k x_\alpha^k \quad (\alpha = 1, 2, 3, 4), \tag{40.2}$$

$$\sum_k A^k = 1 \quad (k = 1, 2, \ldots, N), \tag{40.2'}$$

$$\xi_\alpha^i = \sum_k a^{ik} x_\alpha^k \quad (\alpha = 1, 2, 3, 4), \tag{40.3}$$

$$\sum_k a^{ik} = 0 \quad (i = 1, 2, \ldots, N-1). \tag{40.3'}$$

† We broaden the concept of N-ray stars here to include neutral particles in the set of rays

The operation of translation $T(d)$

$$\hat{T}(d)\, x = x + d\,, \tag{40.4}$$

by use of the conditions (40.2′) and (40.3′), gives

$$\hat{T}(d)\, X = X + d\,, \tag{40.5}$$

$$\hat{T}(d)\, \xi = \xi\,. \tag{40.6}$$

In this manner, under the translation $\hat{T}(d)$, the space $\mathfrak{R}_4(X)$ is displaced by d, and the space $\mathfrak{R}(\xi)$ goes over into itself identically.

For the homogeneous Lorentz transformation, these spaces also transform independently. Let us carry out the Lorentz transformation $\bar{x} = \hat{L}x$, where $\bar{x}$ are the new coordinates, in (40.2) and (40.3). The transformations can be written in greater details as

$$\bar{x}_\alpha^k = L_{\alpha\beta} x_\beta^k\,, \tag{40.7}$$

where $L_{\alpha\beta}$ are the matrix elements of the matrix $\hat{L}$. From (40.7), (40.2) and (40.3), we obtain

$$\bar{X}_\alpha = \sum_{k=1}^{N} A^k L_{\alpha\beta} x_\beta^k = L_{\alpha\beta} \sum_{k=1}^{N} A^k x_\beta^k = L_{\alpha\beta} X_\beta \tag{40.8}$$

and

$$\bar{\xi}_\alpha^i = \sum_{k=1}^{N} a^{ik} L_{\alpha\beta} x_\beta^k = L_{\alpha\beta} \sum_{k=1}^{N} a^{ik} x^k = L_{\alpha\beta} \xi_\beta^i\,, \tag{40.9}$$

or, in short notation,

$$\bar{X} = \hat{L}X\,, \quad \bar{\xi}^i = \hat{L}\xi^i\,. \tag{40.10}$$

In other words, under Lorentz transformation, the spaces $\mathfrak{R}_4(x)$ and $\mathfrak{R}_4(\xi)$ transform independently, each inside of itself. If we now perform the transformation of (40.2) and (40.3) on the variables $x_1, x_2, \ldots, x_N$, then the function s turns out to depend only on the variables ξ:

$$s(x_1, x_2, \ldots, x_N) \to s(\xi_1, \xi_2, \ldots, \xi_{N-1})\,, \tag{40.11}$$

i.e., it is defined in space by the relative variables of $\mathfrak{R}_{4N-4}(\xi)$.

Furthermore, this function which is itself invariant under Lorentz transformation, depends only on the scalar products

$$I^{ik} = \xi^i \xi^k = g_{\alpha\beta} \xi_\alpha^i \xi_\beta^k \qquad (i, k = 1, 2, \ldots, N-1)\,. \tag{40.12}$$

where

$$g_{\alpha x \beta} = \begin{cases} 0, & \text{for} \quad \alpha \neq \beta \\ +1, -1, -1, -1 & \text{for} \quad \alpha = \beta \end{cases}$$

so that we may write

$$\begin{aligned} s(\xi_1, \xi_2, \ldots, \xi_{N-1}) &\equiv s(\ldots, \xi_\iota, \ldots) \\ &= s(I^{12}, I^{13}, \ldots, I^{ik}, \ldots) \\ &\equiv s(\ldots, I^{ik}, \ldots). \end{aligned} \tag{40.13}$$

We must remember here that of all the $N(N-1)/2$ invariants I^{ik}, only $(4N-10)$ are independent.

Thus the function s is defined in the space $\mathfrak{R}_{4N-4}(\xi)$ and it is a function of the scalar products I^{ik}. We now assume that the causal relation characteristic of local theory changes in the vicinity of elementary particles. As was explained in Section 39, in order to define the concept of vicinity, the isotropy of space-time must be violated and this may formally be realized by introducing the function $s(\xi_1, \xi_2, \ldots, \xi_{N-1})$ of time-like vectors n. Such a possibility exists in real "stars" because, by the very fact of its existence, the star violates the isotropy of space and time. In particular, any of the vector n_k $(n_k^2 = 1, n_{k0} > 0)$ which define the direction of the star's rays,

$$n_k = \frac{p_k}{\sqrt{p_k^2}}, \quad \text{for} \quad k = 1, 2, 3, \ldots, N, \tag{40.14}$$

may be taken for the necessary vector n. We could also use their linear combinations – for example, the vector defining the direction of the motion of the star's focus†

$$n = \frac{P}{\sqrt{P^2}}, \quad P = \sum_{k=1}^{N} p_k. \tag{40.15}$$

Taking into account the anisotropy of space-time which results from the anisotropy of the real stars, we may generalize the coefficients of the function $s(\xi_1, \xi_2, \ldots, \xi_{N-1})$ so that their dependence on the directions of the world lines is made explicit – for example,

$$s(\xi_1, \xi_2, \ldots, \xi_{N-1}; n_1, n_2, \ldots, n_{N-1}) \tag{40.16}$$

† Compare with Sections 37 and 39.

or, for a more special case,

$$s(\xi_1, \xi_2, \ldots, \xi_{N-1}; n). \tag{40.17}$$

Thus the following new invariants are introduced

$$R^{ii} = 2(n\xi^i)^2 - \xi^{i2} \geqslant 0 \tag{40.18}$$

and

$$R^{ik} = 2(n\xi^i)(n\xi^k) - \xi^i\xi^k \tag{40.19}$$

or the nondimensional invariant

$$\zeta^{ik} = \frac{R^{ik}}{a^2}, \tag{40.20}$$

where a is the elementary length. Conserving the relativistic invariance of "events" as a whole, using the vector n, we may assume that the function $a(\xi_1, \xi_2, \ldots, \xi_{N-1}; n)$ depends on the invariants $R^{ik}(\xi, n)$ (40.18) and (40.20) as well as the invariant I^{ik} (40.12). We select this dependence in the following manner. Replacing the scalar product (40.12) by a more general one, we set

$$I_a^{ik} = a(z)\, I^{ik} + b(z)\, R^{ik}, \tag{40.21}$$

where

$$z = \tfrac{1}{2}(\zeta^{ii} + \zeta^{kk}), \tag{40.22}$$

and the invariants I^{ik} and R^{ik} are defined by (40.13) and (40.19).

The relativistic invariance of the coefficient function $s(\xi_1, \xi_2, \ldots, \xi_{N-1}; n)$, when the invariants I^{ik} are replaced by the invariants I_a^{ik}, is fully conserved. However, macroscopic causality may be violated. To avoid this happening, we select the functions $a(z)$ and $b(z)$ in such a manner that for $z \to 0$,

$$a(z) \to 0, \quad b(z) \to 1, \tag{40.23}$$

and for $z \to \infty$,

$$a(z) \to 1, \quad b(z) \to 0. \tag{40.23'}$$

Furthermore, note that by "vicinity of an elementary particle", we mean that the quantity $|x_i - x_k|$ is small, and

$$|x_i - x_k| \ll a. \tag{40.24}$$

In the general case, this condition does not correspond at all to the condition that the invariants (40.18)–(40.20), which define the smallness of the quantity $|\xi^i|$, be small

$$|\xi^i| \ll a. \tag{40.25}$$

These conditions may be brought into agreement only by special selection of the variables ξ_i, these being

$$\xi_1 = x_1 - x_2, \quad \xi_2 = x_2 - x_3, \quad \ldots, \quad \xi_{N-1} = x_{N-1} - x_N. \tag{40.26}$$

In satisfying conditions (40.23), (40.23′) and (40.26) for $z \to \infty$, we obtain the usual causal relation because, in this case $I_a^{ik} \to I^{ik}$. However, for $z \to 0$, $I_a^{ik} \to R^{ik}$ so that the usual causal relation is completely destroyed and in place of a succession of events, we have their "lump", which is not ordered in time. In this latter case, in the coordinate system in which $n = (1, 0, 0, 0)$, the invariant I_a^{ik} has the form

$$I_a^{ik} = R^{ik} = \sum_{a=1}^{4} \xi_a^i \xi_a^k, \tag{40.27}$$

which is characteristic of Euclidean geometry. A new symmetry arises in this case. For a given n, for small $|\xi_\alpha^i|$, $|\xi_\beta^k|$ we have four-dimensional, spherical (or elliptical) symmetry. If all the scalar products of the vectors $\xi^i, \xi^k, \ldots$ were defined in terms of the invariants I_a^{ik}, then we could say that in the space $\mathfrak{R}(\xi)$ there is a special geometry which is such that, for $|\xi_i| \gg a$, it is *pseudo Euclidean*, and for $|\xi_i| \ll a$, it is *Euclidean*.

The result of measuring the distance between two points A and B in such a space would depend on the scale on which this length were measured, whether it were larger or smaller than the elementary length. This paradoxical situation should not be suprising, however, because small values of the coordinate ξ are unobservable. Since we restricted ourselves to satisfying the conditions of macroscopic causality in discussing the physically observable matrix elements of the scattering matrix (40.1), this paradox is hidden rather deeply.

41. Stochastic Geometry

A basic assumption of stochastic metric geometry is that, to each pair of elements P, Q of stochastic space $\mathfrak{R}(P)$, it is possible to associate a distribution function $F_{PQ}(z)$ which can be interpreted as the probability that the distance $r(P, Q)$ between the points P and Q is less than z. This function has the following properties

$$\left.\begin{array}{l} 1)\ F_{PQ}(z) = 1 \ \text{for all } z \text{ only when } P = Q; \\ 2)\ F_{PQ}(z) = 0 \ \text{for all } z \leqslant 0; \\ 3)\ F_{PQ} = F_{QP}, \text{ symmetry}; \\ 4)\ \text{if } F_{PQ}(z) = 1 \text{ and } F_{QR}(u) = 1, \text{ then } F_{PR}(z+u) = 1. \end{array}\right\} \quad (41.1)$$

These properties reflect the usual properties of distance in the metric space $\mathfrak{R}_m(P)$:

$$\left.\begin{array}{l} 1)\ r(P, Q) = 0 \ \text{if and only if } P = Q; \\ 2)\ r(P, Q) \geqslant 0; \\ 3)\ r(P, Q) = r(Q, P); \\ 4)\ r(P, Q) + r(Q, R) \geqslant r(P, R). \end{array}\right\} \quad (41.1')$$

This metric follows from the stochastic metric as a special case if

$$F_{PQ}(z) \equiv H[z - r(P, Q)] \quad (41.2)$$

and

$$H(z) = 1 \quad \text{for} \quad z > 0, \quad (41.2')$$

$$H(z) = 0 \quad \text{for} \quad z \leqslant 0 \quad (41.2'')$$

(see [86, 87]). From a physical point of view, the stochastic metric leads to stochastic arithmetization of space. In fact the elements of the space P are arithmetized in a manner such that, to each element P which marks some specific physical event, a number $x(P)$ (coordinate) is assigned which is obtained by measuring the distance $r(P, 0)$ from some arbitrary element (0) (the coordinate "origin").

The stochastic nature of the distance $r(P, Q)$ leads to the fact that if we measure

$$r(P, 0) = L, \quad (41.3)$$

then the actual distance l between the elements P and 0 is equal to L to the probability

$$\mathrm{d}W(L, l) = w(L, l)\,\mathrm{d}l. \quad (41.4)$$

Therefore, the arithmetization of the spatial manifold becomes stochastic and each measured coordinate $x(P)$ of the element P may be assigned an actual coordinate $X(P)$ only to some degree of probability.

In the following, by stochastic space we shall mean that space $\mathfrak{R}_{st}(X)$ which is mapped stochastically into the space of the measurable coordinates $\mathfrak{R}(x)$, and we introduce the probability

$$\mathrm{d}W(X, x) = w(X, x)\,\mathrm{d}x. \tag{41.5}$$

This equation differs from (41.4) only in that the distances L and l are assigned $\pm$ signs so that they must now be considered as coordinates of the point P which is measureable and existent. If our space is uniform, then the probability (41.5) is a function of only the difference $X-x$. In this case, in place of (41.5), we may write

$$\mathrm{d}W(X, x) = w\left(\frac{X-x}{a}\right)\frac{\mathrm{d}x}{a}, \tag{41.6}$$

where a is some scale length which is characteristic of the dispersion of the individual values of the stochastic coordinates x about the mean value $x = X(P)$.

A number of papers [88–92] have been written on stochastic space. We do not intend to give a detailed analysis of these papers in this place. We can summarize their conceptual content in the assumption that the "true" coordinates (x, t) of a physical event P, because of some principal indeterminacy in the instrument used to measure x, t, or in the event P itself, can be defined only to some degree of probability. By this we mean that if we imagine an ensemble of laboratory assistants who all have identical, ideal instruments for measuring the coordinates of an event P, they would all obtain different results for $x = x(P)$ which would be distributed statistically about the average value $\bar{x} = X(P)$

$$\bar{x} = X(P) = \int x w\left(\frac{X-x}{a}\right)\frac{\mathrm{d}x}{a}, \tag{41.7}$$

with the characteristic statistical dispersion equal to

$$\overline{\Delta x^2} = \overline{(X-x)^2} = \int (X-x)^2\, w\left(\frac{X-x}{a}\right)\frac{\mathrm{d}x}{a} = a^2, \tag{41.8}$$

with the normalization condition

$$\int w\left(\frac{X - x}{a}\right)\frac{dx}{a} = 1. \tag{41.9}$$

Going into greater detail, it is possible to give at least three possible cases which lead to a stochastic space $\mathfrak{R}_{st}(X)$:

(a) the ideal instrument for measuring distances has an irremovable indeterminacy $\Delta x \approx a$;
(b) the objects "particles" with the aid of which points in space are fixed, have finite dimensions a, which leads to $\Delta x \approx a$;
(c) the medium in which the signal used for measuring propagates is turbulent so that the signal velocity fluctuates.

The main difference between the physical theory of stochastic space-time and the mathematical theory discussed in [86, 87], results from the fact that physical space has an indefinite metric, whereas the metric described by the distribution $F_{PQ}(z)$ (41.1) is assumed to be positive definite. Because the metric of physical space-time is indefinite, Eq. (41.6) cannot be generalized to real space-time simply by increasing the number of dimensions. In fact the invariance of the probability dW of the Lorentz transformation requires that the probability density ω be a function of an invariant. The possible invariant is the interval

$$s^2 = (X - x)^2 = (T - t)^2 - (X - x)^2.$$

However, the probability

$$dW = w\left(\frac{s^2}{a^2}\right)\frac{d^4x}{a^4} \tag{41.6'}$$

is not normalizable and cannot represent the distribution of the stochastic variable $x(P)$ which is concentrated about the point P (and not about the light cone $s^2 = 0$).

In order to obtain a concentration of the distribution about P, we must have a time-like vector n as well as the four-dimensional vector $(X - x)$. It is possible to construct an invariant that we met earlier, $I(n, X - x)$, with such a vector [see (39.15)], and to write the positive-definite distance R as

$$R = +\sqrt{2(n, X - x)^2 - s^2} \geqslant 0. \tag{41.10}$$

The probability $\mathrm{d}W$ may now be written as a function of this distance

$$\mathrm{d}W(X-x,n)=w\left(\frac{R}{a}\right)\frac{\mathrm{d}^4x}{a^4}. \tag{41.11}$$

This probability may be concentrated about the point $X(P)$ and it may limit the value of the rms deviation

$$\overline{\Delta x^2}=\int (X-x)^2\,w\left(\frac{R}{a}\right)\frac{\mathrm{d}^4x}{a^4}\approx a^2. \tag{41.8'}$$

Inserting the vector n into the equation for the probability $\mathrm{d}W$, we again face the problem of choosing this vector (see Section 39). In accordance with the possibilities (a) and (b)† given in that section, we shall first discuss the former.

(a) The vector n is tied to the reference frame of the observer so that the direction of the vector n corresponds to the direction of the world line of the observer (compare with [50, 91]). In this case averaging physical quantities which depend on the stochastic coordinates x of the world point P will lead to results which depend on the observer.

Let $P(x)$ be a physical quantity, a function of the stochastic variable $x=X-\xi$. Its observed average value will be

$$\begin{aligned}\bar{F}(X)=\overline{F(x)}&=\int F(X+\xi)\,w(\xi,n)\,\mathrm{d}^4\xi\\&=\int \tilde{F}(Q)\,\mathrm{e}^{\mathrm{i}QX}\,\tilde{w}(Q,n)\,\mathrm{d}^4Q,\end{aligned} \tag{41.12}$$

where $\tilde{F}(Q)$ is the Fourier transform of $F(x)$, and $\tilde{\omega}(Q,n)$ is the Fourier transform of the probability $\omega(\xi,n)$. The vector n has components $n=(1,0,0,0)$ for all observers. Therefore, the quantity $\tilde{\omega}(Q,n)$ will not be invariant under changes in coordinate systems. This is particularly evident in the following physical reasoning. Let $F(x)$ be a plane wave. In one reference frame, it may have a frequency ω such that $\omega a\ll 1$ (a is a measure of the dispersion of the stochastic coordinate x) and, therefore, its mean value $\bar{F}(x)$ is practically the same as its true value $F(x)$. In the same reference frame but for a second observer, because of the Doppler effect, the value of ωa may be $\gg 1$ and, therefore, $|\bar{F}(X)|\ll|F(x)|$.

† We shall return to possibility (c) in Section 42.

Thus in a theoretical scheme which is defined by a vector n that is tied to the observer, special care must be given to quantities whose invariance is not in doubt. Such quantities are, for example, the total probability of forming a specific number and type of particles. This quantity could depend on the reference frame only in the case that a special reference frame exist (for example, the system related to physical vacuum, taken in the old sence of some type of medium such as "ether").

(b) We now turn to the second possibility in which the stochastic nature of the coordinates $x(P)$ depends on the particle dimensions being finite. In this case a physical event P is identified by the presence of particles with finite dimensions a at this point. The coordinates $x(P)$ can now be stochastic because it is not possible to define which point inside the particle (to within the limits of $\Delta x \approx a$) scattered the signal, as if it were not localized accurately in space-time.†

We can imagine that space-time is filled with particles which do not differ from each other, as is assumed in the Lagrangian method of describing fluid motion. In this case the time-like vector n in (41.11) can be identified with a four-dimensional particle velocity u. Relativistic invariance of the probability $\mathrm{d}W(X-x, n)$ under changes from the reference frame of one observer to that of another will be observed because the vector u now satisfies the Lorentz transformation. The assumption that the variables u and $\xi = X - x$ commute,

$$[u, \xi] = 0. \tag{41.13}$$

is implicit in the assumption that the probability depends on these variables. Therefore it is natural to consider the variable ξ as one which describes the internal structure of a particle.

In particular, in reference [89] the rest mass of a particle m is assumed to be represented by the operator m which depends on the internal coordinates ξ and the four-dimensional velocity vector u of the particle. That is,

$$\hat{m}(\xi, u) = m_0 + \alpha\left[\frac{\partial^2}{\partial\xi_\mu\,\partial\xi_\mu} + 2\left(u_\mu\frac{\partial}{\partial\xi_\mu}\right)^2 + \xi_\mu\xi_\mu - (u_\mu\xi_\mu)^2\right]. \tag{41.14}$$

† This assumption of accurate localization of a signal is equivalent in this case to the assumption that the instrument used to measure the particle coordinates is ideal.

By changing to the system in which the particle is at rest, $n=(1, 0, 0, 0)$, it is easy to show that the metric in the space $\mathfrak{R}(\xi)$ is the Euclidean metric $I(\xi, \xi)=\xi_1\xi_1+\xi_2\xi_2+\xi_3\xi_3+\xi_4\xi_4$.† The wave function of the free particles which have a momentum P has the form

$$\psi_P(x, \xi) = \varphi_P(\xi)\, e^{iRx/h} \tag{41.15}$$

Placing this in Dirac's Equation (14.1) gives

$$[i\gamma_\mu P_\mu + \hat{m}(\xi, P)]\, \varphi_P(\xi) = 0\,, \tag{41.16}$$

where $\hat{m}(\xi, P)$ is the operator (41.14), in which the speed u is expressed in terms of the total momentum $u = P/(P^2)^{1/2}$. This curious theoretical scheme has, however, not been investigated fully yet.

42. Discrete Space-Time

In this section we shall analyse the ideas concerning discrete, "quantized" space-time. The concept of "quantized" space-time itself is not sufficiently well defined. We shall use this concept here in relation to theoretical schemes in which some discreteness of space-time is assumed which does not result from any material motion and has a purely geometric character. In its simplest form, the idea of discrete quantization of space-time is contained in the assumption that the points in a space-time manifold do not form a continuum, but represent some discrete, denumerable, or even finite set. In particular we can assume that it is represented by points on some lattice (see [94–97]).

In assuming some discreteness to the space-time manifold, we encounter two problems, namely, (a) the problem of transforming such a manifold from one coordinate system to another moving one (relativity), and (b) the problem of the limiting transition from discreteness to the continuum (correspondence).

These problems may be illustrated by an example of a four-dimensional manifold $D_4(x)$ in which each point X is defined by the coordinates $x_1=n_1a$, $x_2=n_2a$, $x_3=n_3a$, $t=\nu a$ where n_1, n_2, n_3 and ν are whole numbers and the length a is the grid constant. Let $\psi(\mathbf{x}, t)$ be a field that describes some physical reality which can propagate in the space $D_4(x)$.

† This metric, which is based on condition (41.13), was derived in [93] in which relativistic gyroscope theory is treated.

We use the word "propagate" here to mean that (a) there exists some relationship between the value of the field ψ at a point $x(\mathbf{x}, t)$ and at its neighboring points, and (b) we assume that it is possible to divide the manifold $D_4(x)$ into space, $D_3(x)$, and time, $D_1(t)$ (see Section 3). Furthermore, we assume that the field ψ is linear, and using "close" to mean neighboring points, i.e. points with the coordinates

$$x_1 \pm a, x_2, x_3, t; \qquad x_1, x_2 \pm a, x_3, t;$$
$$x_1, x_2, x_3 \pm a, t; \qquad x_1, x_2, x_3, t \pm a;$$

we may write the equation for ψ as

$$\begin{aligned}
&\alpha\psi(x_1, x_2, x_3, t) \\
&\quad + \beta_1 [\psi(x_1 + a, x_2, x_3, t) + \psi(x_1 - a, x_2, x_3, t)] \\
&\quad + \beta_2 [\psi(x_1, x_2 + a, x_3, t) + \psi(x_1, x_2 - a, x_3, t)] \\
&\quad + \beta_3 [\psi(x_1, x_2, x_3 + a, t) + \psi(x_1, x_2, x_3 - a, t)] \\
&\quad + \gamma [\psi(x_1, x_2, x_3, t + a) + \psi(x_1, x_2, x_3, t - a)] = 0.
\end{aligned} \tag{42.1}$$

If the manifold $D_4(x)$ is homogeneous, then the coefficients α, β_1, β_2 β_3, γ do not depend on $\mathbf{x}$ and t. If it is also isotropic in the variables x_1, x_2, x_3, then $\beta_1 = \beta_2 = \beta_3$. Separating out time, we assume that the signs of the coefficients ρ and γ are opposite. Then Equation (42.1) may be written in the form

$$(\alpha + \sigma\beta + 2\gamma)\psi + \beta[\Delta_1\psi + \Delta_2\psi + \Delta_3\psi] + \gamma\Delta_t\psi = 0, \tag{42.2}$$

where Δ_1, Δ_2, Δ_3, Δ_t are the second finite differences of the variables x_1, x_2, x_3 and t respectively. By definition,

$$\Delta_i\psi \equiv \psi(\ldots, x_i + a, \ldots) + \psi(\ldots, x_i - a, \ldots) - 2\psi(\ldots, x_i, \ldots), \tag{42.3}$$

where the dots denote the arguments which do not undergo the displacement $\pm a$.

In order to abbreviate our studies of Equation (42.3), we shall restrict ourselves to the special case of $\alpha + \sigma\beta + 2\gamma = 0$, $\beta = -\gamma$ so that in place of (42.2), we have

$$\Delta_1\psi + \Delta_2\psi + \Delta_3\psi - \Delta_t\psi = 0. \tag{42.4}$$

This equation has eigenfunctions that have the form of plane waves

$$\psi = \text{const. } e^{i(\Pi x - \varepsilon t)}, \tag{42.5}$$

where the momentum $\mathbf{\Pi}(\Pi_1, \Pi_2, \Pi_3)$ and energy ε are related by the dispersion relation for these waves, which is obtained by substituting (42.5) into (42.4) in the form

$$\sin^2\left(\frac{\Pi_1 a}{2}\right) + \sin^2\left(\frac{\Pi_2 a}{2}\right) + \sin^2\left(\frac{\Pi_3 a}{2}\right) - \sin^2\left(\frac{\varepsilon a}{2}\right) = 0. \tag{42.6}$$

Equation (42.4), its solution (42.5), and the dispersion relation (42.6) all have the symmetry of a cubic lattice in the space $\mathfrak{R}_3(\mathbf{\Pi})$ and, therefore, they are invariant with respect to the group symmetry of this lattice. Thus transformation in one of the planes (Π_k, ε) just interchanges one of the coordinates x_k and time t, and one of the momentum components Π_k is simultaneously interchanged with the energy ε.

At first glance, these discrete and finite transformations do not have anything in common with the Lorentz transformations. However, for $|\mathbf{\Pi} a| \ll 1$, the dispersion relation (42.6) has the form

$$\Pi_1^2 + \Pi_2^2 + \Pi_3^2 - \varepsilon^2 = O(\Pi a^4), \tag{42.7}$$

where $O(\Pi a^4)$ denotes the remainder that is of the order of $(\Pi a)^4$. The finite differences in (42.4) may simultaneously be replaced by the second derivatives so that in place of (42.4), we have

$$\frac{\partial^2 \psi}{\partial x_1^2} + \frac{\partial^2 \psi}{\partial x_2^2} + \frac{\partial^2 \psi}{\partial x_3^2} - \frac{\partial^2 \psi}{\partial t^2} = O(\Pi a^4)\, \psi. \tag{42.8}$$

Thus for sufficiently small momenta $|\mathbf{\Pi}|$, Lorentz invariance still holds. For larger momenta, $|\mathbf{\Pi}| \sim \Pi/a$, the situation changes essentially and only discrete transformations, characteristic of the lattice, remain valid. Disregarding such a correspondence with the usual theory, we cannot avoid the existence of a preferable coordinate system in this theoretical scheme, this being the system of the lattice $D_4(x)$. This system has the peculiar feature that the usual relations of the special theory of relativity are valid for particles that have a small momentum in this system.

We now turn to another approach, to the idea of a discrete space-time based on the coordinate x_μ as the displacement operator $\hat{x}_\mu$ in a curved momentum space $\mathfrak{R}_4(p)$ [98]. The most direct way of constructing such operators goes back to the concept of bounding the square of the momentum†:

$$p^2 = p_4^2 - p_1^2 - p_2^2 - p_3^2 \leqslant \frac{1}{a^2}, \tag{42.9}$$

where a is some characteristic length for this theory (Planck's constant $\hbar$ and the velocity of light c are taken to be equal to 1 here). Condition (42.9) physically means that the square of the mass of the particle is bounded; $m^2 = 1/a^2$.

We now introduce the homogeneous coordinates $\eta_0, \eta_1, \eta_2, \eta_3$ and η_4 such that

$$p_\mu = \frac{1}{a}\frac{\eta_\mu}{\eta_0} \qquad (\mu = 1, 2, 3, 4). \tag{42.10}$$

In terms of these coordinates, Equation (42.7), for the limiting values of p, becomes

$$\eta_4^2 - \eta_1^2 - \eta_2^2 - \eta_3^2 - \eta_0^2 = 0. \tag{42.11}$$

Together with this quadratic form, we consider the five-dimensional hypersphere

$$\eta_4^2 - \eta_1^2 - \eta_2^2 - \eta_3^2 - \eta_0^2 = -a^2 \tag{42.12}$$

and we shall define the operators $\hat{x}_\mu$ as operators of infinitely small rotations which leave the hypersphere invariant. These operators are

$$\hat{x}_\alpha = ia\left(\eta_0 \frac{\partial}{\partial \eta_\alpha} - \eta_\alpha \frac{\partial}{\partial \eta_0}\right), \qquad \alpha = 1, 2, 3, \tag{42.13}$$

$$\hat{x}_4 = ia\left(\eta_0 \frac{\partial}{\partial \eta_4} + \eta_4 \frac{\partial}{\partial \eta_0}\right). \tag{42.13'}$$

where $\hat{x}_4 = \hat{t}$ is the operator of time. These operators may be supplemented by the angular momentum operators

$$\hat{L}_\alpha = i\left(\eta_\beta \frac{\partial}{\partial \eta_\gamma} - \eta_\gamma \frac{\partial}{\partial \eta_\beta}\right), \qquad \alpha, \beta, \gamma = 1, 2, 3, \tag{42.14}$$

† We may also consider a second limitation, namely $p^2 \gg -1/a^2$ (see [100]).

$$\hat{M}_\alpha = i\left(\eta_4 \frac{\partial}{\partial \eta_\alpha} + \eta_\alpha \frac{\partial}{\partial \eta_4}\right), \qquad \alpha = 1, 2, 3. \tag{42.14'}$$

It is easy to show that the nontrivial permutation laws give

$$[\hat{x}_\alpha, \hat{x}_\beta] = \mathrm{i}a^2 \hat{L}_\gamma, \tag{42.15}$$
$$[\hat{x}_4, \hat{x}_\alpha] = \mathrm{i}a^2 \hat{M}_\gamma, \tag{42.15'}$$
$$[\hat{x}_\alpha, \hat{p}_\beta] = \mathrm{i}(\delta_{\alpha\beta} + a^2 \hat{p}_\alpha \hat{p}_\beta), \tag{42.16}$$
$$[\hat{x}_4, \hat{p}_4] = \mathrm{i}(1 - a^2 \hat{p}_4^2), \tag{42.16'}$$
$$[\hat{x}_\alpha, \hat{p}_4] = \mathrm{i}a^2 \hat{p}_\alpha \hat{p}_4. \tag{42.16''}$$

In the same manner, from (42.14) and (42.14′), we obtain

$$\hat{L}_\alpha = \hat{x}_\beta \hat{p}_\gamma - \hat{x}_\gamma \hat{p}_\beta, \tag{42.17}$$
$$\hat{M}_\alpha = \hat{x}_\alpha \hat{p}_4 + \hat{x}_4 \hat{p}_\alpha, \tag{42.17'}$$

i.e., the operators $\hat{L}_\alpha$ and $\hat{M}_\alpha$ are components of the usual four-dimensional angular momentum operator. In a similar manner we can also obtain the operators $\hat{x}_\alpha$, $\hat{x}_4 \equiv \hat{t}$ in the momentum representation

$$\hat{x}_\alpha = \mathrm{i}\left(\frac{\partial}{\partial p_\alpha} + a^2 p_\alpha p_\mu \frac{\partial}{\partial p_\mu}\right), \qquad \alpha = 1, 2, 3, \tag{42.18}$$

$$\hat{t} \equiv \hat{x}_4 = \mathrm{i}\left(\frac{\partial}{\partial p_4} - a^2 p_4 p_\mu \frac{\partial}{\partial p_\mu}\right). \tag{42.18'}$$

These operators act in the space $\mathfrak{R}_4(p)$ in which the restriction (42.9) holds. The eigenfunctions of these operators which satisfy the requirements of being finite and single valued, are obtained from the equations

$$\hat{x}\psi_x(p) = x\psi_x(p), \tag{42.19}$$
$$\hat{t}\psi_t(p) = t\psi_t(p), \tag{42.19'}$$

where x and t are the characteristic values of the operators $\hat{x}$ (we dropped the subscript α) and $\hat{t}$. It is easy to show that

$$\psi_x(p) = \exp[im \operatorname{arctg} \alpha(p)], \tag{42.20}$$

where

$$\alpha(p) = ap_1 \sqrt{1 - a^2 p^2}, \tag{42.20'}$$

and m is an integer.

In a similar manner, we obtain

$$\psi_t(p) \approx \exp\left[\mathrm{i}\tau\beta(p)\right], \tag{42.21}$$

where

$$\beta(p) = \ln \frac{1 + ap_4}{\sqrt{1 - a^2p^2}}. \tag{42.21'}$$

Thus the spectrum of characteristic values of the variable x is discrete: $x = ma$, $m = 0, \pm 1, \pm 2$, and the spectrum of the variable $t = \tau a$ is continuous. Such a 'quantized' space was first obtained by Snyder [98].

The geometry of the space $\mathfrak{R}_4(p)$ in which the operators discussed above and their eigenfunctions are defined may be completely represented by the geometry of a manifold of points which arise as a projection of the hypersphere (42.12) from its center onto the plane $\eta_0 = +a$ [99, 100]. Such a three-dimensional case is shown in Figure 45. By making

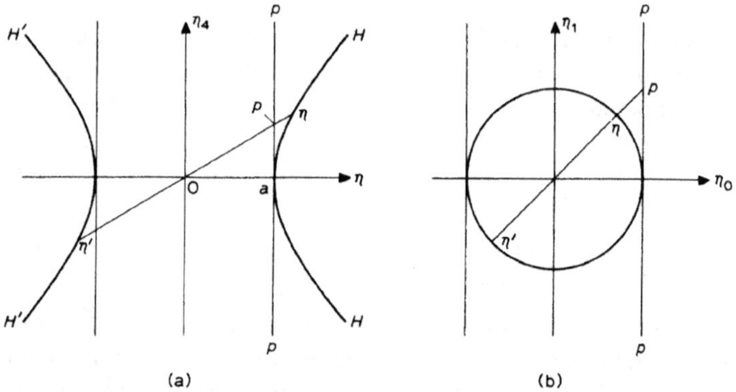

(a) (b)

Fig. 45. (a) Projection of the hypersphere $\eta_4^2 - \eta_1^2 - \eta_0^2 = -a^2$ on the plane $\eta_0 + a$. HH and $H'H'$ are two branches of the hypersphere; pp is the plane $\eta_0 = +a$; η and η' are two points on the hypersphere whose projections on the surface pp are the point p [in the space $\mathfrak{R}_3(p)$]. (b) The same projected onto the plane (η_1, η_0).

such a projection, it is easy to show that we arrive at Equation (42.10) where†

$$\eta_0 = \frac{a}{\sqrt{1 - a^2p^2}}. \tag{42.22}$$

† Other possible versions of the geometry of the space $\mathfrak{R}_4(p)$ may be obtained by displacing the center of the projection along the η_0 axis (cf. [100]).

From (42.10) and (42.12) we obtain the equation for the square of the length element in the space $\mathfrak{R}_4(p)$:

$$d\sigma^2 = (1 - a^2p^2)^{-1}\left[dp^2 + (1 - a^2p^2)^{-1}(p\,dp)^2\right], \qquad (42.23)$$

where

$$dp^2 = dp_4^2 - dp_1^2 - dp_2^2 - dp_3^2, \qquad (42.24)$$

$$p\,dp = p_4\,dp_4 - p_1\,dp_1 - p_2\,dp_2 - p_3\,dp_3. \qquad (42.24')$$

The volume element corresponding to this metric is equal to

$$d\Omega(p) = \sqrt{-g}\,dp_4\,dp_3\,dp_2\,dp_1$$
$$= (1 - a^2p^2)^{-5/2}\,dp_4\,dp_3\,dp_2\,dp_1, \qquad (42.25)$$

where $\sqrt{-g} = (1 - a^2p^2)^{-5/2}$ is the determinant constructed from the metric tensor $g_{\mu\nu}$ which is defined by (42.23). The operators for $\hat{x}_\alpha$, $\hat{x}_4$, $\hat{L}_\alpha$, $\hat{M}_\alpha$ introduced earlier are Hermitian in the space $\mathfrak{R}_4(p)$ with the volume element $d\Omega(p)$.

We may now define the operation of translation in the space $\mathfrak{R}_4(p)$. Translations in the space $\mathfrak{R}_4(p)$ represent rotations of the hypersphere (42.12). Therefore, these translations do not commute. Displacing the vector p to the vector k [denoted by $p(+)k$] is not identical to displacing the vector k to the vector p [denoted by $k(+)p$].† The transformation of the wave function $\varphi(p)$ in the space $\mathfrak{R}_4(p)$ for infinitely small displacements of the space to the vector k may be written as

$$\varphi(p(+)k) = 1 - i\hat{x}_\mu k_\mu \varphi(p). \qquad (42.26)$$

The calculations show that the operators $\hat{x}_\mu$ in (42.26) correspond to the operators (42.18) and (42.18′), as is expected from the fact that displacements in the space $\mathfrak{R}_4(p)$ are equivalent to rotations in the space $\mathfrak{R}_5(\eta)$. The fact that momentum addition is not associative causes the usual laws of energy-momentum conservation to become approximations in Snyder's system.

Practical applications of this refined geometrical theory to quantum field theory has led to serious difficulties [101, 102]. These difficulties are related to what are called the angular divergences which arise in the Feynman diagram calculations in momentum space. These divergences

† For details, see [100].

are not important in the ordinary theory because they may be removed by rotating the integration path in the plane p_4, $p \equiv |\mathbf{p}|$,† through $\pi/2$. Under such rotations, the indefinite metric in the plane (p_0, p) changes to the definite, Euclidean metric. In the case of Snyder's geometry in momentum space, the only modification that arises in computing the Feynman diagrams results from the fact that the factor $(-g)^{1/2}$ appears in the volume element (42.25). From (42.15) it follows that this factor has a branch cut in the complex plane (p_0, p) for $p^2 = p_0^2 - \mathbf{p}^2 = 1/a^2$. This branch cut does not allow a rotation through $\pi/2$. In references [101, 102] attempts were made to first perform the rotation, then to introduce Snyder's metric and then, using analytic continuation, to return to the physical region of the indefinite metric. Such an approach proved successful in certain cases.‡

43. Quasi-particles in Quantized Space

An original equation, equivalent to that of Tomonaga-Schwinger but more convenient for use in curved momentum space, was formulated in reference [104].

Let $\sigma = nx$ be a space-like plane, where n is a unit time-like vector $(n^2 = 1)$, directed towards the future $(n_0 > 0)$. According to (28.14) we have

$$\mathrm{i}\,\frac{\mathrm{d}\hat{U}(\sigma, -\infty)}{\mathrm{d}\sigma} = \hat{W}(\sigma)\,\hat{U}(\sigma, -\infty), \tag{43.1}$$

where $\hat{W}(\sigma)$ is the interaction energy (in the "interaction representation").‡ According to (28.18), the $\hat{S}$-matrix is equal to

$$\hat{S} = 1 + \mathrm{i}\hat{T} = \hat{U}(+\infty, -\infty). \tag{43.2}$$

We write the operator $\hat{U}(\sigma, -\infty)$ in the form of the integral

$$\hat{U}(\sigma, -\infty) = 1 + \frac{1}{2\pi}\int \frac{\mathrm{d}\tau}{\tau - \mathrm{i}\varepsilon}\,\hat{R}(n\tau)\,\mathrm{e}^{\mathrm{i}\sigma\tau}, \tag{43.3}$$

† Note that the analytical properties of the matrix elements in the usual, local theory allow such rotations.

‡ Reference [103] must also be mentioned here. In this work, a geometry of momentum space with variable curvature is used. The difficulties discussed here may perhaps be overcome in such a geometry.

‡ $\hbar = 1$ in (43.1).

where $\hat{R}(n\tau)$ is some new operator and τ is an invariant parameter, a conjugate of σ: $\varepsilon \to +0$. From (43.2) and (43.3), for $\sigma \to +\infty$, it follows that

$$\hat{S} = 1 + \mathrm{i}\hat{R}(0), \tag{43.4}$$

$$\hat{R}(0) = \hat{T}. \tag{43.5}$$

We now set

$$\hat{W}(\sigma) = \int \hat{\tilde{W}}(n\tau)\, \mathrm{e}^{\mathrm{i}\sigma\tau}\, \mathrm{d}\tau \tag{43.6}$$

and substitute (43.3) and (43.6) into (43.1). It is easy to show that this equation will be satisfied if the operator $\hat{R}(n\tau)$ satisfies the condition†

$$\hat{R}(n\tau) = \hat{\tilde{W}}(n\tau) + \frac{1}{2\pi} \int_{-\infty}^{+\infty} \hat{\tilde{W}}(n\tau - n\tau') \frac{\mathrm{d}\tau'}{\tau' - \mathrm{i}\varepsilon} \hat{R}(n\tau'). \tag{43.7}$$

We shall consider the special case of a self-interacting scalar field $\varphi(x)$ with a cubic interaction‡:

$$\hat{W}(x) = g : \hat{\varphi}(x)^3 :. \tag{43.8}$$

Assuming

$$\hat{\varphi}(x) = \frac{1}{(2\pi)^{3/2}} \int \mathrm{e}^{\mathrm{i}kx} \tilde{\varphi}(k)\, \mathrm{d}^4 k, \tag{43.9}$$

we obtain, using (43.6), (43.8) and (43.9),

$$\hat{\tilde{W}}(n\tau) = \frac{g}{\sqrt{2\pi}} \int \delta^4(n\tau - k_1 - k_2 - k_3) : \tilde{\varphi}(k_1) \times \tilde{\varphi}(k_2)\, \tilde{\varphi}(k_3) : \mathrm{d}^4 k_1\, \mathrm{d}^4 k_2\, \mathrm{d}^4 k_3, \tag{43.10}$$

where the vectors k_1, k_2, k_3 denote the particle momenta.

Integrating (43.7) gives the operator $\hat{R}$ in the form of a series in terms of the interaction constant g:

$$\hat{R}(n\tau) = \sum_{s=1}^{\infty} \hat{R}_s(n\tau), \tag{43.11}$$

† A more direct and detailed derivation of (43.7) is given in Appendix XV.

‡ This special case is easily generalized and is only given here for simplicity.

where

$$\hat{R}_s(n\tau) = \frac{1}{(2\pi)^{s-1}} \int \hat{\tilde{W}}(n\tau_1) \frac{d\tau_1}{\tau - \tau_1 + i\varepsilon} \hat{\tilde{W}}(n\tau_2 - n\tau_1) \times \cdots$$
$$\times \frac{d\tau_{s-1}}{\tau - \tau_{s-1} - i\varepsilon} \hat{\tilde{W}}(n\tau_{s-2} - n\tau_{s-1}). \tag{43.12}$$

Rewriting this equation in normal form gives

$$\hat{R}_s(n\tau) = \sum_{m=1}^{3s} \int \delta^4(n\tau - k_1 - k_2 - \cdots - k_m)$$
$$\times S_m^{(s)}(\tau, k_1, k_2, \ldots, k_m) : \tilde{\varphi}(k_1)\, \tilde{\varphi}(k_2) \cdots$$
$$\times \tilde{\varphi}(k_m) : d^4k_1\, d^4k_2 \cdots d^4k_m. \tag{43.13}$$

Here the functions $S_m^{(s)}$ are integrals of the products of the functions $\Delta_+(k) = \theta(k_4)\,\delta(k^2 - m^2)$ and $g(\tau_r) = 1/(\tau - \tau_r - i\varepsilon)$. Because of the properties of the factor $\Delta_+(k)$ in this theory, all particles in the intermediate states are real particles for which $k^2 = m^2$ and $k^2 > 0$. The virtual particles in the intermediate states that are characteristic of the usual perturbation theory are expressed in this new treatment by the appearance of additional integrations over the variables $\tau_1, \tau_2, \ldots, \tau_{s-1}$, that are contained in the function $g(\tau_r)$. These additional integrations must be taken into account in constructing diagrams to represent the functions $R_s(n\tau)$.

The necessary generalization of the Feynman diagram method can be carried out as follows. We compare the continuous outer lines, as usual, to the operators of the field $\tilde{\varphi}(k)$, and for $\tau \neq 0$, we complete their outer, dotted lines in such a manner that the conservation law

$$k_1 + k_2 + \cdots + k_m - n\tau = 0. \tag{43.14}$$

is satisfied. We compare the solid inner lines to the functions $\Delta_+(k)$, and we represent the propagation function by dotted lines. These lines join all the vertices. Some of the simpler diagrams are shown in Figure 46.

Note that in going over the physical $\hat{S}$-matrix, we must set $\tau = 0$. Therefore the simple diagram (a) does not describe a real physical process, because it would describe either the disintegration of one of the particles into two, or the opposite process of synthesis, and both of these would violate the energy-momentum conservation laws. The two dia-

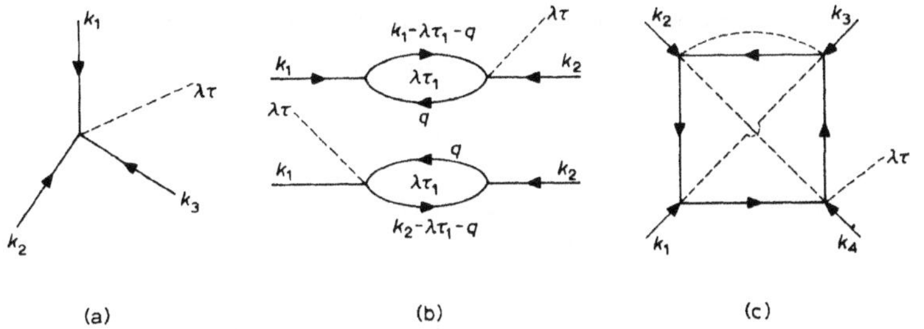

Fig. 46 Some of the simpler diagrams in the theory of quantized space. (a) The vertex for three real particles k_1, k_2, k_3 and one "quasi-particle" $\lambda\tau$; (b) diagram of the proper energies of particles taking into account the possible emission or absorption of "quasi-particles" $\lambda\tau$; (c) diagram for the scattering of real particles k_1, k_2, k_3, k_4; $\lambda\tau$ is the "quasi-particle".

grams in Figure 46b are diagrams of the characteristic energy. Integration over the variables q and τ is assumed in these diagrams. The more complicated diagram in Figure 46c describes particle scattering. On the whole, the complete mechanism of real-particle interactions can be described as an exchange of real particles in which "quasi-particles" with momenta $n\tau_1, n\tau_2, \ldots, n\tau_{s-1}$ (dotted lines) particinate, and the variables $\tau_1, \tau_2, \ldots, \tau_{s-1}$ have values between $-\infty$ and $+\infty$.

It appears that all divergences in these calculations arise in the integration over these variables, i.e., over the momenta of the quasi-particles in the intermediate states. Thus the disagreements may be avoided by restricting the region over which the variable τ is integrated. This may be done by changing over to the curved p-space and assuming that the component points of the momentum are not situated on the direct path from $+\infty$ to $-\infty$, but on a circle of radius $\hbar/a$, where a is some elementary length which relates the distance between diametrally opposite points on the circle. In this case the integration is restricted to within the limits

$$-\frac{\pi\hbar}{2a} \leqslant n\tau \leqslant \frac{\pi\hbar}{2a}. \tag{43.15}$$

For $a \to 0$, the points lie on the direct path and the limits of integration for τ extend from $-\infty$ to $+\infty$.

The group of translations along the direct path, and the group of translations on a circle on which the diametrically opposite points are identified, are isomorphic. The differences in transformation lead to substantial changes in the addition of momenta which, on the circle, are

$$p_1 (+) p_2 = p_1 + p_2 \quad \text{to modulo} \quad \frac{2\pi\hbar}{a}. \tag{43.16}$$

In other words, momentum addition has to be carried out within the limits $-\pi\hbar/a$, $+\pi\hbar/a$. Then Equation (43.7) may be rewritten in the form

$$\hat{R}(n\tau) = \hat{\tilde{W}}(n\tau) + \frac{1}{2\pi} \int_{-\pi}^{+\pi} \hat{\tilde{W}}(n \mid \tau (-) \tau') \frac{d\tau'}{\tau' - i\varepsilon} \hat{R}(n\tau'). \tag{43.7'}$$

From this equation we obtain the unitarity condition

$$\hat{R}(0) - R^{+}(0) = i\hat{R}(0)\,\hat{R}^{+}(0) \tag{43.17}$$

and the relationship between the matrices $\hat{T}$ and $\hat{R}$;

$$\hat{T} = \hat{R}(0), \tag{43.5'}$$

Other relationships, characteristic of the theory for $a=0$, may be similarly obtained.

However the transformations in going over to coordinate space change in the obvious way. All functions of the variable τ in this new scheme must be periodic functions of τ with a period $2\pi\hbar/a$. Therefore, the fourier transform in τ, i.e., along the vector n, must have the form

$$f(\sigma) = \int_{-\pi\hbar/a}^{+\pi\hbar/a} e^{i\sigma\tau} \tilde{f}(\tau)\, d\tau. \tag{43.18}$$

This transformation has meaning only in the case when a spatial variable $\sigma = nx$ is added to τ, where this variable is discrete and has values which are multiples of the elementary length, namely,

$$0, \pm a, \pm 2a, \ldots, \pm ma. \tag{43.19}$$

In this manner, the assumption that this new group of transformations exists leads to a quantization of space-time in the direction of the vector n. On the basis of the new law of addition (43.16), the inner dotted lines may represent real quasi-particles with momentum $2\pi\hbar/a$. Thus the diagram in Figure 46 for $n\tau = 2\pi\hbar/a$ will represent a real process in which quasi-particles with large momenta take part. These processes are analogous to the well-known "umklapp-processes" in crystal lattice theory in which momentum conservation for particles that move in the lattice has meaning only to modulo $2\pi\hbar/a$ (where a is the lattice spacing). The momentum $2\pi\hbar/a$ is transferred in this case to the infinitely heavy lattice. The usual energy-momentum conservation law does not have any meaning in this scheme. This is because, when the vector n assumes the form of a discrete lattice, time in the direction n becomes nonuniform. Physically, this violation may be interpreted as an absorption or emission of quasi-particles with momentum $2\pi\hbar/a$.

The scheme described above of a quantized space-time posesses simplicity and elegance. A fine point which requires special attention is the dependence of physical quantities (momenta of real quasi-particles and scattering matrix elements) on the unit vector n. This vector specifies the direction in which space-time has a nonuniform structure, i.e., the *special direction* in which the isotropy of the space-time continuum that is characteristic of Einstein-Minkowski geometry, is violated. This situation is similar to that of the unit time vector n with which the localness of acausality is ensured (see Sections 36 and 37).

The following three interpretations are possible: (a) the vector n is "external" (see Section 38). For example, it may be related to the motion of the hypothetical "ether" or galaxies [2]; (b) the vector n is a vector characteristic of the observer so that it has the value $n=(1, 0, 0, 0)$ in all reference frames. In this case we have the situation discussed in Section 39; (c) the vector n is an arbitrary unit, time-like vector. In this case the square of the matrix elements $T_{fi}(n)$, the quantities $|T_{fi}(n)|^2$, would be proportional to the probability that, in the process $f \rightarrow i$, quasi-particles, both real and virtual, having momenta parallel to the vector n, take part.

44. Fluctuations of the Metric

In the regions of space in which the motion of matter is turbulent and is accompanied by significant, random changes in the density or velocity of matter, the metric tensor $g_{\mu\nu}$ (8.1), becomes a random quantity. Similarly the temporal, t_{PQ}, and spatial, x_{PQ}, intervals which seperate two events P and Q become random quantities. In this case it is possible to speak only of the probability that $t_{PQ}=t$ and $x_{PQ}=1$.†

We assume that the energy-momentum tensor of matter may be written in the form of a sum

$$T_{\mu\nu} = T^0_{\mu\nu} + \delta T_{\mu\nu}, \tag{44.1}$$

where $T^0_{\mu\nu}$ describes the total motion of matter characterized by large time periods T and large scale lengths L, whereas the additional term $\delta T_{\mu\nu}$ results from turbulent material motions and is characterized small time periods $\tau \ll T$ and small scale lengths $\lambda \ll L$. The average value of $\delta T_{\mu\nu}$ in time intervals $\sim T$ and scales $\sim L$ is taken to be equal to zero

$$\overline{\delta T_{\mu\nu}} = 0, \tag{44.2}$$

where the bar denotes the average over turbulent motions. We expand the metric tensor in a similar manner:

$$g_{\mu\nu} = g^0_{\mu\nu} + h_{\mu\nu}, \tag{44.3}$$

$$\bar{h}_{\mu\nu} = 0. \tag{44.3'}$$

By using linear approximations for the quantities $\delta T_{\mu\nu}$ and $h_{\mu\nu}$, we obtain from Einstein's metric tensor Equation (8.3) an equation that holds for small $\delta T_{\mu\nu}$ and $h_{\mu\nu}$ (see, e.g., [13] or [14]), namely,

$$-\tfrac{1}{2}\Box^2 h_{\mu\nu} = \varkappa t_{\mu\nu}, \tag{44.4}$$

where

$$\Box^2 = g_0^{\alpha\beta} \frac{\partial^2}{\partial x_\alpha\, \partial x_\beta}, \tag{44.5}$$

$$t_{\mu\nu} = \delta T_{\mu\nu} - \tfrac{1}{2} g^0_{\mu\nu}\, \delta T, \tag{44.6}$$

$$\delta T = g_0^{\alpha\beta}\, \delta T_{\alpha\beta}, \tag{44.7}$$

$\varkappa = 8\pi\gamma/c^2$, $\gamma = 6.7\times 10^{-8}$ cm^3 gm^{-1} s^{-2} is Newton's gravitational constant. From Equation (44.4), we obtain

$$h_{\mu\nu}(x) = -\, 2\varkappa\, \Box_x^{-2} t_{\mu\nu}(x), \tag{44.8}$$

where $\Box_x^{-2}$ is the inverse of the operator $\Box_x^2$ (the subscript x indicates that the operator acts on the variable x). Using this expression for $h_{\mu\nu}(x)$,

† This section may be considered as a discussion of the third possible cause of stochastic nature of space and time which was noted in point (c), Section 41. We follow reference [106] in this section.

we write the statistical correlation of the quantity $h_{\mu\nu}(x)$, taken at two points in space-time x and x', as

$$\overline{h_{\mu\nu}(x)\, h_{\alpha\beta}(x')} = 4\varkappa^2\, \Box_x^{-2}\, \Box_{x'}^{-2} C_{\mu\nu\alpha\beta}(x, x'), \tag{44.9}$$

where the quantity

$$C_{\mu\nu\alpha\beta}(x, x') = \overline{t_{\mu\nu}(x)\, t_{\alpha\beta}(x')} \tag{44.10}$$

is the statistical correlation of the tensor component $t_{\mu\nu}$ taken at two points in space-time x and x'.

In Section 8 we showed that the space-time interval between two events P and Q is equal to

$$s_{PQ} = \int_Q^P \sqrt{\pm g_{\mu\nu}\, dx_\mu\, dx_\nu}\,. \tag{44.11}$$

For small $h_{\mu\nu}$, we obtain

$$s_{PQ} = s_{PQ}^0 + \int_Q^P \frac{h_{\mu\nu}(x)\, dx_\mu\, dx_\nu}{\sqrt{\pm g_{\mu\nu}^0\, dx_\mu\, dx_\nu}}, \tag{44.12}$$

where s_{PQ}^0 is the value of the interval s_{PQ} for $h_{\mu\nu}=0$. From (44.2) and (44.8) it follows that

$$\Delta s_{PQ} = \overline{s_{PQ} - s_{PQ}^0} = 0\,. \tag{44.13}$$

Furthermore, from (44.12) the equation for the rms $\overline{\Delta s_{PQ}^2} = \overline{(s_{PQ} - s_{PQ}^0)^2}$ is

$$\overline{\Delta S_{PQ}^2} = \int_Q^P \int_Q^P \frac{dx_\mu\, dx_\nu}{\sqrt{\pm g_{\mu\nu}^0\, dx_\mu\, dx_\nu}}\, \frac{dx'_\alpha\, dx'_\beta}{\sqrt{\pm g_{\alpha\beta}^0\, dx'_\alpha\, dx'_\beta}}\, \overline{h_{\mu\nu}(x)\, h_{\alpha\beta}(x')} \tag{44.14}$$

or, substituting the expression for the correlation, (44.9) into this equation gives

$$\overline{\Delta s_{PQ}^2} = 4\varkappa^2 \int_Q^P \int_Q^P \frac{dx_\mu\, dx_\nu}{\sqrt{\pm g_{\mu\nu}^0\, dx_\mu\, dx_\nu}} \times \frac{dx'_\alpha\, dx'_\beta}{\sqrt{\pm g_{\alpha\beta}^0\, dx_\alpha\, dx_\beta}}\, \Box_x^{-2} \Box_{x'}^{-2} C_{\mu\nu\alpha\beta}(x, x')\,. \tag{44.15}$$

In the special case when the direction of PQ is along one of the coordinate axes, then Equation (44.15) simplifies to

$$\overline{\Delta s^2_{PQ}} = 4\varkappa^2 \int\limits_{x_\mu(Q)}^{x_\mu(P)} \frac{\mathrm{d}x_\mu}{\sqrt{\pm\, g_{\mu\mu}(x)}} \times \int\limits_{x'_\mu(Q)}^{x'_\mu(P)} \frac{\mathrm{d}x'_\mu}{\sqrt{\pm\, g^0_{\mu\mu}(x')}}\, \Box_x^{-2}\, \Box_{x'}^{-2}\, C_{\mu\mu\mu\mu}(x, x'). \qquad (44.16)$$

From this equation we see that the problem is reduced to one of computing the correlation $C_{\mu\mu\mu\mu}(x, x')$ and the quantity

$$M_{\mu\mu\mu\mu}(x, x') = \varkappa^2 \Box_x^{-2} \Box_{x'}^{-2} C_{\mu\mu\mu\mu}(x, x'). \qquad (44.17)$$

We now turn to the microworld. In the microworld, the statistical nature of the metric may be a result of field fluctuations in a physical vacuum. The existence of such fluctuations, just as the fluctuations of electric charge in vacuum, may be considered as proven on the basis of the discovery and measurement of the Lamb-shift of the electron level in hydrogen.†

In order to avoid unnecessary complications, we shall consider the fluctuations of a scalar field $\varphi(x)$ in which the particles have a rest mass of m. The Lagrangian in this case has the form

$$\hat{L} = \frac{1}{2}\left(g^{\alpha\beta} \frac{\partial \hat{\varphi}}{\partial x_\alpha} \frac{\partial \hat{\varphi}}{\partial x_\beta} - m^2 \hat{\varphi}^2\right), \qquad (44.18)$$

and the corresponding energy-momentum tensor is

$$\hat{T}_{\mu\nu} = :\frac{\partial \hat{\varphi}}{\partial x_\mu} \frac{\partial \hat{\varphi}}{\partial x_\nu} - g_{\mu\nu} \hat{L}:, \qquad (44.19)$$

where $\hat{\varphi}(x)$ is the scalar field operator and products of the form $(\partial\varphi/\partial x_\alpha)$ $(\partial\varphi/\partial x_\beta)$ are assumed to be normalized (see Appendix VII).

The vacuum average of $\hat{T}_{\mu\nu}(x)$ is equal to zero – i.e.,

$$\langle 0|\, \hat{T}_{\mu\nu}(x)\, |0\rangle = 0, \qquad (44.20)$$

† The possibility of this was first shown by the author (see [143] and [144]).

and so in this case $\delta T_{\mu\nu} \equiv T_{\mu\nu}$. Using (44.6), we find it easy to show that

$$\hat{t}_{\mu\nu}(x) = :\frac{\partial\hat{\varphi}(x)}{\partial x_\mu}\frac{\partial\hat{\varphi}(x)}{\partial x^\nu} - g_{\mu\nu}h: \tag{44.21}$$

Therefore,

$$M_{\mu\mu\mu\mu}(x, x') = \varkappa^2 \Box_x^{-2} \Box_{x'}^{-2} \left\langle :\left[\frac{\partial\hat{\varphi}(x)}{\partial x_\mu}\frac{\partial\hat{\varphi}(x)}{\partial x_\mu} - \tfrac{1}{2}g_{\mu\mu}m^2\hat{\varphi}^2(x)\right]: \right.$$
$$\left. :\left[\frac{\partial\hat{\varphi}(x')}{\partial x'_\mu}\frac{\partial\hat{\varphi}(x')}{\partial x'_\mu} - \tfrac{1}{2}g_{\mu\mu}m^2\hat{\varphi}^2(x')\right]:\right\rangle. \tag{44.22}$$

In computing this quantity we must remember that the vacuum average of the products of the normal products is defined by the equation

$$\langle 0| :\hat{\varphi}(x)\,\hat{\varphi}(y):\ :\hat{\varphi}(z)\,\hat{\varphi}(u): |0\rangle$$
$$= -\Delta_-(x-z)\,\Delta_-(y-u) + \Delta_-(x-u)\,\Delta_-(y-z), \tag{44.23}$$

where $\Delta_-(x)$ is a singular function that is defined in Appendix XIII.†
From (44.3) it follows that

$$\left\langle :\frac{\partial\hat{\varphi}(x)}{\partial x_\alpha}\frac{\partial\hat{\varphi}(x)}{\partial x_\beta}:\ :\frac{\partial\hat{\varphi}(y)}{\partial y_\rho}\frac{\partial\hat{\varphi}(y)}{\partial y_\sigma}:\right\rangle = -2\frac{\partial^4\Delta_-^2(x-y)}{\partial x_\alpha\,\partial x_\beta\,\partial y_\rho\,\partial y_\sigma}. \tag{44.24}$$

If we place the function $\Delta_-(x)$ by its Fourier transform $\tilde{\Delta}_-(p)$ we obtain

$$\Delta_-(x) = \int \hat{\Delta}_-(p)\,e^{ipx}\,d^4p = \int_C \frac{e^{ipx}\,d^4p}{p^2+m^2} \tag{44.25}$$

(the contour C is shown in Figure 49), and using (44.23) and (44.24), from (44.22) we obtain

$$M_{\mu\mu\mu\mu}(\xi) = \alpha\frac{\varkappa h^2}{c^2}\int\frac{\tilde{\Delta}_-(p)\,\tilde{\Delta}_-(q)\,e^{i(p+q,\,\xi)}}{(p+q)^4}$$
$$\times\left\{p_\mu^2 q_\mu^2 + g_{\mu\mu}m^2(p_\mu^2+q_\mu^2) + \tfrac{1}{4}g_{\mu\mu}^2 m^4\right\} d^4p\,d^4q, \tag{44.26}$$

where $\xi = x' - x$ and α is a numerical coefficient of order 1. It is easy to show that this integral diverges at the upper limit $|p_\mu|\ |q_\mu| \approx K \to \infty$. For

† In order to obtain this equation, all particle production operators had to be transposed on the left of the anihilation operators in the product of the fields $\varphi(x)\,\varphi(y)\,\varphi(z)\,\varphi(u)$.

$\xi=0$, it reduces to

$$M_{\mu\mu\mu\mu}(0) = \beta \frac{\varkappa^2 \hbar}{c^2} K^4, \tag{44.27}$$

and for $\xi=t$, $Kt \gg 1$, to

$$M_{\mu\mu\mu\mu}(t) = \gamma \frac{\varkappa^2 \hbar^2}{c^2} \frac{K^2}{t^2} \cos Kt. \tag{44.27'}$$

where β and γ are numerical coefficients. Because of this divergence, there is no point in calculating $M_{\mu\mu\mu\mu}(x)$ in detail. However, Equation (44.27) and (44.27′) are useful in the sense that they show that the fluctuations in the metric tensor are due to fluctuations in the vacuum field and are important only in regions with scales of the order of

$$1/K \approx \Lambda_g, \tag{44.28}$$

where Λ_g is the length

$$\Lambda_g = \sqrt{\frac{\hbar\varkappa}{c}} = \sqrt{\Lambda_c r_g} = 0.82 \times 10^{-32}\ \text{cm}, \tag{44.28'}$$

where $\Lambda_g = \hbar/mc$ is the Compton wavelength of the particles and $r_g = \varkappa m$ is the gravitational radius of the particle. It is significant that the fluctuations do not depend on the mass of the particles of the vacuum field insofar as this mass $m \ll \hbar/\Lambda_g c \approx 10^{-5}$ gm.† Field fluctuations on a scale of the order of $\Lambda_g = \hbar/mc$ give a negligible contribution to the metric tensor.

From (44.28′) we can see that the critical scale Λ_g is the geometric average of the gravitational radius of a particular r_g, and its Compton wavelength $\hbar/mc$. The length Λ_g is significantly greater than the gravitational radius of the particle, but it is less than the length that is characteristic for weak particle interactions, $\Lambda_F \sim 10^{-10}$ cm. Thus if weak interactions should play the defining role in the structure of elementary particles, then gravitational phenomena are either in general insignificant in the world of elementary particles, or they are buried at such depths that they can be reached only in dreams.

† See reference [107] where "maximon" particles with limiting mass $m_g = \hbar/\Lambda_g c$ are considered.

45. Nonlinear Fields and the Quantization of Space-Time

In this section we shall consider the problem of quantizing a nonlinear field $\varphi(x, t)$ which is described by the Born-Infeld equation in two-dimensional space-time $\mathfrak{R}_2(x)$. This equation is obtained from the variational principle

$$\delta S = 0, \quad S = -\int L \, \mathrm{d}x \, \mathrm{d}t, \tag{45.1}$$

where the Lagrangian is

$$L = -(1 + \varphi_x^2 - \varphi_t^2)^{1/2}, \tag{45.1'}$$

where $\varphi_x = \partial\varphi/\partial x$, $\varphi_t = \partial\varphi/\partial t$. This function is a special case of the Lagrangian that was considered earlier in Section 7. The scale of the field which is a measure of its nonlinearity is taken to be unity in (45.1′)

From the variational principle we obtain the equation

$$(1 - \varphi_t^2)\,\varphi_{xx} + 2\varphi_x\varphi_t\varphi_{xt} - (1 - \varphi_x^2)\,\varphi_{tt} = 0, \tag{45.2}$$

where

$$\varphi_{xx} = \frac{\partial^2\varphi}{\mathrm{d}x^2}, \qquad \varphi_{xt} = \frac{\partial^2\varphi}{\partial x\partial t}, \qquad \varphi_{tt} = \frac{\partial^2\varphi}{\partial t^2}$$

For $\varphi_0 \to \infty$, this corresponds to φ_x, $\varphi_t \to 0$, and (45.2) reduces to the linear wave equation

$$\varphi_{xx} - \varphi_{tt} = 0. \tag{45.2'}$$

If we set $z = \varphi(x, t)$, then this equation may be considered as the equation of the minimal surface in space having the metric $\mathrm{d}s^2 = \mathrm{d}t^2 - \mathrm{d}x^2 - \mathrm{d}z^2$. In fact the integral S in the variational principle is equal to the area of such a surface. Note that the equation which results from the Lagrangian

$$L = -(1 + \varphi_x^2 + \varphi_y^2)^{1/2}, \tag{45.3}$$

would give the minimal surface in the usual Euclidean space with the metric $\mathrm{d}s^2 = \mathrm{d}x^2 + \mathrm{d}y^2 + \mathrm{d}z^2$. A soap film stretched on a given contour is an example of such a surface.

At first glance, the problem of quantizing a nonlinear field does not appear to have anything in common with quantization of space-time.

However, as was shown in [108, 109], there is a direct relationship between these problems.† In order to demonstrate this relationship, we turn to Cauchy's problem for Equation (45.2). We take the initial conditions for $t=0$ to be

$$\varphi(x, 0) = a(x), \qquad \varphi_t(x, 0) = b(x) \tag{45.4}$$

and assume that this function satisfies the condition that it be hyperbolic:

$$1 + a^2(x) - b^2(x) > 0. \tag{45.5}$$

In order to solve Cauchy's problem, we introduce the variables α, β which are analogous to the variables $\alpha = x - t$, $\beta = x + t$ that are characteristic of the linear equation (45.2′), and we consider the variables x, t and $z \equiv \varphi$ as functions of the new variables α and β:

$$\begin{aligned} &x = x(\alpha, \beta), \qquad t = t(\alpha, \beta), \\ &\varphi = \varphi(t(\alpha, \beta), x(\alpha, \beta)) \equiv z(\alpha, \beta). \end{aligned} \tag{45.6}$$

Introducing the vector $\mathbf{r} = \mathbf{r}(\alpha, \beta)$ with components t, x and z and defining the scalar product of $\mathbf{r}_1$ and $\mathbf{r}_2$ as

$$(\mathbf{r}_1 \mathbf{r}_2) = t_1 t_2 - x_1 x_2 - z_1 z_2, \tag{45.7}$$

we can rewrite Equation (45.2) in the form

$$\mathbf{r}_\alpha^2 D_{\beta\beta} - 2(\mathbf{r}_\alpha \mathbf{r}_\beta) D_{\alpha\beta} + \mathbf{r}_\beta^2 D_{\alpha\alpha} = 0, \tag{45.8}$$

where

$$\mathbf{r}_\alpha = \frac{\partial \mathbf{r}}{\partial \alpha}, \quad \mathbf{r}_\beta = \frac{\partial \mathbf{r}}{\partial \beta}, \quad \mathbf{r}_{\alpha\alpha} = \frac{\partial^2 \mathbf{r}}{\partial \alpha^2}, \quad \mathbf{r}_{\alpha\beta} = \frac{\partial^2 \mathbf{r}}{\partial \alpha\, \partial \beta} \tag{45.8′}$$

and $D_{\rho\sigma}$ is the determinant

$$D_{\rho\sigma} = \begin{vmatrix} t_{\rho\sigma} & x_{\rho\sigma} & z_{\rho\sigma} \\ t_\alpha & x_\alpha & z_\alpha \\ t_\beta & x_\beta & z_\beta \end{vmatrix} \tag{45.8″}$$

Here the subscripts ρ and σ have two values, α and β. The condition that the equation be hyperbolic gives

$$(\mathbf{r}_\alpha \mathbf{r}_\beta)^2 - \mathbf{r}_\alpha^2 \mathbf{r}_\beta^2 > 0, \tag{45.9}$$

† We follow these works in the following.

and the characteristics of Equation (45.8) are given by

$$\mathbf{r}_\alpha^2 = 0, \qquad \mathbf{r}_\beta^2 = 0. \tag{45.10}$$

Equations (45.10) and (45.8) together form a system of equations from which the three unknown functions $t(\alpha, \beta)$, $x(\alpha, \beta)$, and $z(\alpha, \beta)$ may be obtained. From (45.10) and (45.8) we have

$$D_{\alpha\beta} = 0, \tag{45.11}$$

and therefore

$$\mathbf{r}_{\alpha\beta} = A\mathbf{r}_\alpha + B\mathbf{r}_\beta, \tag{45.12}$$

where A and B are arbitrary functions of α and β. From (45.10) and (45.12) we have

$$(\mathbf{r}_{\alpha\beta}\mathbf{r}_\alpha) = (\mathbf{r}_\alpha\mathbf{r}_\beta)\, B = \frac{1}{2}\frac{\partial}{\partial\beta}\mathbf{r}_\alpha^2 = 0, \tag{45.13}$$

$$(\mathbf{r}_{\alpha\beta}\mathbf{r}_\beta) = (\mathbf{r}_\alpha\mathbf{r}_\beta)\, A = \frac{1}{2}\frac{\partial}{\partial\alpha}\mathbf{r}_\beta^2 = 0 \tag{45.13$'$}$$

and as $(\mathbf{r}_\alpha, \mathbf{r}_\beta) \neq 0$, $A = B = 0$. In this manner we arrive at the final system of equations

$$\mathbf{r}_\alpha^2 = 0, \qquad \mathbf{r}_\beta^2 = 0, \qquad \mathbf{r}_{\alpha\beta} = 0. \tag{45.14}$$

From the third equation in (45.14), it immediatly follows that

$$\mathbf{r}(\alpha, \beta) = \mathbf{r}_1(\alpha) + \mathbf{r}_2(\beta), \tag{45.15}$$

where $\mathbf{r}_1$ and $\mathbf{r}_2$ are two arbitrary vectors. The first equation in (45.14) can now be written as

$$\left(\frac{\partial \mathbf{r}_1(\alpha)}{\partial\alpha}\right)^2 = 0. \qquad \left(\frac{\partial \mathbf{r}_2(\beta)}{\partial\beta}\right)^2 = 0. \tag{45.16}$$

Making use of the fact that the variables α and β are arbitrary, we select them in such a manner that for $t = 0$, $\alpha = \beta = x$. Then the initial conditions (45.4) can be written in the form

$$t(\alpha, \alpha) = t_1(\alpha) + t_2(\alpha) = 0, \tag{45.17}$$

$$x(\alpha, \alpha) = x_1(\alpha) + x_2(\alpha) = \alpha, \tag{45.17$'$}$$

$$z(\alpha, \alpha) = z_1(\alpha) + z_2(\alpha) = \alpha(\alpha). \tag{45.17$''$}$$

We now write the derivative $\partial\varphi(x, t)/\partial t$ in terms of derivatives of t, x and z with respect to α and β:

$$\frac{\partial\varphi}{\partial t} = \frac{\begin{vmatrix} x_\alpha & z_\alpha \\ x_\beta & z_\beta \end{vmatrix}}{\begin{vmatrix} x_\alpha & t_\alpha \\ x_\beta & t_\beta \end{vmatrix}} = \frac{\begin{vmatrix} x_1'(\alpha) & z_1'(\alpha) \\ x_2'(\beta) & z_2'(\beta) \end{vmatrix}}{\begin{vmatrix} x_1'(\alpha) & t_1'(\alpha) \\ x_2'(\beta) & t_2'(\beta) \end{vmatrix}}, \tag{45.18}$$

and for $\alpha=\beta$ we obtain

$$\frac{\partial\varphi}{\partial t} = b(\alpha) = \frac{\begin{vmatrix} x_1'(\alpha) & z_1'(\alpha) \\ x_2'(\alpha) & z_2'(\alpha) \end{vmatrix}}{\begin{vmatrix} x_1'(\alpha) & t_1'(\alpha) \\ x_2'(\alpha) & t_2'(\alpha) \end{vmatrix}}. \tag{45.18'}$$

Further, from (45.14) for $\alpha=\beta$, we have

$$t_1'(\alpha)^2 - x_1'(\alpha)^2 - z_1'(\alpha)^2 = 0, \tag{45.19}$$

$$t_2'(\alpha)^2 - x_2'(\alpha)^2 - z_2'(\alpha)^2 = 0. \tag{45.19'}$$

We now write the desired result in the form

$$\mathbf{r}(\alpha, \beta) = \tfrac{1}{2}[(\boldsymbol{\rho}(\alpha) + \boldsymbol{\rho}(\beta)] + \tfrac{1}{2}\int_\alpha^\beta \boldsymbol{\pi}(\lambda)\, d\lambda. \tag{45.20}$$

From (45.17)–(45.17″), it follows that the vector $\boldsymbol{\rho}(\alpha)$ is equal to

$$\boldsymbol{\rho}(\alpha) \equiv [0, \alpha, a(\alpha)], \tag{45.21}$$

and the vector $\boldsymbol{\pi}(a)$

$$\boldsymbol{\pi}(\alpha) = [\pi_t(\alpha), \pi_x(\alpha), \pi_z(\alpha)] \tag{45.22}$$

is defined from (45.12), (45.19) and (45.19′). A few calculations lead to the result

$$\pi_t(\alpha) = -\frac{1 + a'(\alpha)^2}{L}, \quad \pi_x(\alpha) = \frac{-a'(\alpha)\, b(\alpha)}{L}, \quad \pi_z(\alpha) = -\frac{b(\alpha)}{L}, \tag{45.23}$$

where

$$L(\alpha) = [1 + \alpha'(\alpha)^2 - b(\alpha)^2]^{1/2}. \tag{45.23'}$$

The quantities π_t, π_x and π_z have simple physical meanings. For $t=0$

$$\pi_z = \pi(x, 0) = \frac{\partial L}{\partial \varphi_t} \tag{45.24}$$

is the momentum which is the canonical conjugate of the field φ. Then

$$\pi_x = G(x, 0) = -\pi(x, 0)\, a'(x) = \pi(x) \frac{\partial L}{\partial \varphi_x} \tag{45.25}$$

is the momentum density of the field φ and finally

$$\pi_t = H(x, 0) = \frac{\partial L}{\partial \varphi_t} \varphi_t - L = \pi_z \varphi_t - L \tag{45.26}$$

is the energy density of the field. Equation (45.20) may now be written as

$$t(\alpha, \beta) = \frac{\beta - \alpha}{2} + \frac{1}{2} \int_\alpha^\beta [H(\lambda) - 1]\, \mathrm{d}\lambda, \tag{45.27}$$

$$x(\alpha, \beta) = \frac{\beta + \alpha}{2} + \frac{1}{2} \int_\alpha^\beta G(\lambda)\, \mathrm{d}\lambda, \tag{45.28}$$

$$z(\alpha, \beta) = \tfrac{1}{2}[a(\alpha) + a(\beta)] + \tfrac{1}{2} \int_\alpha^\beta \pi(\lambda)\, \mathrm{d}\lambda. \tag{45.29}$$

From these solutions it follows that the variables t, x, and z, when considered as functions of α and β, depend also on the dynamic variables H, G, π, taken on the spatial surface $t=0$. The dependence of t and x on these dynamic variables is a result of the fact that in the case of the non-linear equation, the characteristic lines are not straight lines $\alpha = x - t$, $\beta = x + t$, but are curved, and this warping depends on the magnitude of the field φ and its derivatives φ_x, φ_t.

Equations (45.27)–(45.29) may be rewritten in the form

$$t(\xi, \tau) = \tau + \tfrac{1}{2} \int_{\xi - \tau}^{\xi + \tau} [H(\lambda) - 1]\, \mathrm{d}\lambda, \tag{45.27'}$$

$$x(\xi, \tau) = \xi + \tfrac{1}{2} \int_{\xi - \tau}^{\xi + \tau} G(\lambda)\, \mathrm{d}\lambda, \tag{45.28'}$$

$$z(\xi, \tau) = \tfrac{1}{2}[a(\xi + \tau) + a(\xi - \tau)] + \tfrac{1}{2} \int_{\xi-\tau}^{\xi+\tau} \pi(\lambda)\, d\lambda, \qquad (45.29')$$

where

$$\beta = \xi + \tau, \quad \alpha = \xi - \tau, \quad \tau = \tfrac{1}{2}(\beta - \alpha), \quad \xi = \frac{\beta + \alpha}{2}. \qquad (45.30)$$

We shall now consider the field $\varphi(\xi, \tau) \equiv z(\xi, \tau)$ as the Heisenberg operator $\hat{\varphi}(\xi, \tau)$ in the space $\mathfrak{R}_2(\xi, \tau)$. For $\tau = 0$, the quantization law for scalar fields now gives

$$[\hat{\varphi}(\xi, 0), \hat{\pi}(\xi, 0)] = \frac{\hbar}{i}\, \delta(\xi' - \xi). \qquad (45.31)$$

The quantization of the field $\hat{\varphi}$ leads to the quantization of the quantities t and x which become operators. Then x and t no longer commute with the field $\hat{\varphi}$. In order to show this, we form the Poisson brackets $[\hat{\varphi}(\xi, \tau), t(\xi', \tau)]$ and $[\hat{\varphi}(\xi, \tau), x(\xi', \tau)]$ and compute them to the first non-vanishing terms of order $1/\varphi_0$. In this approximation, the operators $\hat{H}(\alpha)$ and $\hat{G}(\alpha)$ in (45.27′), (45.28′) and (45.29′) may be taken in the linear approximation

$$\hat{G}(\alpha) = \frac{\partial a}{\partial \alpha}\, \hat{\pi}(\alpha), \qquad (45.25')$$

$$\hat{H}(\alpha) - 1 = \frac{1}{2}\left\{\left(\frac{\partial a}{\partial \alpha}\right)^2 + \pi(\alpha)^2\right\}. \qquad (45.26')$$

Then we obtain

$$[\hat{\varphi}(\xi, \tau), t(\xi', \tau)] = \frac{1}{4\varphi_0^2} \int_{\xi'-\tau}^{\xi'+\tau} [a(\xi + \tau) + a(\xi - \tau), \hat{H}(\lambda)]\, d\lambda, \qquad (45.31')$$

$$[\hat{\varphi}(\xi, \tau), x(\xi', \tau)] = \frac{1}{4\varphi_0^2} \int_{\xi'-\tau}^{\xi'+\tau} [a(\xi + \tau) + a(\xi - \tau), \hat{G}(\lambda)]\, d\lambda, \qquad (45.32)$$

where $\hat{G}(\alpha)$ and $\hat{H}(\alpha)$ are taken from (45.25′) and (45.26′), and the dimension of the equation is easily obtained by introducing explicitly the scale of the gradient of the field φ_0.

Using the commutation law (45.31) and noting that $a(\xi \pm \tau)$ is obtained from $a(\xi)$ by simply displacing the coordinates ξ by $\pm\tau$, we obtain

$$[\hat{\varphi}(\xi, \tau), t(\xi', \tau)] = \frac{1}{2\varphi_0^2}\frac{\hbar}{\mathrm{i}}\int_{\xi'-\tau}^{\xi'+\tau} \pi(\lambda)\,\{\delta[\lambda = (\xi + \tau)] + \delta[\lambda - (\xi - \tau)]\}\,\mathrm{d}\lambda = \frac{1}{2\varphi_0^2}\frac{\hbar}{\mathrm{i}}\{\pi(\xi + \tau) + \pi(\xi - \tau)\}, \tag{45.33}$$

$$[\hat{\varphi}(\xi, \tau), x(\xi', \tau)] = \frac{1}{2\varphi_0^2}\frac{\hbar}{\mathrm{i}}\int_{\xi'-\tau}^{\xi'+\tau} \frac{\partial a(\lambda)}{\partial\lambda}\{\delta[\lambda - (\xi + \tau)] + \delta[\lambda - (\xi - \tau)]\}\,\mathrm{d}\lambda = \frac{1}{2\varphi_0^2}\frac{\hbar}{\mathrm{i}}\left\{\frac{\partial a(\xi + \tau)}{\partial\xi} + \frac{\partial a(\xi - \tau)}{\partial\xi}\right\}, \tag{45.34}$$

where the right-hand sides of (45.33) and (45.34) have to be set to zero if ξ lies in the interval

$$\xi' - 2\tau < \xi < \xi' + 2\tau\,. \tag{45.34'}$$

If two physical quantities can be represented by linear Hermitian operators $\hat{A}$ and $\hat{B}$ and if their commutator is $\hat{C} = [\hat{A}, \hat{B}]$, then it is well known that the rms deviations $\overline{\Delta A^2}$, $\overline{\Delta B^2}$ of A and B satisfy the inequality

$$\overline{\Delta A^2}\;\overline{\Delta B^2} \geqslant \tfrac{1}{4}\,|\overline{C^2}| \tag{45.35}$$

in any state (see, e.q., [110]). On the basis of this generalized uncertainty relationship, from (45.33) and (45.35), we obtain the following relationships for the rms values $\overline{\Delta\varphi^2}$, $\overline{\Delta t^2}$ and $\overline{\Delta x^2}$.

$$\overline{\Delta\varphi(\xi, \tau)^2}\;\overline{\Delta t(\xi, \tau)^2} \geqslant \frac{\hbar^2}{4\varphi_0^4}\,\overline{|\pi(\xi + \tau) + \pi(\xi - \tau)|^2}\,, \tag{45.36}$$

$$\overline{\Delta\varphi(\xi, \tau)^2}\;\overline{\Delta x(\xi, \tau)^2} \geqslant \frac{\hbar^2}{4\varphi_0^4}\,\overline{\left|\frac{\partial\varphi(\xi + \tau)}{\partial\xi} + \frac{\partial\varphi(\xi - \tau)}{\partial\xi}\right|^2}. \tag{45.37}$$

From this it follows that for $\tau > 0$, there does not exist any specific value of the field φ at a given point with coordinates x and t. The field φ is not defined in the space $\mathfrak{R}_2(x, t)$, but in the space $\mathfrak{R}_2(\xi, \tau)$. Because of this, for $\tau > 0$, there does not exist any point in the space $\mathfrak{R}_2(x, t)$ which would correspond to a given point in the space $\mathfrak{R}_2(\xi, \tau)$.†

From the commutative laws it is also evident that "quantization" of the space-time $\mathfrak{R}_2(x, t)$ can only arise in the regions of the space-time $\mathfrak{R}_2(\xi, \tau)$ where the nonlinearity is large. For

$$\frac{1}{\varphi_0}\pi(\xi) \to 0 \qquad \text{and} \qquad \frac{1}{\varphi_0}\frac{\partial \varphi}{\partial \xi} \to 0,$$

the variables φ, t and x commute because t and x become C-numbers again. The example of the nonlinear field discussed in this section is interesting in that the quantization of space-time is not introdiced externally, but it results from the dynamics of the nonlinear field.

† In the approximation we considered, however, that x and t do commute.

CHAPTER VIII

EXPERIMENTAL QUESTIONS

46. Concluding remarks about the theory

The main idea of this monograph is based on the assumption that the $\hat{S}$-matrix, which asymptotically describes microphenomena, will retain its meaning in future theories. This assumption does not imply that the methods of constructing this matrix will not undergo great changes and differ significantly from present-day methods, be they based on the traditional concept of Lagrangian interaction of local fields, or those based on the more abstract concepts characteristic of the axiomatic approach.

A second basic assumption used is that modern free-particle theory is valid. In other words, we assumed that modern theory asymptotically correctly reflects physical reality for large separations between particles. These assumptions served as the basis for considering possible deviations from present-day theory in regions with very small scales in space and time. We considered two main classes of possible modifications to modern theory which, by convention were characterized using the vector n "internal" or "external" (see Section 39).

We now turn to the first class. The current concept of uniformity and isotropy of space-time is conserved in this class. Moreover the asymptotic momentum space $\mathfrak{R}_3(p)$ (see Section 31), and the space-time $\mathfrak{R}_4(x)$ are conserved in the form in which they are familiar to us in modern theory. The possible deviations from modern local field theory themselves have a *local* character, i.e., they are concentrated in small regions of space-time (with volume $\sim a^4$ where a is the "elementary length") in the neighborhood of the interacting particles. Using the schemes we had considered for local violation of microcausality (see Sections 36–40), we attempted to describe a "magic" circle – a region in space-time within which the situation could perhaps be described using only an essentially new concept that arises from the limits of our present-day concepts. Thus the theoretical schemes we considered may be considered as models. This model character is expressed in the fact that, when we localized acausality, we used the spacetime description with the variables x and t which are defined for free-point particles. The "equality in rights" of these variables for small separations in space and time is subject to question.

However the logical consequence of these models is supported by the fact that we used the space-time description only as a device for guaranteeing that macroscopic causality would not be violated. Then we returned to momentum space and in this manner constructed an acausal $\hat{S}$-matrix which was similar to the original causal matrix $\hat{S}$. The fact that the acausal schemes we considered may be treated as models makes the experimental verification of facts which could be used to show the inadequacies of the usual local theory expecially important.

47. Experimental Consequences of Local Acausality

An important consequence of *local acausality* is the changes that occur in the analytic properties of the local scattering amplitude $T(\omega)$. These, in turn, cause changes in the dispersion relations and those between asymptotic cross sections (see Section 30 and 38). These consequences of acausal theory can be investigated experimentally. In order to be able to verify the dispersion relations however, they must contain only experimentally observable quantities.

The dispersion relation for the forward elastic-scattering amplitude for the scattering of π-mesons off of nucleons has these necessary properties. Therefore, only this type of scattering is suitable for experimental verification of such an essential part of the theory as microcausality. The corresponding dispersion relations were given in Section 30. Experiment and theory can only be compared for the dispersion relation for charged mesons, i.e., for the amplitudes $D_1(\omega)$ and $D_2(\omega)$ [see Section 30, Equations (30.32′) and (30.32″)].

The point ω_0 at which the computations are usually made is the meson rest energy, so that $\omega_0 = m$, $D_1(\omega) = D_1(m)$. If the absorption part of the amplitudes A_1 and A_2 in Equations (30.32′) and (30.32″) are expressed using the optical theorem [see (XIII.4)] in terms of the total cross sections σ_+ and σ_-, then these equations may be written in the form

$$D_1(\omega) = D_1(m) + \frac{f^2}{M} \frac{k^2}{\omega^2 - (m^2/2M)^2} \frac{1}{1 - (m/2M)^2}$$

$$+\frac{k^2}{4\pi}\int_m^\infty \frac{d\omega'}{k'}\frac{\omega'}{\omega'^2-\omega^2}(\sigma_+(\omega')+\sigma_-(\omega')), \tag{47.1}$$

$$D_2(\omega)=\frac{f^2}{M}\frac{2k^2}{\omega^2-(m^2/2M)^2}\frac{\omega M}{m^2}\frac{1}{1-(m/2M)^2}$$

$$+\frac{k^2\omega}{4\pi}\int_m^\infty \frac{d\omega'}{k'}\frac{1}{\omega'^2-\omega^2}(\sigma_+(\omega')-\sigma_-(\omega')). \tag{47.1'}$$

where $k^2=\omega^2-m^2$, where k is the meson momentum and the constant $f^2=(g^2/4)\,(m^2/M^2)=0.081$ and is measured experimentally from the scattering of low energy π^+-mesons.

The biggest problem in comparing theory with experiment is that of extrapolating the total cross sections $\sigma_\pm(\omega)$ to very high energies. This extrapolation is necessary for the evaluation of the integrals in (47.1) and (47.1′). The extrapolation equation is of the form

$$\sigma_\pm=a+\frac{b}{k^n}; \tag{47.2}$$

The constants a and b and the index n are selected experimentally (see [111–114]). The first experiments to determine the real part of the scattering amplitude for very small angles $D(\omega)$ were made in Dubna [115, 116]. The experimental data was analyzed and compared to the theoretical dispersion relations (47.1) and (47.1′) using (47.2) in references [111-114].

The most complete set of data to date is given in [114]. Figure 47 shows the experimental data (dotted line) for the relationship between the real and imaginary parts of the amplitude $\alpha=D(\omega)/A(\omega)$ as a function of the energy of the mesons k. The solid line is the theoretical curve obtained using Equations (47.1), (47.1′) and the experimental formula (47.2). This figure shows that the dispersion relation obtained experimentally agrees well with the theoretical result obtained on the basis of local (causal) theory up to values of the π-meson momentum of $k\approx 20$-26 GeV s^{-1}.

We now turn to the additional terms that result from the local acausal amplitude (see Section 38). According to Equations (38.13) and (38.13′), we can write

$$D_{a1}(\omega)=D_1(\omega)+\psi_1(\omega), \tag{47.3}$$

$$D_{a2}(\omega) = D_2(\omega) + \psi_2(\omega), \tag{47.3'}$$

where $D_{a1}(\omega)$ and $D_{a2}(\omega)$ are the acausal amplitudes; the amplitudes $D_1(\omega)$, $D_2(\omega)$ are given by (47.1) and (47.1′); and ψ_1 and ψ_2 denote the additional acausal terms.† The necessary transformations of Equations

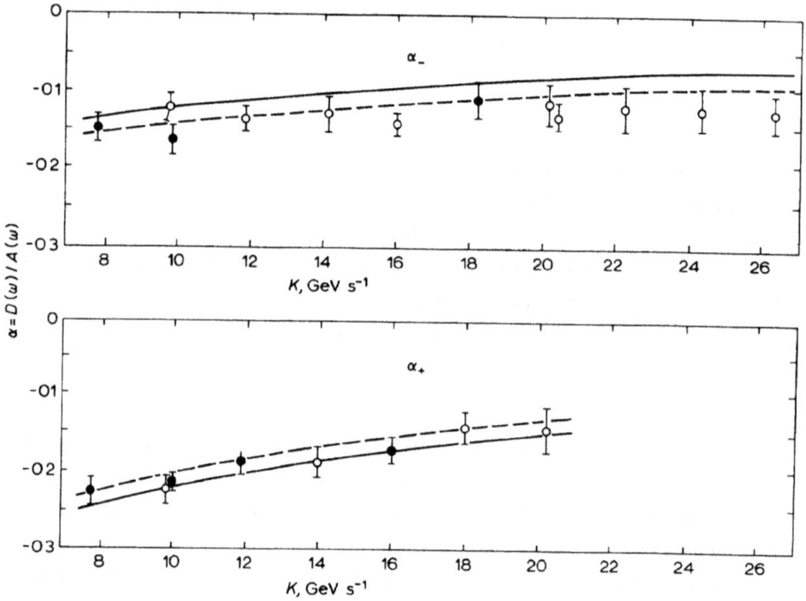

Fig. 47. The ratio α of the real and imaginary parts of the scatttering amplitude. The solid curves were calculated from the dispersion relations (47.1) and (47.1′) using Equation (47.2).

(38.13) and (38.13′) are carried out in Appendix XVI [Equations (XVI.11′) and (XVI.12)] and the result is

$$\psi_1(\omega) = \frac{\omega^2 - m^2}{4\pi\sqrt{M^2 + m^2}}\,\tilde{\rho}(\omega)\,(\sigma_+(M) + \sigma_-(M)), \tag{47.4}$$

$$\psi_2(\omega) = -\frac{\omega^2 - m^2}{M^2 + m^2}\,\tilde{\rho}(\omega)\,\frac{\omega}{M}\,(D_+(M) + D_-(M)), \tag{47.4'}$$

† Note that ω denotes the meson energy in the laboratory reference frame.

where $\sigma_{\pm}(M)$ is the total cross section for $\omega = M$, and $D_{\pm}(M)$ is the real part of the amplitude at the point $\omega = M$.

For $\omega \gtrsim M$, these corrections which were added to the real part of the scattering amplitude $D(\omega)$ become very important and, therefore, they would be observed in experiments if the elementary length $a = 1/M$ was greater than 10^{-15} cm.

We now turn to the experimental verification of the equality of the total asymptotic cross sections (see Section 30). The total cross sections for $\bar{p}p$, pp, $\pi^{\pm}p$, $K^{\pm}p$ collisions from references [117, 118] are shown in Figure 48. From this figure, it is obvious that the equality predicted in

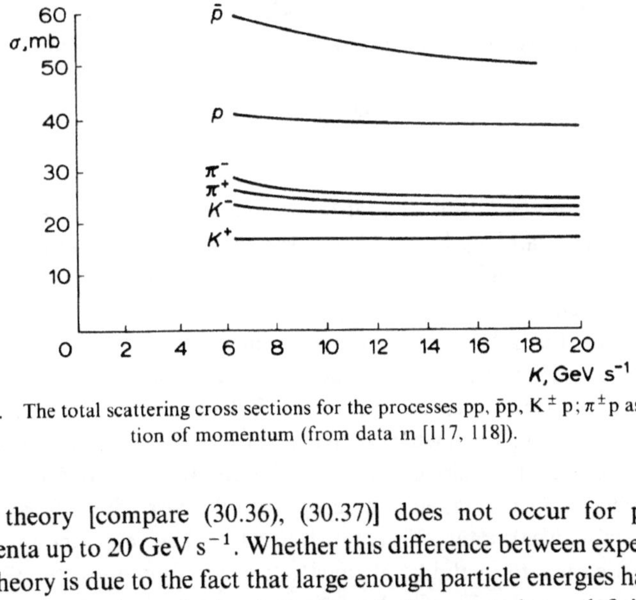

Fig. 48. The total scattering cross sections for the processes pp, $\bar{p}p$, $K^{\pm}p$; $\pi^{\pm}p$ as a function of momentum (from data in [117, 118]).

local theory [compare (30.36), (30.37)] does not occur for particle momenta up to 20 GeV s^{-1}. Whether this difference between experiment and theory is due to the fact that large enough particle energies have not yet been reached, or whether it ought to be interpreted as a deficiency in local theory remains an open question. Some of the curves ($K^{\pm}p$) seem to indicate that the curves tend to different limits. Further investigation of the limits of the cross sections represents an important problem for future accelerators.

Quantum electrodynamics is the classic example of a local theory. Therefore, in the past few years much effort was put into searching for deviations from the predictions of this theory. One of the more important ways of checking the validity of quantum electrodynamics is that of measuring the Landé g factor which defines the relationship between the magnetic moment of a particle and its spin. For the electron and $\mu^{\pm}$-mesons, this factor is close to 2. The experimental data for μ-mesons may be summed up by the Equation [121]

$$\tfrac{1}{2}(g-2)_{\text{exp}} = \tfrac{1}{2}(g-2)_{\text{theor}} \pm 0{,}6\left(\frac{a}{\pi}\right)^2, \tag{47.5}$$

where $\alpha^2/\hbar c$ is the fine-structure constant. The divergence given by (47.5) from the values predicted by local quantum electrodynamics may be formally stated in terms of changes in the μ-meson distribution function $\tilde{D}_c(q)$ (where q is the momentum transfered). We first replace the local function $\tilde{D}_c(q)$ by the nonlocal function $\tilde{D}_c(q, a)$ using the equation

$$\tilde{D}_c(q) = \frac{1}{q^2 - m_\mu^2} \to \tilde{D}_c(q, a) = \tilde{D}_c(q) - \frac{1}{q^2 - m_\mu^2 - 1/a^2}. \tag{47.6}$$

Nonlocal nature is introduced into this equation by the last term which contains the small length $a\,(a \ll 1/m_\mu)$ where m_μ is the μ-meson mass.† In order for theory and experiment to agree, the length a would have to be $< 10^{-14}$ cm.

Experimental data show that the Landé g factor is 2 for electrons and this corresponds, to a high degree of accuracy, to the value for μ-mesons. It is only in terms of the order of $(\alpha/\pi)^3$ that differences appear. However, in the region in which such high-order terms become important, quantum electrodynamics ceases to be a closed theory and strong interaction effects must be taken into account. In the case when intermediate bosons exist, weak interactions must also be taken into account [121]. In investigating the validity of quantum electrodynamics, (e^+e^-) and $(\mu^+\mu^-)$ pair production for large momentum transfers were studied [121], as well as elastic scattering of electrons and mesons off of protons (with a momentum transfer q of up to several GeV/sec) [145]. The basic results of these experiments showed that the predictions of local quantum electrodynamics hold up to values of $a < 10^{-14}$ cm.

It is significant that within these limits, there do not appear to be any

† This estimate was introduced in [122] and is arbitrary because (47.6) can only be used for estimating the possible nonlocal effects.

differences in properties of μ-mesons and electrons, not counting the great difference in their mass ($m_\mu/m_e \sim 200$) [145]. This fact indicates that the mass difference of the μ-meson and electron must be related to space-time regions on the order of 10^{-15} cm or smaller. Essentially this estimate of the scale of the nonlocal nature of a, which is based on studies of πp- scattering given in [144], leads to values of $a = 1/M \approx 10^{-16}$ cm. Therefore, it becomes probable that the mass of the particle is defined, not only by strong electromagnetic interactions, but also by weak interactions which evidently play a fundamental role [144–146]. We recall that the length a that is characteristic of weak interactions (see Section 22) is equal to $a = \Lambda_F \approx 10^{-16}$ cm.

Modern local theory leads to the well-known *CPT* theorem according to which the use of the operations of charge conjugation $\hat{C}$, spatial parity $\hat{P}$ and time reversal $\hat{T}$ does not change the state of elementary particles. Formally, this is written as

$$\hat{C}\hat{P}\hat{T} = 1. \tag{47.7}$$

The operation $\hat{C}\hat{P}$ is called the operation of *combined parities*. This important relationship between the three operations was already shown in the fact that in the Feynman diagram technique, antiparticles could be considered as particles that move in the opposite direction in time [127]. A deeper basis for Equation (47.7) lies in the proper relation of the statistics of a particle to its spin [128]. In the axiomatic approach to quantum field theory, the *CPT* theorem can be derived from local causality, from the positiveness of particle energy and Lorentz invariance [129].

If the *CPT* theorem were violated, this would imply that the basic principles of local theory had been violated and this could be related to possible changes in the principles of causality in the microworld and, more generally, in the geometry of the microworld.† However as we have seen, the invariance of the *CPT* theory has been verified to great accuracy. One consequence of this invariance is the fact that the mass of a particle m and its antiparticle $\tilde{m}$ are equal. The difference in mass of the K_L- and K_S-mesons has been measured very accurately, and these measurements show that

$$\left|\frac{\tilde{m}_K - m_K}{m_K}\right| \leqslant 10^{-14}.$$

† Compare B. A. Arbusov. Preprint 71–26, J. Ph. H. E. Serpuhov (1971).

This result shows that Equation (47.7) is valid to a high degree of accuracy.

Another of the consequences of the *CPT* theorem is that the lifetime of a particle τ, and that of the antiparticle $\tilde{\tau}$ are equal. The mass of unstable particles is a complex quantity ; $m = m_0 + i\lambda/2$ where λ is the disintegration constant. Therefore the fact that τ and $\tilde{\tau}$ are equal follows from the *CPT* theorem in the same way as the equality of the masses m and $\tilde{m}$. The data obtained to date on the disintegration times of π^+- and π^--mesons gives

$$\frac{\tau^+}{\tau^-} - 1 = 0.064 \pm 0.069,$$

i.e., the lifetimes are equal to the degree of accuracy that is presently attainable [149].

Another piece of data that supports the validity of the *CPT* theorem is the Landé g factor, or rather the quantity $(g-2)$ for μ^-- and μ^+-mesons [145]. These give $(g-2)_+ = (1162+5) \times 10^{-6}$ and $(g-2)_- = (1165+3) \times \times 10^{-6}$. Thus we may consider $\hat{C}\hat{P}\hat{T} = 1$ for the time being. The combined partities $\hat{C}\hat{P}$ pose a more complicated problem. A few years ago it was discovered that this parity was not conserved during the disintegration of long lived K_L-mesons (K_2^0-mesons) into two π-mesons [123, 124]. If K^0 denotes the K^0-meson state and $\tilde{K}^0$ denotes the anti-K^0-meson state, then the short-lived K-meson state (K_1^0) is

$$K_S = \frac{1}{\sqrt{2}} (K^0 + \tilde{K}^0), \tag{47.8}$$

and the long-lived state (K_2^0) is

$$K_L = \frac{1}{\sqrt{2}} (K^0 - \tilde{K}^0). \tag{47.8'}$$

The first of these states is even with respect to the operation $\hat{C}\hat{P}$, and the second is odd. On the other hand the state of a pair of mesons $(\pi^+\pi^-)$ is even with respect to $\hat{C}\hat{P}$. In fact the parity inversion $\hat{P}$ changes the sign of the π-meson state because it is pseudoscalar and it changes the direction of the radius vector which joins the two mesons. Because of this, the wave function which describes the state of the meson pair acquires the factor $(-1)^l$ (here l is the orbital momentum). Under charge conjugation $\hat{C}$, the interchange $\pi^+ \rightleftarrows \pi^-$ takes place and, therefore, a second factor $(-1)^l$ is required.

In this manner the parity of the state $(\pi^+\pi^-)$ is proved. Therefore, the disintegration process

$$K_L \to \pi^+ + \pi^- \tag{47.9}$$

is forbidden if $\hat{C}\hat{P}$-parity is to be conserved. In the experiment described in [123] in 1964, disintegrations of the type (47.9) were shown to be possible, even though the probability of this occurring is

$$\frac{K_L \to \pi^+ + \pi^-}{K_S \to \pi^+ + \pi^-} \approx 2 \times 10^{-3}. \tag{47.10}$$

Since then these results have been supported and recently a violation of $\hat{C}\hat{P}$-parity was discovered during the disintegration of a K-meson into a π-meson and leptons in the reactions [126]

$$K_L \to \pi^- + \mu^+ + \nu, \tag{47.11}$$

$$K_L \to \pi^+ + \mu^- + \tilde{\nu}, \tag{47.11'}$$

and

$$K_L \to \pi^- + e^+ + \nu, \tag{47.12}$$

$$K_L \to \pi^+ + e^- + \tilde{\nu}. \tag{47.12'}$$

The asymmetry was discovered in the fact that reactions (47.11) and (47.12) appeared to be prefered to the reactions (47.11′) and (47.12′). However, because of the symmetry of the K_L-meson with respect to the operation $\hat{C}\hat{P}$, reactions (47.11) and (47.11′), and correspondingly (47.12) and (47.12′), ought to have occured with equal frequencies. The violation of $\hat{C}\hat{P}$-parity could either indicate that the $\hat{C}\hat{P}\hat{T}$ theorem is violated, or that the physical phenomena are not invariant with respect to time reversal operations $\hat{T}$, i.e., the future is not symmetric with the past. This asymmetry is well known in macroscopic phenomena and in everyday life, but it is most unusual in mechanics or in field theory [130].

The extent to which $\hat{C}\hat{P}$-parity (and, consequently, the asymmetry of time) violation may be related to the problem of causality and geometry in the microworld depends to a great extent on the degree of generality of the violation of $\hat{C}\hat{P}$ invariance.† So far no phenomena, except for

† See [132, 133].

the disintegration of K-mesons, have been discovered in which $\hat{C}\hat{P}$-parity is not conserved. This fact hinders the investigation of $\hat{C}\hat{P}$ non-conservation from the geometrical point of view.

48. Experimental Results of Models with the "External" Vector

We now turn to the other class of models which, in our classification, we refered to as the class of models having an "external" vector n. In these models, in contrast to the local acausal case, Lorentz invariance is violated and it becomes in some sense an approximation. This departure from Lorentz invariance depends either on the selection of a coordinate system, as in some of the models of stochastic space (see Section 41), or on the violation of the homogeneity and isotropy of space-time which is characteristic of models with a discrete or quantized space-time (see Section 42).

In Section 44 we considered zero-point fluctuations of the scalar field $\varphi(x)$ which gave rise to fluctuations in the metric tensor $g_{\mu\nu}$. In the usual understanding of vacuum, these fluctuations as well as those in the vacuum charge do not violate the homogeneity or isotropy of space-time because, at any point in the space-time continuum $\mathfrak{R}_4(x)$, they occur in an identical manner. However, we may also regard physical vacuum in the older way according to which the vacuum is a reference frame singled out in our Universe. Although these assertions may seem improbable, they may not be logically excluded from these considerations.

In particular, it was noted in [105] that as the scale size increases, the number of reference frames which can serve as inertial frames decreases significantly. For particles moving within the atomic nucleus for example, the acceleration reaches values of 10^{32} cm s^{-2}, for macroscopic systems, the value is of the order of 10^8 cm s^{-2}, for planets, 1 cm s^{-2} and for stars or galaxies only about 10^{-8} cm s^{-2}. Therefore, although all macroscopic systems may appear to be inertial for atomic systems, it is difficult to find inertial frames for large systems, other than those of the stars or galaxies. Remembering the well-known expanding universe, it would be more correct to consider reference frames as locally at rest, but changing slowly as one goes over to more distant galaxies. It is highly probable that the "most inertial" reference frame could be that which is at rest with respect to the "relic" radiation that was recently found in the universe.

Let us now return to the microworld, and for simplicity consider elastic scattering. In the usual local theory, this scattering is completely defined by the scattering amplitude $A(s, t)$ which depends on two invariants, $a = P^2$ and $t = q^2$ (P is the total momentum of the particles and q is the momentum transfered during scattering). If there exists some time-like "outer", unit vector n, then the scattering amplitude A may also depend on the two invariants $\alpha = a(n, P)$ and $\beta = a(n, q)$, where a is the elementary length. Therefore,

$$A = A(s, t, \alpha, \beta). \tag{48.1}$$

We now assume that the laboratory reference frame is almost at rest with respect to some prefered reference frame (prefered because of inhomogeneities in space-time, "most inertiality", etc.), so that in the laborabory frame $n = (1, 0, 0, 0)$ and in the other, $\alpha = aE$, $\beta = aq_0$. We shall now consider elastic scattering in this reference frame. In particular, if the center of mass of the colliding particles is at rest, then $\alpha = a[(E_0 + m)m/2]^{1/2}$, $\beta = 0$ and m is the particle rest mass. Thus scattering in the laboratory frame and in the center-of-mass frame can be different. This difference can of course only become observable for very high particle energies (or sufficiently large momentum transfer). In the opposite case, $\alpha \approx \beta \approx 0$ and $A \approx A(s, t, 0, 0)$.

But is such an experiment possible? Very accurate data exist for the reaction $K_2^0 \rightarrow \pi^+ + \pi^-$ for two energies. According to the local theory, the probability of the disintegration $\mathrm{d}W$ is a function of the two invariants $s = P^2$ and $r = Pp_1 = P^2 - Pp_2$, where P is the four-dimensional momentum of the K_2^0-meson and p_1 and p_2 are the momenta of the $\pi^{\pm}$-mesons:

$$\mathrm{d}W \equiv \mathrm{d}W(s, r). \tag{48.2}$$

This probability can be converted in the usual manner from one reference frame to another and in particular, if it is known for the meson at rest, it can be computed for moving mesons by simple recalculation of the invariants s and r. The criterion used to show the validity of such a transformation, based on the usual kinematics of the theory of relativity, is the invariance of the rest mass of the K^0-meson. Experiments performed at energies of the K_2^0-meson of $E_0 = 1$ GeV [134] and $E_0 = 10.7$ GeV [135] show the spread Δm_k of the possible values of the meson rest mass m_k in the first case is $\sim 0.7\%$ and in the second is $\sim 1\%$ for the same average value of m_k. Thus these experiments show that relativistic kinematics in the region of a few GeV gives good agreement to an accuracy of $\sim 1\%$.

In the physics of the last century, conservation laws had a fundamental meaning. In present-day theory however, symmetry and group properties have a more fundamental meaning. From this point of view, the conservation laws are defined in terms of symmetry. Thus the conservation laws of energy, momentum and angular momentum are consequences of the homogeneity and isotropy of space-time. Violations of the homogeneity and isotropy would lead to violations of the conservation laws. We recall here that in the general theory of relativity in which the inhomogeneous Riemann space is used, the local conservation laws of energy and momentum do not exist.

It would be interesting to determine to what degree of accuracy and reliability the laws of energy and momentum conservation apply to the world of elementary particles and, in particular, to regions of high energies. This proves a difficult question to answer because the significance of these fundamental laws is assumed to be obvious and no experiments have been set up to verify them. One must remember that the possible violations of laws could be a result of violating homogeneity and isotropy in the microworld space-time, but there is hardly any basis for making the concept of homogeneity and isotropy of space-time a subject of faith in physics.

And what information is it possible to obtain from modern high energy experiments? It turns out that the most accurate results are obtained from studies of elastic scattering of protons. The results show that the relativistic conservation laws of energy end momentum for elastic proton scattering

$$p_1 + p_2 = p_1' + p_2', \tag{48.3}$$

$$E(p_1) + E(p_2) = E(p_1') + E(p_2') \tag{48.3'}$$

(where p_1, p_2 are the proton momenta before impact, p_1', p_2' are the momenta after impact, and $E = (p^2 + m^2)^{1/2}$ is the proton energy) are satisfied in the region $E = 2$–10 GeV to an accuracy of about 3%. The corresponding wavelength in the laboratory reference frame is $\lambda \approx 10^{-15}$ cm. Furthermore, we can consider as sufficiently well determined that the sharp discontinuous violations of kinematics (48.3) and (48.3′) do not exist at the level of 10%, but that they could exist at the level of 3% [136]. However, there are as yet no experimental data on this subject.

If any inhomogeneity exists in space-time on the ultrasmall scale of order a, then the expected violations of the conservation laws would be significant, $\Delta E \sim \hbar c/a$, $\Delta p \sim \hbar/a$, although they could be rare. Such a situation exists in the theory of "quasi-particles" that was discussed in Section 43. Obviously, it would be difficult to distinguish between such violations and those which result from processes in which neutral particles take part. In the case when some momentum and energy is lost, it would be natural to interpret this as being probably due to the formation of a neutral particle. Thus it would be more interesting to observe a spontaneous increase in energy and momentum under the conditions of sufficient shielding from external neutral radiation.

APPENDICES

APPENDIX I (TO SECTIONS 14 AND 16)

Singular Functions in Field Theory

The function $D(x)$ in (14.4) and (16.3) is derived from the four-dimensional integral

$$I(x) = \frac{1}{(2\pi)^4} \int \frac{e^{ikx}}{k^2 + m^2} d^4k, \tag{I.1}$$

where $d^4k = dk_0\ d^3k$, $kx = \mathbf{kx} - k_0 t$, $k^2 = \mathbf{k}^2 - k_0^2$ for bosons. For fermions we shall use the vector p instead of k.

The integrand in (I.1) has poles at the points

$$k_0 = \pm \sqrt{\mathbf{k}^2 + m^2} = \pm\, \omega \tag{I.2}$$

or for fermions, at the points $k_0 = \pm(\mathbf{p}^2 + m^2)^{1/2} = \pm E$. Therefore, the result obtained by integrating (I.1) over dk_0 depends on the choice of integration contour in the complex plane. The various possibilities are shown in Figure 49. The function $D(x)$ corresponds to the contour C_0 in this figure.

We now list the various singular functions that can be obtained from the integral (I.1) depending on the integrating contour chosen.

1. The function $D(x)$ – "commutative function": integration contour is C_0 and the function has the form

$$D(x) = \frac{1}{(2\pi)^3} \int e^{ikx} \frac{\sin \omega t}{\omega} d^3k$$
$$= \frac{1}{2\pi}\, \varepsilon(t) \left\{ \delta(x^2) + \frac{m^2}{2} \frac{J_1(m\sqrt{x^2})}{m\sqrt{x^2}}\, \theta(x^2) \right\}, \tag{I.3}$$

where $x^2 = t^2 - \mathbf{x}^2 = t^2 - r^2$ (with $c = 1$), $J_1(z)$ is the Bessel function, and

$$\begin{aligned} \theta(\xi) &= 1, \quad \xi > 0, \quad \varepsilon(\xi) = +1, \quad \xi > 0, \\ \theta(\xi) &= 0, \quad \xi < 0, \quad \varepsilon(\xi) = -1, \quad \xi < 0. \end{aligned} \tag{I.4}$$

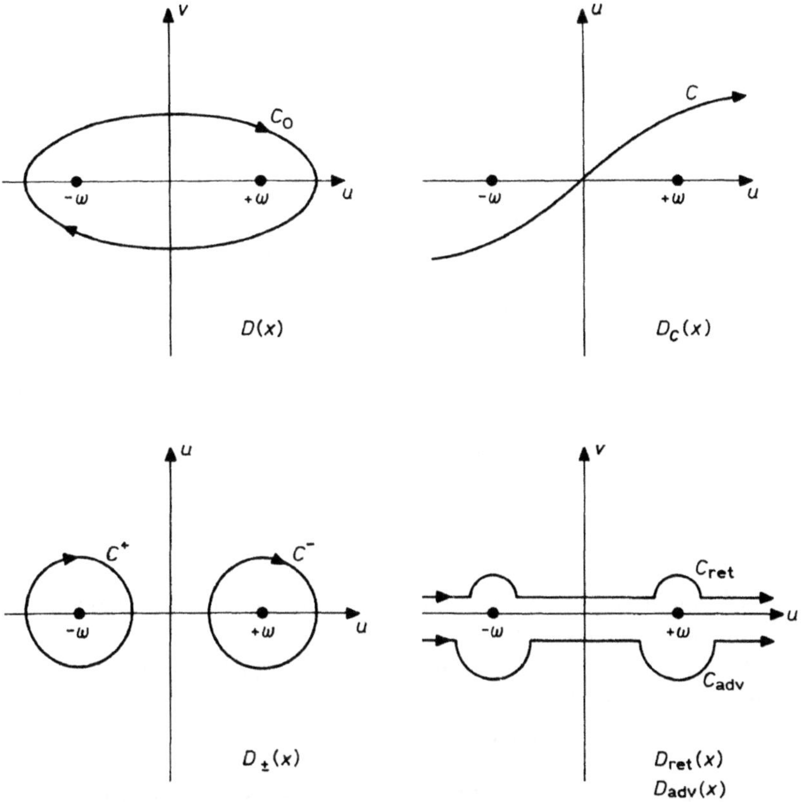

Fig. 49. The various integration contours in the complex plane $k_0 = u + iv$ that can be used to compute the singular functions D, D_c, $D_\pm$, D_{ret}, D_{adv}.

2. The function $D_c(x)$ is "the causal function"; integration contour is C. This function has the form

$$D_c(x) = \frac{1}{4\pi}\,\delta(x^2) - \frac{m}{8\pi\sqrt{x^2}}\,\theta(x^2)\,[J_1(m\sqrt{x^2}) - iN_1(m\sqrt{x^2})] + \theta(x^2)\,\frac{im}{4\pi^2\sqrt{x^2}}\,K_1(m\sqrt{x^2})\,. \qquad \text{(I.5)}$$

Here $J_1(z)$, $N_1(z)$, $K_1(z)$ are the usual Bessel functions. For $m=0$, the function D_c becomes

$$D_c(x) = \frac{1}{4\pi}\,\delta^+(x^2), \tag{I.6}$$

where

$$\delta^+(z) = \frac{1}{2\pi}\int_0^\infty e^{i\omega(z+i\varepsilon)}\,d\omega = \frac{1}{2\pi}\frac{1}{z+i\varepsilon} \tag{I.7}$$

for $\varepsilon \to \infty$. Similarly

$$\delta^-(z) = \frac{1}{2\pi}\int_0^\infty e^{-i\omega(z-i\varepsilon)}\,d\omega = \frac{1}{2\pi}\frac{1}{z-i\varepsilon} \tag{I.8}$$

for $\varepsilon \to \infty$. This equation is equivalent to

$$\delta^\pm(z) = \frac{1}{2}\left[\delta(z) \pm \frac{i}{\pi}\frac{1}{z}\right], \tag{I.9}$$

which is more useful for integrating along the real axis if the term in $1/z$ is the most important in the integral near $z=0$.

3. The functions $D_\pm(x)$, which are obtained by integrating along the contours C^+ and C^-, have the form

$$D_\pm(x) = \tfrac{1}{2}\{D(x) \pm iD_1(x)\}, \tag{I.10}$$

where $D_1(x)$ is the function obtained by integrating over the contour C_1:

$$D_1(x) = \begin{cases} \dfrac{m^2}{2\pi^2}\dfrac{K_1(m\sqrt{x^2})}{m\sqrt{x^2}} & \text{for } x^2 > 0, \quad \text{(I.11)} \\[2ex] \dfrac{m^2}{2\pi^2}\dfrac{N_1(m\sqrt{x^2})}{m\sqrt{x^2}} & \text{for } x^2 < 0. \quad \text{(I.11')} \end{cases}$$

4. Lastly, the functions $D_{\text{ret}}(x)$ and $D_{\text{adv}}(x)$ are obtained by integrating along the contours C_{ret} and C_{adv}. These are equivalent to the Green's functions for the inhomogeneous equations

$$(\Box^2 - m^2)\,D_{\text{ret}}(x) = -\,\delta(x), \tag{I.12}$$

$$(\Box^2 - m^2)\,D_{\text{adv}}(x) = -\,\delta(x), \tag{I.12'}$$

$$D_{\text{ret}}(x) = \begin{cases} D(x) & \text{for} \quad t > 0, \\ 0 & \text{for} \quad t < 0, \end{cases} \qquad \begin{matrix} (\text{I.13}) \\ (\text{I.13}') \end{matrix}$$

$$D_{\text{adv}}(x) = \begin{cases} 0 & \text{for} \quad t > 0, \\ D(x) & \text{for} \quad t < 0; \end{cases} \qquad \begin{matrix} (\text{I.14}) \\ (\text{I.14}') \end{matrix}$$

$D_{\text{ret}}(x)$ is the *retarded* Green's function and D_{adv} is the *advanced* Green's function.

The corresponding functions for a four-dimensional field are obtained by operating on the function $D(x)$ with the operator $(\gamma_\mu \partial/\partial x_\mu - m)$. In particular, the *commutative* function for a fermion field $S(x)$ is equal to

$$S(x) = -\left(\gamma_\mu \frac{\partial}{\partial x_\mu} - m\right) D(x). \qquad (\text{I.15})$$

The expression for the *causal* function for a fermion field is

$$S_c(x) = -\left(\gamma_\mu \frac{\partial}{\partial x_\mu} - m\right) D_c(x). \qquad (\text{I.16})$$

Finally the *retarded* function $S_{\text{ret}}(x)$ and the *advanced* function $S_{\text{adv}}(x)$ for the fermion field are

$$S_{\text{ret}}(x) = -\left(\gamma_\mu \frac{\partial}{\partial x_\mu} - m\right) D_{\text{ret}}(x), \qquad (\text{I.17})$$

$$S_{\text{adv}}(x) = -\left(\gamma_\mu \frac{\partial}{\partial x_\mu} - m\right) D_{\text{adv}}(x) \qquad (\text{I.17}')$$

and they satisfy the inhomogeneous equations

$$\left(\gamma_\mu \frac{\partial}{\partial x_\mu} - m\right) S_{\text{ret}}(x) = -\delta(x), \qquad (\text{I.18})$$

$$\left(\gamma_\mu \frac{\partial}{\partial x_\mu} - m\right) S_{\text{adv}}(x) = -\delta(x). \qquad (\text{I.18}')$$

The derivations of these equations may be found in textbooks on quantum field theory [37, 38].

APPENDIX II (TO SECTION 15)

The components of the spinor $u^r_\alpha(\mathbf{p})$ for $E>0$ may be written as

$$\left.\begin{array}{ll} r=1 & r=2 \\ u_1 = N, & u_1 = 0, \\ u_2 = 0, & u_2 = N, \\ u_3 = \dfrac{p_z}{m+E} N, & u_3 = \dfrac{\Pi^*}{m+E} N, \\ u_4 = \dfrac{\Pi}{m+E} N, & u_4 = \dfrac{p_z}{m+E} N, \end{array}\right\} \tag{II.1}$$

where

$$N = \frac{1}{\sqrt{2}} \left(1 + \frac{m}{E}\right)^{1/2}, \quad \Pi = p_z + \mathrm{i} p_y.$$

From this we obtain

$$\begin{aligned} \frac{\partial u^*}{\partial p_z} \frac{\partial u}{\partial p_z} = \sum_{\alpha=1}^{4} \frac{\partial u_\alpha^*}{\partial p_z} \frac{\partial u_\alpha}{\partial p_z} = \frac{1}{8} \left(1 + \frac{m}{E}\right)^{-1} \frac{m^2 p_z^2}{E^6} + \\ + \frac{1}{2} \left(1 + \frac{m}{E}\right)^{-1} \frac{1}{E^2} \left[1 + \frac{p_z^4}{(1+m/E)^2 E^4} + \frac{1}{4} \right. \\ \times \frac{m^2 p_z^4}{(1+m/E)^2 E^6} - \frac{2 p_z^2}{(1+m/E) E^2} - \frac{m p_z^2}{(1+m/E) E^3} \\ \left. + \frac{m p_z^4}{(1+m/E)^2 E^5} \right] + \frac{1}{2} \frac{(E^2 - m^2 - p_z^2) p_z^2}{E^6} \\ \times \left(1 + \frac{m}{E}\right)^3 \left(1 + \frac{1}{2} \frac{m}{E}\right)^2 . \end{aligned} \tag{II.2}$$

Note that

$$\int p_z^2 \, \mathrm{d}\Omega = \tfrac{4}{3} \pi p^2, \qquad \int p_z^4 \, \mathrm{d}\Omega = \tfrac{4}{5} \pi p^4, \tag{II.3}$$

and so

$$M = \frac{1}{4\pi} \int \frac{\partial u^*}{\partial p_z} \frac{\partial u}{\partial p_z} \, \mathrm{d}\Omega = \begin{cases} \dfrac{1}{4m^2} & \text{for} \quad p \ll mc, \qquad \text{(II.4)} \\ \dfrac{1}{3} \dfrac{1}{p^2} & \text{for} \quad p \gg mc \qquad \text{(II.4')} \end{cases}$$

or

$$M\left(\xi, \frac{m}{p_0}\right) = \begin{cases} \dfrac{1}{4m^2} & \text{for} \quad \xi \ll \dfrac{m}{p_0}, \qquad \text{(II.5)} \\[2ex] \dfrac{1}{4m^2}\dfrac{4}{3}\dfrac{m^2}{p_0^2}\dfrac{1}{\xi^2} & \text{for} \quad \xi \gg \dfrac{m}{p_0}. \qquad \text{(II.5')} \end{cases}$$

The integral of $M(\xi)$ has the form

$$I_2\left(\frac{m}{p_0}\right) = \int_0^\infty f^2(\xi)\,\xi^2\,\mathrm{d}\xi\; M\left(\xi, \frac{m}{p_0}\right). \tag{II.6}$$

For $m/p_0 \to \infty$ this integral is equal to

$$I_2\left(\frac{m}{p_0}\right) = \frac{1}{4m^2}\int_0^\infty f^2(\xi)\,\xi^2\,\mathrm{d}\xi = \frac{\alpha}{4m^2}. \tag{II.7}$$

and for $m/p_0 \to 0$, to

$$I_2\left(\frac{m}{p_0}\right) = \frac{1}{4m^2}\int_0^{m/p_0} f^2(\xi)\,\xi^2\,\mathrm{d}\xi + \frac{1}{4m^2}\frac{4}{3}\frac{m^2}{p_0^2}\int_{m/p_0}^\infty f^2(\xi)\,\mathrm{d}\xi. \tag{II.8}$$

The first integral in (II.8) tends to zero as $(m/p_0)^3$. The second integral in (II.8) tends to unity and gives a contribution of the order of $(1/p_0)^2$ to I_2. This is combined with the contribution from the integral I_1 in (15.9).

APPENDIX III (TO SECTION 16)

The example referred to in Section 16 is as follows. Let the field $\varphi(x)$ (16.13) have an amplitude $A(\mathbf{k})$ that is equal to

$$A(\mathbf{k}) = \frac{c_1}{\omega}\exp\left(-\frac{(\mathbf{k}-\mathbf{k}_1)^2}{2b^2}\right) + \frac{c_2}{\omega}\exp\left(-\frac{(\mathbf{k}-\mathbf{k}_2)^2}{2b^2}\right). \tag{III.1}$$

We compute $\omega(\mathbf{x}, 0)$. On the basis of (16.16) we have

$$\varphi(\mathbf{x}, 0) = c_1 \int \frac{\exp(-(\mathbf{k}-\mathbf{k}_1)^2/2b^2 + i\mathbf{k}\mathbf{x})}{\omega}\,\mathrm{d}^3k$$

$$+ c_2 \int \frac{\exp(-(\mathbf{k}-\mathbf{k}_2)^2/2b^2 + i\mathbf{kx})}{\omega} \mathrm{d}^3 k$$

$$= \frac{c_1}{\omega_1} \exp\left(-\frac{b^2 x^2}{2} + i\mathbf{k}_1\mathbf{x}\right)$$

$$+ \frac{c_2}{\omega_2} \exp\left(-\frac{b^2 x^2}{2} + i\mathbf{k}_2\mathbf{x}\right), \qquad \text{(III.2)}$$

where we assumed that $|\mathbf{k}_1 - \mathbf{k}_2| \gg b$. Therefore,

$$\hat{\omega}\varphi = c_1 \exp\left(-\frac{b^2 x^2}{2} + i\mathbf{k}_1\mathbf{x}\right) + c_2 \exp\left(-\frac{b^2 x^2}{2} + i\mathbf{k}_2\mathbf{x}\right). \qquad \text{(III.3)}$$

Using (16.24) we have

$$\omega(\mathbf{x}, 0) = \tfrac{1}{2}(\hat{\omega}\varphi^*\varphi + \varphi^*\hat{\omega}\varphi)$$

$$= e^{-b^2x^2} \left\{\frac{|c_1|^2}{\omega_1} + \frac{|c_1 c_2|}{\omega_1} \cos(\Delta\mathbf{kx} + \alpha)\right.$$

$$\left. + \frac{|c_1 c_2|}{\omega_2} \cos(\Delta\mathbf{kx} + \alpha) + \frac{|c_2|^2}{\omega_2}\right\}, \qquad \text{(III.4)}$$

where $\Delta\mathbf{k} = \mathbf{k}_2 - \mathbf{k}_1$, $\alpha = \arg(c_2/c_1)$. Assuming that $\omega_2 \gg \omega_1$ we get

$$\omega(\mathbf{x}, 0) = \mathrm{e}^{-b^2x^2} \frac{|c_1|^2}{\omega_1} \left[1 + \left|\frac{c_2}{c_1}\right| \cos(\Delta\mathbf{kx} + \alpha)\right]. \qquad \text{(III.5)}$$

For $|c_2/c_1| > 1$, the quantity $\omega(\mathbf{x}, 0)$ oscillates. Note that, because of the assumption that $|\mathbf{k}_2 - \mathbf{k}_1| \gg b$, the density is not strongly localized (in the small). The oscillations have a period of the order of $1/|k_2 - k_1|$ in space, and of the order of $1/|\omega_2 - \omega_1|$ in time.

APPENDIX IV (TO SECTION 20)

The Absorbing Diaphragm

We investigate the solution of Dirac's equation for a particle of mass m and energy E which moves in the field of a strongly absorbing diaphragm. We must assume that the interaction between the particles in the diaphragm and those we are considering is covariant. For definiteness, we assume that this interaction is of a vector nature.

Let the fourth component of this interaction be V. Then Dirac's equation gives

$$(\boldsymbol{\alpha}\mathbf{p} + \beta m)\,\psi = (E - V)\,\psi\,. \tag{IV.1}$$

Assuming $\psi = \begin{Bmatrix}\varphi\\ \chi\end{Bmatrix}$, we obtain for φ

$$\left\{(\boldsymbol{\sigma}\hat{\mathbf{p}})\frac{1}{(E - V + m)}(\boldsymbol{\sigma}\hat{\mathbf{p}})\right\}\varphi = (E - V - m)\,\varphi\,, \tag{IV.2}$$

where $\boldsymbol{\sigma}$ is the Pauli matrix and

$$\boldsymbol{\alpha} = \begin{pmatrix}0 & \boldsymbol{\sigma}\\ \boldsymbol{\sigma} & 0\end{pmatrix}, \qquad \beta = \begin{pmatrix}I & 0\\ 0 & -I\end{pmatrix}. \tag{IV.3}$$

In regions in which $V = 0$ (outside of the screen but inside of the aperture) we have

$$\hat{p}^2\varphi = (E^2 - m^2)\,\varphi\,. \tag{IV.4}$$

For simplicity we shall restrict ourselves to two dimensions, and we assume the beam is directed along the Oy axis. Noting that

$$p^2 = -\nabla^2 = -\left(\frac{\partial^2}{\partial x^2} + \frac{\partial^2}{\partial y^2}\right),$$

φ can be written in the form

$$\varphi \sim \exp(\mathrm{i}\,(k_x x + k_y y))\,, \tag{IV.5}$$

$$E^2 = k_x^2 + k_y^2 + m^2 = k^2 + m^2\,, \tag{IV.6}$$

$$E = +\sqrt{k^2 + m^2}\,. \tag{IV.6$'$}$$

In the region outside the screen, where $V \neq 0$, Equation (II.2) has the form

$$\hat{p}^2\varphi + [(E - V)^2 - m^2]\,\varphi = 0 \tag{IV.7}$$

or

$$\nabla^2\varphi + [(E - V)^2 - m^2]\,\varphi = 0\,. \tag{IV.7$'$}$$

If the energy V is real, there are two possible cases. These are:

(a) $V < 0$ (attraction); in this case the quantity $[(E - V)^2 - m^2] = q^2 =$

$=q_x^2+q_y^2>k^2$ and the solution has the form of a wave

$$\varphi \sim \exp(\mathrm{i}(q_x x + q_y y)). \tag{IV.5'}$$

For $E<m$ bound states may occur. If $V<-2m$, then bound states in which electron and positron states become interchangeable can also occur.

(b) $V>0$ (repulsion); in this case the quantity $[(E-V)^2-m^2]$ is positive for sufficiently great values of E and so the solution may again be written in the form (IV.5′), but with a different value for the wave vector q'_x, q'_y. For $(E-V)^2-m^2>0$, the particle passes over the barrier. For $(E-V)^2-m^2=\lambda^2<0$, $\lambda_x^2+\lambda_y^2=\lambda^2$, the wave function has the form

$$\varphi \approx \exp(\pm\lambda_x x \pm \lambda_y y), \tag{IV.5''}$$

Inside of an infinitely large barrier we must take the solution which corresponds to a damped wave φ.

Finally, for large V ($V>2m$), the barrier once again becomes transparent. This corresponds to the overlapping inside of the barrier of levels m and $-m$, i.e., to a sufficiently large polarization of vacuum so that the field may form positron and electron pairs.

We now turn to the case of interest in which the purely imaginary part of the potential is absorbed. Assuming $V=\mathrm{i}W$ we obtain

$$\nabla^2\varphi + [(E^2-m^2) - 2\mathrm{i}EW - W^2]\,\varphi = 0. \tag{IV.8}$$

From this equation it is obvious that for $W\to\infty$ it assumes the form

$$\nabla^2\varphi - W^2\varphi = 0 \tag{IV.8'}$$

and, therefore, we have an exponential solution with real indices λ_x, λ_y, $\lambda_x^2+\lambda_y^2=W^2>0$:

$$\varphi \sim \exp(\pm\lambda_x x \pm \lambda_y y), \tag{IV.9}$$

For large W (larger than λ_x, λ_y), these waves are damped very quickly inside the screen. Note that Klein's paradox does not arise in this case and the solutions $c\pm m$ cannot be interchanged.

APPENDIX V (TO SECTION 22)

A. The space-time structure of a nucleon may be formally defined by the vector Γ_μ [see (22.3)] which depends on two variables, $X=y-\frac{1}{2}(x_1+x_2)$ and $x=x_2-x_1$:

$$\Gamma_\mu = \Gamma_\mu(X, x). \tag{V.1}$$

The function of the two variables $\Gamma_\mu(X, x)$ is derived using perturbation theory. We compute the vertex Γ_μ for the interactions that are described by the Lagrangian

$$\hat{W} = g\{\hat{\bar{\psi}}\gamma_5\tau_1\hat{\psi}\hat{\varphi}_1 + \hat{\bar{\psi}}\gamma_5\tau_2\hat{\psi}\hat{\varphi}_2\} + \mathrm{ie} \times \left\{\hat{\varphi}_1 \frac{\partial\hat{\varphi}_2}{\partial x_\mu} - \hat{\varphi}_2 \frac{\partial\hat{\varphi}_1}{\partial x_\mu}\right\} \hat{A}_\mu + \frac{\mathrm{e}}{2}\hat{\bar{\psi}}(1+\tau_3)\gamma_\mu\hat{\psi}\hat{A}_\mu. \tag{V.2}$$

where the operators $\hat{\bar{\psi}}$ and $\hat{\psi}$ describe a nucleon field ("bare" protons and neutrons), the operators $\hat{\varphi}_1$ and $\hat{\varphi}_2$ describe the pseudoscalar field of charged mesons, and τ_1, τ_2 and τ_3 are the usual isotopic spin matrices. The first term describes the "strong" interaction between the nucleon and meson fields, g is the constant of this interaction. We shall formally also consider it sufficiently small so that perturbation theory may be used. The second term describes the interaction between a meson field and an electromagnetic field. This field is characterised by the vector potential $\hat{A}_\mu$. Finally, the third term describes the electromagnetic interaction of a "bare" nucleon. For a neutron this term is equal to zero because in this case $\frac{1}{2}(1+\tau_3)=0$.

We now compute an element of the scattering matrix for the scattering of a photon on a nucleon to the first order in e, and to the second order in g. We must first calculate the average over meson vacuum of the interaction energy $\hat{W}$, taking into account terms of the order of e, $-$eg., and eg^2. Using the usual Feynman diagram method, we obtain

$$\langle 0|\,\hat{W}\,|0\rangle = \int J_\mu(x)\,A_\mu(x)\,\mathrm{d}^4x + \int J_\mu(x_1, x_2, y)\,A_\mu(y) \times \mathrm{d}^4x_1\,\mathrm{d}^4x_2\,\mathrm{d}^4y, \tag{V.3}$$

where

$$J_\mu(x) = \frac{\mathrm{e}}{2}\bar{\psi}(x)\,(1+\tau_3)\,\gamma_\mu\psi(x), \tag{V.4}$$

$$J_\mu(x_1, x_2, y) = \mathrm{i}eg^2\bar{\psi}(x_2)\left\{\gamma_5 S_c(x_2-x_1)\,\gamma_5\tau_3 \times\right.$$

$$\times \left[D_c(x_2 - y) \frac{\partial D_c(x_1 - y)}{\partial y_\mu} - D_c(x_1 - y) \frac{\partial D_c(x_2 - y)}{\partial y_\mu} \right] \Bigg\} \psi(x_1) . \qquad (V.5)$$

The current (V.4) is due to the charge of the "bare" proton, and the current (V.5) is due to the charge of the meson atmosphere around a nucleon. The latter current corresponds to the Feynman diagram shown in Figure 22b. In this equation $S_c(x_2-x_1)$ is the nucleon causal function defined in (14.3) and (I.16); $D_c(x_2-y)$, $D_c(x_1-y)$ are the meson causal functions. The term in curly brackets in (V.5) is, to an accuracy of ieg^2, the vertex of the function $\Gamma_\mu(x_1, x_2, y)$ to the second order in g. It depends only on the difference of the coordinates x_1, x_2, y and, therefore, it may be rewritten in terms of the variables $X-y$ and x where $X=y-\frac{1}{2}(x_1+x_2)$. Thus from the point of view of field theory, the space-time structure of a nucleon is described by functions of two variables. The difference between the point of entry x_1 and the exit point x_2 takes into account the effect of nucleon recoil (changing its position) in going from the initial state $\psi(x_1)=\psi_i(x_1)$ to the final state $\psi(x_2)=\bar{\psi}_f(x_2)$. This reflects the dynamic character of the nucleon structure described by the vector $\Gamma_\mu(X, x)$. The Fourier transform of this quantity $\tilde{\Gamma}(\alpha, \beta)$ is a function of the matrix γ_μ and the vector α, β. In the following it will be sufficient to limit ourselves to the terms of $\tilde{\Gamma}$ which are diagonal with respect to the spinor indices. It is easy to show that these terms have the form

$$\tilde{\Gamma}_\mu(\alpha, \beta) = \alpha_\mu \Phi(\alpha^2, \alpha\beta, \beta^2) \equiv \Phi_\mu(\alpha^2, \alpha\beta, \beta^2), \qquad (V.6)$$

where Φ is a function that depends on the invariants α^2, $\alpha\beta$, β^2. Note that the quantity $\tilde{\Gamma}_\mu(\alpha, \beta)$ would have exactly the same form if the nucleon were a scalar particle with zero spin.

From particle scattering, using (22.7), we can obtain the value of the function $\tilde{\Gamma}_\mu(\alpha, \beta)$ only for $\alpha=q$ and $\beta=P$:

$$\tilde{\Gamma}_\mu(q, P) = \Phi_\mu(q^2, 0, P^2) = \Phi_\mu(q^2, 0, 4M^2 - q^2). \qquad (V.7)$$

From this we see that the spatial transformation

$$\rho_\mu(\mathbf{X}) = \int \tilde{\Gamma}_\mu(q, P)\, e^{i\mathbf{q}\mathbf{X}}\, d^3q \qquad (V.8)$$

has physical meaning only if $q^2 \ll 4M^2$ ($P^2 \approx 4M^2$). In fact, the variable $\beta\,(=P)$ is coupled to the coordinate $x = x_2 - x_1$, but not to the coordinate $X = y - \frac{1}{2}(x_1 + x_2)$, and from (V.7) we see that the variable β in (V.8) is a function of the variable $\alpha = q$. Therefore, we must write

$$\rho_\mu(\mathbf{X}) = \int \tilde{\Gamma}_\mu(q, 2M)\, \mathrm{e}^{\mathrm{i}\mathbf{qX}}\, \mathrm{d}^3 q \tag{V.9}$$

and restrict ourselves to values of $\mathbf{q}^2 \ll 4M^2$.

From this it follows that the spatial distribution $\rho(\mathbf{X})$ has physical meaning only for heavy, almost immovable nucleons and for small momentum transfers q.

B. In order to simplify (22.10):

$$J_\mu = \bar{u}_f(p_f)\{aq_\mu + bP_\mu + c\gamma_\mu + \mathrm{d}\sigma_{\mu\nu}q_\nu + \mathrm{e}\sigma_{\mu\nu}P_\nu\}\, u_i(p_i), \tag{V.10}$$

note that, from the vectors q, P, and γ_μ, it is possible to construct only the scalars q^2, P^2, $\gamma_\mu q_\mu$, $\gamma_\mu P_\mu$, (the scalar $qP \equiv 0$). Furthermore, the scalar P^2 can be expressed in terms of the scalar q^2 ($P^2 = 4M^2 - q^2$) and the scalars $\gamma_\mu q_\mu$ and $\gamma_\mu P_\mu$ may be removed using the Dirac equations

$$(\gamma_\mu p_\mu - M)\, u(p) = 0, \tag{V.11}$$

$$\bar{u}(p)\,(\gamma_\mu p_\mu + M) = 0, \tag{V.11'}$$

where M is the nucleon mass. It follows from these equations that the coefficients a, b, c, d, e are functions of q^2 only. From these equations we also have

$$(\bar{u}_j|\, \mathrm{i}\sigma_{\mu\nu}P_\nu\, |u_i) = (\bar{u}_j|\, -\, q_\mu\, |u_i), \tag{V.12}$$

$$(\bar{u}_f|\, \mathrm{i}\sigma_{\mu\nu}q_\nu\, |u_i) = (\bar{u}_f|\, 2M\gamma_\mu - P_\mu\, |u_i). \tag{V.12'}$$

These relationships allow us to remove from (V.10) the terms in P_μ and $\sigma_{\mu\nu}P_\nu$. Thus the current J_μ becomes

$$J_\mu = \bar{u}_j(p_f)\{Aq_\mu + B\gamma_\mu + C\sigma_{\mu\nu}q_\nu\}\, u_i(p_i), \tag{V.13}$$

where A, B, C are form-factor functions which depend only on q^2.

Using (22.4″) we now obtain

$$0 = J_\mu q_\mu \equiv \bar{u}_f(p_f)\{Aq^2 + B\gamma_\mu q_\mu + C\sigma_{\mu\nu}q_\mu q_\nu\}\, u_i(p_i). \tag{V.14}$$

The factor $\gamma_\mu q_\mu$ is equal to zero because of (V.11) and (V.11′) and the quantity $\sigma_{\mu\nu} q_\mu q_\nu = 0$ because the matrix $\sigma_{\mu\nu}$ is antisymmetric. Therefore, the form factor $A(q^2) \equiv 0$ and the current may be written in the form

$$J_\mu = \bar{u}_f(p_f) \{B(q^2)\, \gamma_\mu + C(q^2)\, \sigma_{\mu\nu} q_\nu\}\, u_i(p_i). \tag{V.15}$$

APPENDIX VI (TO SECTION 23)

The equation for elastic π-N scattering (23.6) was obtained in the following way. Let $\Phi_s = \Phi_s(\mathbf{k}_1, \mathbf{k}_2, \mathbf{k}_3, \ldots, \mathbf{k}_s)$ be the wave function of a system of s mesons and nucleons (here $\mathbf{k}_1, \mathbf{k}_2, \ldots, \mathbf{k}_s$ are the momenta). Mesons and nucleons can be annihilated or created in this interaction. Thus the wave functions Φ_s with different numbers of particles are related by the equation

$$(E - H_0)\, \Phi_s = \sum_{s' \neq s} W_{ss'} \Phi_{s'}. \tag{VI.1}$$

where H_0 is the Hamiltonian of the noninteracting particles, E is the total energy of the system, and $W_{ss'}$ is a matrix element of the energy of interaction operator. Of all the possible functions Φ_s, we are interested in $\Phi_2(\mathbf{k}_1, \mathbf{k}_2) = \psi(\mathbf{k}, \mathbf{p})$ where $\mathbf{k}_1 = \mathbf{k}$ is the meson momentum and $\mathbf{k}_2 = \mathbf{p}$ is the nucleon momentum. In order to compute these components, we first rewrite the system of Equations (VI.1) in the form

$$(E - H_0)\, \psi = \sum_{s' \neq 2} W_{2s'} \Phi_{s'}, \tag{VI.2}$$

$$(E - H_0)\, \Phi_{s'} = \sum_{s'' \neq s'} W_{s's''} \Psi_{s''} + W_{s'2} \psi. \tag{VI.2′}$$

We introduce the "elementary scattering matrix" $\hat{r}$ (see [41]):

$$\hat{r} = \delta^+ (E - H_0)\, \hat{W}, \tag{VI.3}$$

where $\delta^+(z)$ is the function defined in Appendix I by (I.7). Using this matrix, Equations (VI.2) and (VI.2′) may be written in symbolic form

$$(E - H_0)\, \psi = \hat{\omega} \Phi, \tag{VI.4}$$

$$\Phi = \hat{r} \Phi + \hat{r} \psi; \tag{VI.4′}$$

and by iterating we obtain for the function ψ the equation

$$(E - H_0)\, \psi = \hat{\omega} \frac{\hat{r}}{1 - \hat{r}} \psi + (\omega \hat{r}^N \Phi)_{N \to \infty}. \tag{VI.5}$$

If we assume (a) that divergent terms are excluded from this equation and (b) that the remaining terms tend to zero as $N \to \infty$, then this equation may be written in the form

$$\{E - E(\mathbf{k}, \mathbf{p})\} \psi(\mathbf{k}, \mathbf{p}) = \int \mathfrak{G}(\mathbf{k}, \mathbf{p} \mid \mathbf{k}', \mathbf{p}') \psi(\mathbf{k}', \mathbf{p}') \, d^3k' \, d^3p', \tag{VI.6}$$

where $\mathfrak{G}(\mathbf{k}, \mathbf{p}|\mathbf{k}', \mathbf{p}')$ is the matrix element of the operator

$$\hat{\mathfrak{G}} = \frac{\hat{r}}{1 - \hat{r}} \hat{\omega}, \tag{VI.7}$$

$$E(\mathbf{k}, \mathbf{p}) = E_\pi(k) + E_N(p) \tag{VI.8}$$

is the sum of the π-meson energy $E_\pi(k)$ and the nucleon energy $E_N(p)$. For real processes $E = H_0$ and the matrix r, because the function

$$\delta^+(E - H_0) = \frac{1}{E - H_0} + i\pi\delta(E - H_0) \tag{VI.9}$$

is imaginary, will contribute to the imaginary part of the operator $\hat{\mathfrak{G}}$. Therefore, the operator $\hat{\mathfrak{G}}$ is complex. This is due to the presence of real, inelastic processes.

We now change over to the center of mass system in which $k + p = 0$. Equation (VI.6) may now be written as

$$\{E - E_N(\mathbf{q}) - E_\pi(\mathbf{q})\} \psi(\mathbf{q}) = \int \mathfrak{G}(E, \mathbf{q} \mid \mathbf{q}') \psi(\mathbf{q}') \, d^3\mathbf{q}' \tag{VI.10}$$

or, in the coordinate representation, as

$$\{E - \hat{H}_N^0(\mathbf{x}) - \hat{H}_\pi^0(\mathbf{x})\} \psi(\mathbf{x}) = \int F(E, \mathbf{x} \mid \mathbf{x}') \psi(\mathbf{x}') \, d^3x'. \tag{VI.11}$$

where $\hat{H}_N^0(\mathbf{x})$ and $\hat{H}_\pi^0(\mathbf{x})$ are the Hamiltonians of the unbound nucleon and π-meson respectively. By operating on both sides of this equation by the operators $[E - \hat{H}_N^0(\mathbf{x})] + \hat{H}_\pi^0(\mathbf{x})$ and $[E + \hat{H}_\pi^0(\mathbf{x})] - \hat{H}_N^0(\mathbf{x})$ in turn, we obtain

$$[E_\pi^2 - \hat{K}^2] \psi(\mathbf{x}) = \int U(\mathbf{x}, \mathbf{x}'; E) \psi(\mathbf{x}') \, d^3x', \tag{VI.12}$$

$$[E_{\rm N} - \hat{H}^0_{\rm N}(x)]\,\psi(\mathbf{x}) = \int V(\mathbf{x}, \mathbf{x}'; E)\,\psi(\mathbf{x}')\,\mathrm{d}^3x'. \tag{VI.13}$$

These are the equations given in the text as (23.6) and (23.7).

These equations may also be obtained from the relativistic wave function equations for the system of particles $\varphi(x_1, x_2, \ldots, x_k, \ldots, x_{\rm N})$. Here $x_k = t_k$, $\mathbf{x}_k$ is the time and space coordinate of the k-particle (see [47]). The analysis that most closely approaches this method is that for the simultaneous function $\varphi(t, \mathbf{x}_1, t, \mathbf{x}_2) = \psi(t, \mathbf{x}_1, \mathbf{x}_2)$ given in [42] on the basis of Green's function theory. The authors obtained an equation that they called a *quasipotential*. The advantages of this new approach to equal time equations are that (a) it automatically includes the necessary renormalization, and (b) the potential is local and so approximations may be written in terms of the scattering amplitude.

APPENDIX VII (TO SECTIONS 24 AND 29)

The interaction of fields in modern theory is described by an operator of the Lagrange function $\hat{L}(x)$ which is usually written as the product of the operators of the fields being considered. This product must satsify the requirements of relativistic invariance, some of the symmetry conditions, and the requirements that correspond to those of the classical theory, if any exist. In particular, for the interaction between the electromagnetic field $\hat{A}_\mu$ and the electron-positron field ψ, the Lagrangian $\hat{L}$ has the form

$$\hat{L} = -\frac{1}{\hbar}\,\hat{J}_\mu(x)\,\hat{A}_\mu(x), \tag{VII.1}$$

where the current $\hat{J}_\mu(\mathrm{x})$ is equal to

$$\hat{J}_\mu(x) = e\bar{\psi}(x)\,\gamma_\mu\psi(x). \tag{VII.2}$$

The constant in this case is the elementary charge e. For the case of a scalar, electrically neutral field $\hat{\varphi} = \hat{\varphi}^+$, the Lagrangian is

$$\hat{L} = -\frac{\lambda}{\hbar}\frac{\hat{\varphi}^4}{4}, \tag{VII.3}$$

where λ is some constant.†

† Note that $\hat{L} \sim \hat{\phi}^3$ leads to a dynamically unstable system and that $\hat{L} \sim \hat{\phi}^2$ leads to trivial linear equations.

For a sufficiently small interaction constant and for a completely switched on interaction, the scattering matrix $\hat{S}$

$$\hat{S} = T \exp \mathrm{i} \int \hat{L}(x) \, \mathrm{d}^4 x \tag{VII.4}$$

can be written in the form of a series of functionals [see (28.11), (29.22)]

$$\hat{S} = \sum_{n=0}^{\infty} \frac{\mathrm{i}^n}{n!} \int T[\hat{L}(x_1) \hat{L}(x_2) \cdots \hat{L}(x_n)] \, \mathrm{d}^4 x_1 \, \mathrm{d}^4 x_2 \cdots \mathrm{d}^4 x_n . \tag{VII.5}$$

Because the Lagrangian $\hat{L}(x)$ is the product of the operators of the fields $\hat{\varphi}$ or $\hat{\psi}$ or $\hat{A}$, each term in the series (VII.5) itself represents the sum of the products of the operators of particle creation a_k^+ and the operators of particle annihilation a_k. The order of these operators which transform the initial state of the field Ψ_i into the final state Ψ_f defines the history of the particles (in the Feynman diagram sense). The computations of the matrix element $S_{fi} = \langle \Psi_f | \hat{S} | \Psi_i \rangle$ which describes this transformation may be simplified if the factor in (VII.2) is written in the form in which all the annihilation operators are on the right of the formation operators. Then the annihilation part of the operators that operate on Ψ_i would transform it into a functional of vacuum Ψ_0 and the formation part of the operators, which operate from the left of the functional $\bar{\Psi}_f$ would also transform it into a functional of vacuum $\bar{\Psi}_0$. Calculation of the effect of the "extra" operators reduces to finding their vacuum average.

This process of changing the position of the operators is called the process of transforming the T-products by the normal N-products. The validity of this process is based on the easily verifiable equation (see [37, 38])

$$T\{\varphi(x)\,\varphi(y)\} = N\{\varphi(x)\,\varphi(y)\} + D_c(x-y), \tag{VII.5'}$$

which holds for operators of a boson field $\varphi(x)$, and the equation

$$T\{\psi_\alpha(x)\,\bar{\psi}_\beta(y)\} = N\{\psi_\alpha(x)\,\bar{\psi}_\beta(y)\} + [S_c(x-y)]_{\alpha\beta}, \tag{VII.6}$$

which holds for operators of a fermion field $\bar{\psi}\psi$. Note that, for the combinations of operators such as (ψ, ψ) or $(\bar{\psi}, \bar{\psi})$, the T- and N-products are equal. The *causal functions* $D_c(x)$ and $S_c(x)$ are defined in Appendix I and are also called *contractions*. The vacuum average of the normal product is obviously equal to zero. Thus the nonvanishing contribution to the matrix element S_{fi} consists of the product of singular functions such as $D_c(x-y)$ and $S_c(x-y)$ and of the limiting, or "bounding" operators of particle production and annihilation, which act on the functionals Ψ_f and Ψ_i.† Therefore, in the simple case of a real, scalar field $\hat{\varphi}(x)$

$$\hat{\varphi}(x) = \int d^3k \{\hat{a}(k)\, U_k(x) + \hat{a}^+(k)\, U_k^*(x)\}, \tag{VII.7}$$

where $U_k(x) = e^{ikx}/(2\omega)^{1/2}$, $\omega = (k^2 + m^2)^{1/2}$, the matrix element S_{fi} has the form

$$\begin{aligned} S_{fi} = \langle \Psi_f|\, a^+(k_1)\, a^+(k_2) \cdots a^+(k_m) \int \bar{\Phi}_f(y_1, y_2, \ldots, y_m) \\ \times \hat{S}_{mn}\, \Phi_i(x_1, x_2, \ldots, x_n)\, d^4y_1\, d^4y_2 \cdots d^4y_m\, d^4x_1\, d^4x_2 \cdots \\ \times d^4x_n a(k_1')\, a(k_2') \cdots a(k_m')\, |\Psi_i\rangle, \end{aligned} \tag{VII.8}$$

where $\hat{S}_{mn}$ denotes the sum of the products of all operators $a^+(k)\, U_k^+(x)$ and $\hat{a}(k)\, U_k(x)$ except for the "bounding" operators which are given in (VII.8). The function $\bar{\Phi}_f$ is equal to

$$\bar{\Phi}_f(y_1, y_2, \ldots, y_m) = \sum_P U_{k_1}^*(y_1)\, U_{k_2}^*(y_2) \cdots U_{k_m}^*(y_m), \tag{VII.9}$$

where $\sum_P$ denotes the sum over all combinations of m-particles in the final state. Similarly,

$$\Phi_i(x_1, x_2, \ldots, x_n) = \sum_P U_{k'_1}(x_1)\, U_{k'_2}(x_2) \ldots U_{k'_n}(x_n) \tag{VII.9'}$$

in the symmetrized function of n-particles in the initial state. From the definition of the "bounding" operators, we have

$$\begin{aligned} S_{fi} = \int \bar{\Phi}_f(y_1, y_2, \ldots, y_m)\, s(y_1, y_2, \ldots, y_m \mid x_1, x_2, \ldots, x_n) \\ \times \Phi_i(x_1, x_2, \ldots, x_n)\, d^4y_1\, d^4y_2 \cdots d^4y_m \\ \times d^4x_1\, d^4x_2 \cdots d^4x_n, \end{aligned} \tag{VII.10}$$

† Some of the matrix elements may diverge. We shall not describe the methods of regularization (renormalization) of such elements here. These methods are given in special courses on quantum field theory.

where

$$s(y_1, y_2, \ldots, y_m \mid x_1, x_2, \ldots, x_n) = \langle \Psi_0 | \hat{S}_{nm} | \Psi_0 \rangle \qquad \text{(VII.11)}$$

is the vacuum average of all combinations of particle formation and annihilation that describe the history of the field in the interval $t_1 = -\infty$ to $t_2 = \infty$ in which the state changes from the initial i to the final f.

From this short outline, it follows that the scattering matrix $\hat{S}$ may be written in the form of a series of functions

$$s_n = s(x_1, x_2, \ldots, x_n), \qquad \text{(VII.12)}$$

each of which represents the vacuum average over the n-particle history. The division of the particles into those in the initial state (x) and those in the final state (y) is not important, as we can see from (VII.9). Each function in (VII.11) can represent various physical processes depending on what divisions of the initial (i) and final (f) particle states are possible. The function $s(x_1, x_2, \ldots, x_{n-1})$ can represent the processes

$$\left.\begin{array}{l} s(y_1 \mid x_1, x_2, \ldots, x_{n-1}), \\ s(y_1, y_2 \mid x_1, x_2, \ldots, x_{n-1}, x_{n-2}), \\ \vdots \\ s(y_1, y_2, \ldots, y_{n-1} \mid x_n). \end{array}\right\} \qquad \text{(VII.13)}$$

Not all the possible divisions will describe real, physical processes. For example, the first division in (VII.13) would imply that a particle y is in a resonant state $(n-1)$ of the particles $(x_1, x_2, \ldots, x_{n-1})$, which contradicts the spectral decomposition (VII.7) in which only one type of particle, which is stable and has mass m, is assumed.

APPENDIX VIII (TO SECTION 27)

A. Virtual photons may not be considered separately from their source. We assume that the source is a particle that is localized near the point $\mathbf{x} = \mathbf{0}$, and that a quantum jump takes place at a time close to $t = 0$. The current of particles formed in such a transition is

$$J_\mu^{(x)} = \int \mathrm{d}^3 p_3 \, \mathrm{d}^3 p_1 \exp\left[-\tfrac{1}{2} a^2 \left[(\mathbf{p}_3 - \mathbf{p}_f)^2 + (\mathbf{p}_1 - \mathbf{p}_i)^2\right]\right] \times \exp\left[-\mathrm{i}(\mathbf{kx} - k_0 t)\right] \bar{u}_f(p_3) \, F_\mu u_i(p_1). \qquad \text{(VIII.1)}$$

The field of the packet of virtual photons that are formed in this transition is described by the equation

$$\begin{aligned} A_\mu(\mathbf{x}, t) &= \frac{1}{(\Box^2 + m^2)} J_\mu(x) \\ &= \int \mathrm{d}^3 p_3 \, \mathrm{d}^3 p_1 \frac{\bar{u}_f F_\mu u_i}{(k^2 - k_0^2 + m^2)} \\ &\quad \times \exp\left[-\frac{a^2}{2} [(\mathbf{p}_3 - \mathbf{p}_f)^2 + (\mathbf{p}_1 - \mathbf{p}_i)^2] \right. \\ &\quad \left. - \mathrm{i}(\mathbf{kx} - k_0 t) \right]. \end{aligned} \tag{VIII.2}$$

In order to compute the integrals in (VIII.2), we set

$$\mathbf{p}_1 = \boldsymbol{\xi} + \mathbf{p}_i, \quad \mathbf{p}_3 = \boldsymbol{\eta} + \mathbf{p}_f, \tag{VIII.3}$$

so that

$$\left. \begin{aligned} \mathbf{p}_3 - \mathbf{p}_1 &= (\boldsymbol{\eta} - \boldsymbol{\xi}) + (\mathbf{p}_f - \mathbf{p}_i), \quad \mathbf{k} \approx (\mathbf{p}_f - \mathbf{p}_i), \\ p_{03} - p_{01} &= E(\mathbf{p}_3) - E(\mathbf{p}_1) \\ &= E_f - E_i + \nabla E_f \boldsymbol{\eta} - \nabla E_i \boldsymbol{\xi} + \cdots, \\ k_0 &\approx E_f - E_i, \quad \nabla E_f = \mathbf{u}_f, \quad \nabla E_i = \mathbf{u}_i. \end{aligned} \right\} \tag{VIII.3'}$$

We consider the packet to be small, but $|\mathbf{q}_{fi}| = |\mathbf{p}_f - \mathbf{p}_i| \gg \hbar/a$. Furthermore, for simplicity, we assume that the radiating particle is a point particle, i.e., its mass $m \gg |q_{fi}|$. Then from (VIII.2), the potential of the packet is

$$A_\mu(\mathbf{x}, t) \approx \exp\left[\mathrm{i}(\mathbf{kx} - k_0 t) - \frac{1}{2a^2} ((\mathbf{x} - \mathbf{u}_f t)^2 + (\mathbf{x} - \mathbf{u}_i t)^2) \right] \tag{VIII.4}$$

This expression was used in the text for the case when $u_i = 0$, $u_f = u$.

B. Note that the integral

$$I = \int_{-\infty}^{+\infty}\!\!\int \mathrm{e}^{-\Phi(x,t)} \, \mathrm{d}x \, \mathrm{d}t, \tag{VIII.5}$$

where

$$\Phi(x, t) = A^2 x^2 + Bxt + C^2 t^2 + Dx + Et + G, \tag{VIII.6}$$

for $\Delta^2 = 4A^2C^2 - B^2 > 0$, is equal to

$$I = \frac{2\pi}{\Delta} e^{\Psi}, \tag{VIII.7}$$

where

$$\Psi = \frac{A^2}{\Delta^2} E^2 + \frac{C^2}{\Delta^2} D^2 - \frac{B}{\Delta^2} ED - G. \tag{VIII.8}$$

For our case

$$\Delta^2 = 2\left(\frac{1}{a^2} + \frac{1}{b^2}\right)\left[\frac{\mathbf{v}_f^2 + \mathbf{v}_i^2}{b^2} + \frac{\mathbf{u}^2}{a^2}\right] - \left[\frac{\mathbf{v}_f + \mathbf{v}_i}{b^2} + \frac{\mathbf{u}}{a^2}\right]^2, \tag{VIII.9}$$

where

$$A^2 = \frac{1}{a^2} + \frac{1}{b^2}, \tag{VIII.10}$$

$$C^2 = \frac{1}{2}\left(\frac{\mathbf{v}_f^2 + \mathbf{v}_i^2}{b^2} + \frac{\mathbf{u}^2}{a^2}\right), \tag{VIII.10'}$$

$$\mathbf{B} = -\left(\frac{\mathbf{v}_f - \mathbf{v}_i}{b^2} + \frac{\mathbf{u}}{a^2}\right), \tag{VIII.10''}$$

$$\mathbf{D} = \mathrm{i}(\mathbf{k} - \mathbf{q}) + \frac{1}{a^2}(2\mathbf{X} - \mathbf{u}T), \tag{VIII.10'''}$$

$$E = -\mathrm{i}(k_0 - q_0) + \frac{1}{a^2}[(\mathbf{X} - \mathbf{u}T, \mathbf{u})]. \tag{VIII.10''''}$$

and where q_0 is the difference between the fourth components of the vectors p_3 and p_1. For the particular values of Δ, A, C, $\mathbf{B}$;

$$\Delta^2 = \frac{1}{a^4}[2(\mathbf{v}_f^2 + \mathbf{v}_i^2 + \mathbf{u}^2) + (\mathbf{v}_f + \mathbf{v}_i - \mathbf{u})^2], \tag{VIII.11}$$

$$A^2 = \frac{2}{a^2}, \tag{VIII.11'}$$

$$C^2 = -\frac{1}{a^2}(\mathbf{v}_f^2 + \mathbf{v}_i^2 + \mathbf{u}^2), \tag{VIII.11''}$$

$$\mathbf{B} = -\frac{1}{a^2}(\mathbf{v}_f + \mathbf{v}_i + \mathbf{u}), \tag{VIII.11'''}$$

we obtain

$$\begin{aligned}\Psi = &-\frac{\mathbf{X}^2}{2a^2} - \frac{(\mathbf{X}-\mathbf{u}T)^2}{2a^2} + \frac{C^2}{a^4\Delta^2}(2\mathbf{X}-\mathbf{u}T)^2 \\ &+ \frac{A^2}{a^4\Delta^2}(\mathbf{X}-\mathbf{u}T, \mathbf{u})^2 + \frac{\mathbf{B}}{a^4\Delta^2}(2\mathbf{X}-\mathbf{u}T)(\mathbf{X}-\mathbf{u}T, \mathbf{u}) \\ &- \frac{C^2}{\Delta^2}(\mathbf{k}-\mathbf{q})^2 - \frac{A^2}{\Delta^2}(k_0 - q_0) + \operatorname{Im}\Psi, \end{aligned} \tag{VIII.12}$$

where Im Ψ denotes the imaginary part of Ψ. Because $k_0 = q_0$ Im $\Psi = 0$ for $\mathbf{k} = \mathbf{q}$, then Im Ψ may in general be neglected in (VIII.12).

The real part of Ψ is a positive-definite quadratic function of the variables $\mathbf{X}$ and T. We write it out for the special case for $a = b$, $X_\perp = 0$, $X_u \neq 0$, $T = 0$ [here $X_u = 1/u(\mathbf{Xu})$];

$$\Psi = -\frac{X_\mu^2}{\mathrm{d}^2}, \quad \mathrm{d}^2 = 2a^2\left[1 + \frac{1}{2}\frac{(\mathbf{v}_f + \mathbf{v}_i - \mathbf{u})^2}{(\mathbf{v}_f^2 + \mathbf{v}_i^2 + \mathbf{u}^2)}\right], \tag{VIII.13}$$

for $X_u = 0$, $T = 0$

$$\Psi = -\frac{X_\perp^2}{\mathrm{d}^2}, \tag{VIII.13'}$$

for $X = 0$, $T \neq 0$,

$$\Psi = -\frac{u^2 T^2}{\beta^2}, \tag{VIII.13''}$$

where

$$\beta^2 = 2a^2 \frac{2(\mathbf{v}_f^2 + \mathbf{v}_i^2 + \mathbf{u}^2) + (\mathbf{v}_f + \mathbf{v}_i + \mathbf{u})^2}{2(\mathbf{v}_f^2 + \mathbf{v}_i^2 + \mathbf{u}^2) + \mathbf{u}^2}. \tag{VIII.14}$$

APPENDIX IX (TO SECTION 28)

Let Ψ be a functional of some function which is defined in the space $\mathfrak{R}(x)$

$$\Psi = \Psi\{u(x)\}. \tag{IX.1}$$

We vary the function $u(x)$ about the point x such that

$$u'(x) = u(x) + \delta u(x)$$

where the variation $\delta u(x)$ differs from zero in the neighborhood of x_0. If the new functional can be written in the form

$$\Psi' \equiv \Psi\{u'(x)\} = \Psi\{u(x)\} + \int F\{u(x')\}\,\delta u(x')\,\mathrm{d}x', \qquad \text{(IX.2)}$$

then the limit

$$\frac{\delta\psi}{\delta u(x)} = \lim_{\delta u\, dx \to 0} \frac{\Psi\{u'(x)\} - \Psi\{u(x)\}}{\delta u(x)\,\mathrm{d}x} = F\{u(x)\} \qquad \text{(IX.3)}$$

is called the functional derivative of the functional Ψ with respect to the function $u(x)$ at the point x. Another, more formal definition is

$$\frac{\delta\Psi}{\delta u(x)} = \lim \frac{\Psi\{u'(x)\} - \Psi\{u(x)\}}{\varepsilon}, \qquad \text{(IX.3')}$$

where $\delta u(x') = \varepsilon\delta(x - x')$. In this definition, the localization of the variation about the point x is clearly shown. Obviously, the two definitions (IX.3) and (IX.3′) are equivalent.

If $u(x) = \sigma(x)$ is a space-like surface, then

$$\frac{\delta\Psi}{\delta\sigma(x)} = \lim_{\delta\omega \to 0} \frac{\Psi\{\sigma'(x)\} - \Psi\{\sigma(x)\}}{\delta\omega}, \qquad \text{(IX.4)}$$

where $\delta\omega = \delta\sigma(x)\,\mathrm{d}^3x$ is a four-dimensional volume contained between the surfaces $\sigma'(x)$ and $\sigma(x)$ (see Figure 30).

If the functional $\Psi\{u(x)\}$ is defined by the series

$$\begin{aligned}\Psi\{u\} = \Psi_0(x) &+ \frac{1}{1!}\int \psi(x, x_2)\,u(x_2)\,\mathrm{d}x_2 \\ &+ \frac{1}{2!}\int \psi(x, x_1)\,u(x_1)\,\mathrm{d}x_1 \\ &+ \frac{1}{3!}\int \psi(x, x_2, x_3)\,u(x_2)\,u(x_3)\,\mathrm{d}x_2\,\mathrm{d}x_3, \qquad \text{(IX.5)}\end{aligned}$$

then its functional derivative at the point x will be

$$\begin{aligned}\frac{\delta\Psi\{u\}}{\delta u(x)} = \psi_1(x) &+ \frac{1}{2!}\int \psi(x, x_2)\,u(x_2)\,\mathrm{d}x_2 \\ &+ \frac{1}{2!}\int \psi(x_1, x)\,u(x_1)\,\mathrm{d}x_1\end{aligned}$$

$$+\frac{1}{3!}\int \psi(x, x_2, x_3)\, u(x_2)\, u(x_3)\, dx_2\, dx_3$$
$$+\frac{1}{3!}\int \psi(x_1, x, x_3)\, u(x_1)\, u(x_3)\, dx_1\, dx_3$$
$$+\frac{1}{3!}\int \psi(x_1, x_2, x)\, u(x_1)\, u(x_2)\, dx_1\, dx_2 + \cdots . \tag{IX.6}$$

APPENDIX X (TO SECTION 31)

The Relativistic Wave Function for Two Particles

We shall investigate the wave function for two free particles which are described by the plane waves

$$\psi(\mathbf{x}_1, t_1; \mathbf{x}_2, t_2) = \exp\left[i(\mathbf{p}_1\mathbf{x}_1 - E_1 t_1)\right] \exp\left[i(\mathbf{p}_2\mathbf{x}_2 - E_2 t_2)\right]. \tag{X.1}$$

This equation has physical meaning only when the points $\mathbf{x}_1$, t_1, and $\mathbf{x}_2$, t_2 lie on a space-like plane. We introduce the coordinates

$$\left.\begin{aligned} \mathbf{X} &= \alpha\mathbf{x}_1 + \beta\mathbf{x}_2, & T &= \alpha t_1 + \beta t_2, \\ \mathbf{x} &= \mathbf{x}_1 - \mathbf{x}_2, & t &= t_1 - t_2, \end{aligned}\right\} \tag{X.2}$$

where α, $\beta > 0$, $\alpha + \beta = 1$. Then

$$\left.\begin{aligned} \mathbf{x}_1 &= \mathbf{X} + \beta\mathbf{x}, & t_1 &= T + \beta t, \\ \mathbf{x}_2 &= \mathbf{X} - \alpha\mathbf{x}, & t_2 &= T - \alpha t. \end{aligned}\right\} \tag{X.3}$$

The total momentum is

$$\mathbf{P} = \mathbf{p}_1 + \mathbf{p}_2, \qquad E = E_1 + E_2. \tag{X.4}$$

In terms of these variables, the wave function (X.1) becomes

$$\psi(\mathbf{X}, T, \mathbf{x}, t) = \exp\left[i(\mathbf{P}\mathbf{X} - ET) + i(\mathbf{p}\mathbf{x} - \varepsilon t)\right], \tag{X.5}$$

where

$$\mathbf{p} = \beta\mathbf{p}_1 - \alpha\mathbf{p}_2, \qquad \varepsilon = \beta E_1 - \alpha E_2 \tag{X.6}$$

are the relative momentum and energy.

As the points (x_1, t_1) and (x_2, t_2) lie on a space-like plane, we may set $t_1 = t_2$. Then $t = 0$ and T is the time common to both particles. In the center of mass system, $\mathbf{P}=0$ and so, for $t_1=t_2$, we have

$$\psi(\mathbf{X}, T, \mathbf{x}, 0) = \exp\left[\mathrm{i}(\mathbf{px} - ET)\right]. \tag{X.7}$$

where the quantity

$$E = E(p) = \sqrt{m_1^2 + \mathbf{p}^2} + \sqrt{m_2^2 + \mathbf{p}^2} \tag{X.8}$$

is the total energy of the two particles in the center-of-mass system. We now change notation and use t instead of T, x as the relative coordinate of the particle and t as the common time. In this notation, the expansion of the function ψ in terms of spherical waves has the form

$$\psi(\mathbf{x}, t) = \sum_{l=1}^{\infty} (2l+1)\, P_l(\cos\vartheta)\, R_l(pr)\, \mathrm{e}^{-\mathrm{i}E(p)t} \tag{X.9}$$

but that for zero-spin particles does not differ from the nonrelativistic case. In this equation r is the distance between the particles in the center-of-mass system

$$R_l(pr) = \frac{J_{l+1/2}(pr)}{\sqrt{pr}}, \tag{X.10}$$

$J_{l+1/2}(z)$ is the Bessel function and $P_l(\xi)$ the Legendre polynomial. The asymptotic form of equation (X.9) is

$$\psi(\mathbf{x}, t) = \sum_{l=0}^{\infty} (2l+1)\, P_l(\cos\upsilon)\left[a_l(E)\exp(\mathrm{i}(\mathbf{pr} - Et)) + \right.$$
$$\left. + b_l(E)\exp(-\mathrm{i}(\mathbf{pr} - Et))\right], \tag{X.11}$$

APPENDIX XI (TO SECTION 34)

Resonance States. Nonrelativistic Theory

We shall consider the scattering of two particles that can have a resonant interaction. For definiteness, we assume that the resonance takes place in the s-state.† We write the wave function of the s-state in the form

$$\varphi(r, t) = \frac{1}{r}\psi(kr)\exp\left(-\mathrm{i}\frac{Et}{\hbar}\right). \tag{XI.1}$$

† The calculations for other states are analogous.

The function ψ satisfies the equation

$$\frac{d^2\psi}{dr^2} + [k^2 - k^2(r)]\,\psi = 0, \tag{XI.2}$$

where

$$k^2 = \frac{2mE}{\hbar^2}, \qquad k^2(r) = \frac{2mV(r)}{\hbar^2}; \tag{XI.3}$$

E is the total energy, k is the wave vector, m is the reduced mass of the interacting particles: $1/m = 1/m_1 + 1/m_2$, and V is the potential energy of interaction in the s-state.

We define a special form of $V(r)$ in the following manner (R is the radius of the sphere of influence):

$$k^2(r) = 0 \qquad \text{for} \quad R < r, \tag{XI.4}$$

$$k^2(r) = k_0^2 = \frac{2mV_0}{\hbar^2} \qquad \text{for} \quad a < r < R, \tag{XI.4'}$$

$$k^2(r) = 0 \qquad \text{for} \quad 0 < r < a \tag{XI.4''}$$

(see Figure 50). This form of $V(r)$ allows for the existence of resonance levels in the continuous spectrum.

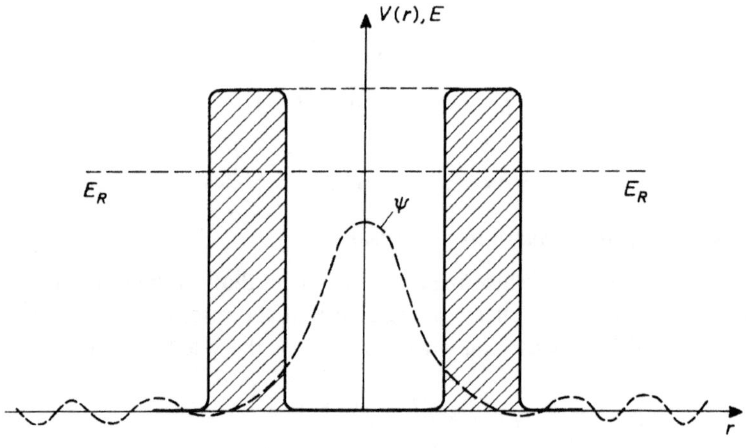

Fig. 50. The form of the potential $V(r)$ that gives a resonance for an energy of $E = E_R$. The dotted line represents the wave function ψ for $E = E_R$.

The solution of (XI.2) may be written in the form

$$\psi_{\mathrm{I}} = a\mathrm{e}^{\mathrm{i}k(r-R)} + b\mathrm{e}^{-\mathrm{i}k(r-R)} \qquad \text{for} \quad R < r, \tag{XI.5}$$

$$\psi_{\mathrm{II}} = \alpha\mathrm{e}^{+\lambda(r-R)} + \beta\mathrm{e}^{-\lambda(r-R)} \qquad \text{for} \quad a < r < R, \tag{XI.5$'$}$$

$$\psi_{\mathrm{III}} = \gamma \sin kr \qquad \text{for} \quad 0 < r < a. \tag{XI.5$''$}$$

Matching the solutions on the boundaries $r=R$ and $r=a$ gives

$$\left.\begin{aligned} a + b &= \alpha + \beta, \\ \frac{\mathrm{i}k}{\lambda}(a - b) &= \alpha - \beta \end{aligned}\right\} \quad \text{for} \quad r = a, \tag{XI.6}$$

$$\left.\begin{aligned} \alpha\mathrm{e}^{-\Lambda} + \beta\mathrm{e}^{\Lambda} &= \gamma \sin ka, \\ \alpha\mathrm{e}^{-\Lambda} - \beta\mathrm{e}^{\Lambda} &= \frac{k}{\lambda}\gamma \cos ka \end{aligned}\right\} \quad \text{for} \quad r = R, \tag{XI.6$'$}$$

where $\Lambda=\lambda(R-a)>0$. For $\Lambda\to\infty$ $(R\to\infty)$, we obtain from (XI.6) and (XI.6′) the condition

$$D_+(k) = 1 + \frac{k}{\lambda}\operatorname{ctg} ka = 0, \tag{XI.7}$$

which defines the discrete, steady-state levels for the case of an infinitely wide barrier.

For finite values, quasi-stationary resonance levels arise [28]. In order to compute these levels, we write the system of equations (XI.6) and (XI.6′) in the form

$$2\alpha = Aa + A^*b, \qquad Ma + M^*b = \tfrac{1}{2}\gamma \sin ka, \tag{XI.8}$$

$$2\beta = A^*a + Ab, \qquad Na + N^*b = \tfrac{1}{2}\gamma \frac{k}{\lambda}\cos ka, \tag{XI.8$'$}$$

where

$$A = 1 + \frac{\mathrm{i}k}{\lambda}, \quad M = A\mathrm{e}^{-\Lambda} + A^*\mathrm{e}^{\Lambda}, \quad N = A\mathrm{e}^{-\Lambda} - A^*\mathrm{e}^{\Lambda}. \tag{XI.9}$$

From (XI.8) and (XI.8′) we have

$$\frac{a}{b} = \frac{A}{A^*}\,\frac{\dfrac{A^*}{A}\mathrm{e}^{-2\Lambda}D_- - D_+}{\dfrac{A^*}{A}\mathrm{e}^{-2\Lambda}D_- + D_+}, \tag{XI.10}$$

where

$$D_-(k) = 1 - \frac{k}{\lambda}\operatorname{ctg} ka. \tag{XI.10'}$$

Let k_R be one of the roots of Equation (XI.7). Then

$$D_+(k) = \left[\frac{\mathrm{d}D(k)}{\mathrm{d}k}\right]_{k=k_R}(k - k_R) + \cdots, \tag{XI.11}$$

$$\left[\frac{\mathrm{d}D(k)}{\mathrm{d}k}\right]_{k=k_R} = -\frac{1}{k_R}\Delta(k_R), \tag{XI.11'}$$

where

$$\Delta(k_R) = 1 + \frac{k_R^2}{\lambda_R^2} + \frac{k_R a}{\lambda_R}\left(1 + \frac{\lambda^2}{k_R^2}\right) > 0. \tag{XI.12}$$

Therefore, close to the resonance Equation (XI.10) takes on the form

$$\frac{a}{b} = -\frac{A}{A^*}\frac{[k - (k_R + \delta) - \mathrm{i}\Gamma]}{[k - (k_R + \delta) + \mathrm{i}\Gamma]}, \tag{XI.13}$$

where

$$\Gamma = \operatorname{Im}\left[\frac{A}{A^*}\mathrm{e}^{-2\Lambda}k_R\frac{D_-(k_R)}{\Delta}\right], \tag{XI.14}$$

$$\delta = \operatorname{Re}\left[\frac{A}{A^*}\mathrm{e}^{-2\Lambda}ka\frac{D_-(k_R)}{\Delta}\right]. \tag{XI.15}$$

Note that close to the resonance, the amplitude of the wave function in the inner region γ tends to α.

In fact, close to the resonance

$$\frac{\gamma}{a} = \frac{4(A^{*2} - A^2)}{\sin k_R a\,[M^* + N^*]} = \frac{4(A^{*2} - A^2)}{\sin k_R a}2A^*\mathrm{e}^{\Lambda}, \tag{XI.16}$$

so that for $\Lambda \to \infty$ and for $E \approx E_R$, $\gamma/a \to \infty$. Particles accumulate inside the potential barrier, in the region $0 < \mathrm{r} < a$.

We now rewrite the solution $\psi_1(kr)$ for $r > R$ in the form of the superposition of the initial s-waves $\psi_\mathrm{I}^0 = (\sin kr)/kr$ and the scattered waves $u = f\mathrm{e}^{\mathrm{i}kr}/r$:

$$\psi_\mathrm{I} = \psi_\mathrm{I}^0 + u. \tag{XI.17}$$

Simple computations using (XI.5), (XI.9) and (XI.13) give

$$f = \frac{e^{-2ikR}}{2ik} \frac{k - i\lambda_R}{k + i\lambda_R} \frac{k - \bar{k}_R - i\Gamma_R}{k - \bar{k}_R + i\Gamma_R}, \tag{XI.18}$$

where

$$\bar{k}_R = k_R + \delta_R .$$

The scattering matrix S which transforms the incident wave e^{-ikr} into the scattered wave e^{ikr} has, from (XI.17) and (XI.18), the form

$$S(k) = e^{-2ikR} \frac{k - i\lambda_R}{k + i\lambda_R} \frac{k - \bar{k}_R - i\Gamma_R}{k - \bar{k}_R + i\Gamma_R}. \tag{XI.19}$$

The first factor contains the factors that violate causality. This violation is due to the fact that the radius of the sphere of influence of the potential $V(r)$ is finite. The second factor describes the resonance about the point $k = \bar{k}_R$.

(a) *The analytic properties of the solution in the complex plane E.* By considering $k^2 = 2mE/\hbar^2$ and assuming, for simplicity, that $2m/\hbar^2 = 1$, we may write the solution ψ_1 [see (XI.5)] in the form

$$\psi_1(kr) = a \exp[-\sqrt{-E}(r - R)] + b \exp[\sqrt{-E}(r - R)] \tag{XI.20}$$

We set

$$E = \varepsilon e^{i\varphi}, \qquad \varepsilon = |E|, \qquad \varepsilon^{1/2} = |E^{1/2}|. \tag{XI.21}$$

On the first, "physical," sheet of the Riemann surface, we define

$$\sqrt{-E} = \varepsilon^{1/2} \exp[\tfrac{1}{2} i(\varphi - \pi)],$$

$$\mathrm{Re}\sqrt{-E} = \varepsilon^{1/2} \sin\tfrac{1}{2}\varphi \geqslant 0. \tag{XI.22}$$

On the second, "nonphysical" sheet of this surface, we define E by the equations

$$\sqrt{-E} = \varepsilon^{1/2 i(\varphi + \pi)},$$

$$\mathrm{Re}\sqrt{-E} = -\varepsilon^{1/2} \sin\tfrac{1}{2}\varphi \leqslant 0 \tag{XI.22'}$$

and we make a branch cut from $E=0$ to $E+\infty$ which separates the two sheets (see Figure 51). In accordance with (XI.5), (XI.20)-(XI.22), the wave vector k is defined thus:

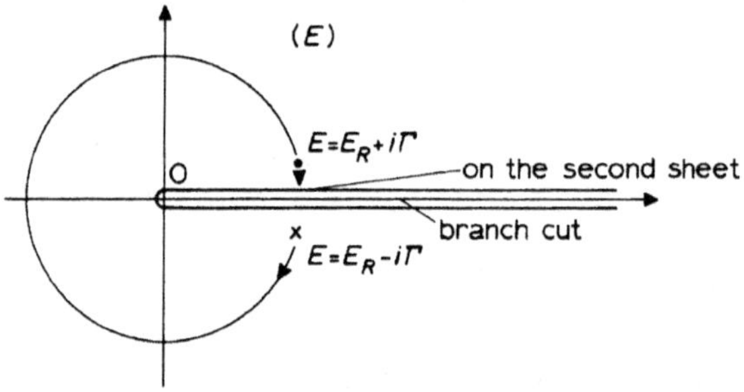

Fig. 51. The complex plane (E). This figure shows the path from a point on the first sheet (the cross) to a point on the second sheet (the black dot).

on the physical sheet

$$ik = -\sqrt{-E} = +\,i\varepsilon^{1/2}\exp\left(i\frac{\varphi}{2}\right), \tag{XI.23}$$

$$k = \varepsilon^{1/2}\exp\left(i\frac{\varphi}{2}\right), \qquad \operatorname{Im} k > 0; \tag{XI.23'}$$

and on the nonphysical sheet

$$ik = \mp\sqrt{-E} = -\,i\varepsilon^{1/2}\exp\left(i\frac{\varphi}{2}\right), \tag{XI.24}$$

$$k = -\,\varepsilon^{1/2}\exp\left(i\frac{\varphi}{2}\right), \qquad \operatorname{Im} k < 0, \tag{XI.24'}$$

Thus the physical sheet of the surface (E) is mapped onto the upper half-plane of the variable k, and the non-physical sheet is mapped onto the lower half-plane. Note that in the stationary state, the component of the current that is directed along the radius vector must be equal to zero. Therefore,

$$b(E) = a(E^*). \tag{XI.25}$$

Furthermore, from the symmetry of the upper and lower half-planes of

(E), we have

$$a(E^*) = a^*(E). \tag{XI.26}$$

We first consider the stable levels which, in the preceeding example can occur for $\Lambda = 0$ $(R \to \infty)$. On the first sheet we have

$$\psi(r, t) = [a(E) \exp(-\varepsilon^{1/2} r) + b(E) \exp(\varepsilon^{1/2} r)] \exp(-\mathrm{i}Et). \tag{XI.27}$$

The second solution increases without limit for $r \to \infty$ and, therefore, we must impose the requirement [the characteristic solution of (XI.2)] that

$$b(E) = 0. \tag{XI.28}$$

This equation defines the stable levels which are defined by zeros in the functions on the surface (E) which lie on the real poles $E < 0$.

Note that on the second sheet, for $E < 0$, we would have

$$\psi(r, t) = [a(E) \exp(\varepsilon^{1/2} r) + b(E) \exp(-\varepsilon^{1/2} r)] \exp(-\mathrm{i}Et), \tag{XI.29}$$

so that the levels would be defined by the condition

$$a(E) = 0. \tag{XI.28'}$$

Because of (XI.25), Equations (XI.28′) and (XI.28) give the same result.

(b) *Unstable levels.* We consider the levels that are possible for complex values of E when Re $E > 0$. We again require that

$$b(E) = 0. \tag{XI.30}$$

Let

$$E = E_R - \mathrm{i}\Gamma, \tag{XI.31}$$

so that

$$\exp(-\mathrm{i}Et) = \exp(-\mathrm{i}E_R t - \Gamma t). \tag{XI.32}$$

On the physical sheet we have

$$\begin{aligned} \mathrm{i}k &= -\sqrt{-E} = -\varepsilon^{1/2} \exp[\tfrac{1}{2}\mathrm{i}(\gamma + 2\pi - \pi)] \\ &= -\varepsilon^{1/2} \exp[\tfrac{1}{2}\mathrm{i}(\gamma + \pi)] = -\mathrm{i}\varepsilon^{1/2} \exp\left(\mathrm{i}\frac{\gamma}{2}\right), \end{aligned} \tag{XI.33}$$

where

$$\gamma = -\operatorname{arctg}\frac{\Gamma}{E_R} + 2\pi n, \qquad n = 0, 1\,, \tag{XI.34}$$

from which

$$k = -k' - i\varkappa'\,, \tag{XI.35}$$

$$k' = \varepsilon^{1/2}\cos\frac{\gamma}{2}, \qquad \varkappa' = -\varepsilon^{1/2}\sin\frac{\gamma}{2}, \tag{XI.35$'$}$$

so that $k'>0$, $\varkappa'<0$.

Thus on the physical sheet, the function would have the form

$$\begin{aligned}\psi(r, t) = a(E)\exp\left[-ik'(r-R) + \varkappa'(r-R)\right] \\ \times\exp(-iE_R t - \Gamma t)\,.\end{aligned} \tag{XI.36}$$

This solution contradicts the continuity equation because $\psi(r, t)$ decreases as $r(\varkappa'<0)$ and t increase.

We now turn to the solution on the second, nonphysical sheet. In this case

$$ik = -(-E)^{1/2} = -i\varepsilon^{1/2}e^{i\varphi/2}, \qquad k = -\varepsilon^{1/2}e^{i\varphi/2}$$

and $\varphi=\gamma+2\pi$, where $\gamma<0$, so that $k=k''+i\varkappa''$

$$k'' = \varepsilon^{1/2}\cos\frac{\gamma}{2}, \qquad \varkappa'' = \varepsilon^{1/2}\sin\frac{\gamma}{2}, \tag{XI.37}$$

$$k'' > 0\,, \qquad \varkappa'' < 0\,, \tag{XI.37$'$}$$

For $r>R$, the wave function has the form

$$\psi(r, t) = a(E)\exp(ik''r - \varkappa''r)\exp(-iE_R t - \Gamma t)\,. \tag{XI.38}$$

This solution increases as $r(\varkappa''<0)$ increases, and decreases as t increases. By assuming $-\varkappa''=+|\varkappa''|$, we obtain

$$\psi(r, t) = a(E)\exp\left[i(k''_R r - E_R t)\right]\exp(\varkappa'' r - \Gamma t)\,. \tag{XI.38$'$}$$

This solution is compatible with the continuity equation and is, therefore, considered correct [138].

This crossing over from the first to the second sheet of the plane E is easily illustrated using Figure 51. The zero point $E=E_R-i\Gamma$, indicated by a cross, lies below the branch cut. If we remain on the first sheet, we can only get from the point $E=E_R+i\Gamma$ to the point $E=E_R-i\Gamma$ by going around the branch cut from the left. The phase of the wave vector changes by $e^{i\pi}$ when we do this, and the converging and diverging waves change places. In order to satisfy condition (XI.30) and the earlier definitions of the converging and diverging waves, we must cross over onto the second sheet directly across the branch cut OE.

(c) *Space-time properties of the resonance levels.* We define the initial state ψ_I^0 in the form of a localized wave packet

$$\psi_I^0 = \int c(k)\frac{\sin kr}{kr}\exp[-iE(k)t]\,dk, \qquad \text{(XI.39)}$$

which is concentrated around the value $k=k_R$.† For definiteness we set

$$\left.\begin{array}{ll} c(k)=c & \text{for}\quad k_R-\Delta<k<k_R+\Delta, \\ c(k)=0 & \text{outside of this intervals.} \end{array}\right\} \qquad \text{(XI.40)}$$

The wave packet (XI.39) now has the form

$$\psi_I^0(r,t) = \frac{c}{2ik_Rr}\exp[i(k_Rr-E_Rt)]\,f(r-vt)$$
$$-\frac{c}{2ik_Rr}\exp[-i(k_Rr+E_Rt)]\,f(r+vt), \qquad \text{(XI.41)}$$

where the function $f(r+vt)$ differs significantly from zero near

$$r\pm vt = \frac{\hbar}{\Delta}. \qquad \text{(XI.42)}$$

We assume $\Delta>R$. Then the collision occurs at a time close to

$$t\approx\frac{\hbar}{\Delta v}. \qquad \text{(XI.43)}$$

We now turn to the scattered wave $u_s(kr,t)$. According to (XI.17), (XI.18) and (XI.39), it has the form

† We assume that k_R is not too close to the origin.

$$u_s(kr, t) = \frac{c}{2ik_R r} \int_{k_R - \Delta}^{k_R + \Delta} c(k) \exp[i(kr - Et)]\, g(k) \times \frac{k - \bar{k}_R - i\Gamma}{k - \bar{k}_R + i\Gamma}\, dk. \tag{XI.44}$$

By replacing the continuous functions, $g(x)$, by their values at the point $k = k_R$, we obtain

$$u_s(kr, t) = \frac{c \exp[i(k_R r - E_R t) - \Gamma t]}{2ik_R \varepsilon}\, g(k_R) \times \int_{-\Delta}^{+\Delta} \exp[i(r - vt)\,\xi]\, \frac{\xi - i\Gamma}{\xi + i\Gamma}\, d\xi. \tag{XI.44'}$$

For $\Delta \to \infty$, we close the integration contour in the lower half-plane for $r - vt \leqslant 0$. We obtain

$$\begin{aligned} u_s(kr, t) &= \frac{2\pi i c g(k_R)}{2ik_R r} \exp[i(k_R r - E_R t)](-2i\Gamma) \times \exp[-i(r - vt)\, i\Gamma] \\ &= \frac{4\pi \Gamma c g(k_R)}{2ik_R} \exp[i(k_R r - E_R t)] \exp[\Gamma(r - vt)]. \end{aligned} \tag{XI.45}$$

If $r - vt > 0$, then we can close the integration contour in the upper half-plane and obtain

$$u_s(kr, t) = 0, \qquad \text{for} \quad r - vt > 0, \tag{XI.45'}$$

which agrees with the results of Section 34.

APPENDIX XII (TO SECTION 35)

The unitarity relation for the scattering matrix $\hat{S} = \hat{I} + i\hat{T}$ may be written as

$$\hat{S}\hat{S}^{+} = \hat{I} + i\hat{T} - i\hat{T}^{+} + \hat{T}\hat{T}^{+} = 1 \tag{XII.1}$$

or

$$\operatorname{Im} \hat{T} = \tfrac{1}{2} \hat{T}\hat{T}^{+}. \tag{XII.2}$$

This operator equation may be written in matrix form as

$$\operatorname{Im} T_{\alpha\beta} = \tfrac{1}{2} \sum_{\gamma} T_{\alpha\gamma} T_{\alpha\beta}^{+} .$$

In particular, noting that $T_{\alpha\gamma} = T^*_{\gamma\alpha}$, for the diagonal term $\alpha = \beta$ we have

$$\operatorname{Im} T_{\alpha\alpha} = \tfrac{1}{2} \sum_{\gamma} |T_{\alpha\gamma}|^2 . \tag{XII.3}$$

We apply this equation to the scattered particle. Then $T_{\alpha\alpha}$ denotes the "forward-scattering" amplitude (i.e., the scattering in which the particle does not change state), and the sum of terms $|T_{\alpha\gamma}|^2$ denotes the total probability of a transition from an initial state α to all other states γ. Therefore $\sum |T_{\alpha\gamma}|^2$ is proportional to the total scattering cross section σ_π. A more detailed calculation gives the result

$$\operatorname{Im} T(0) = \frac{k}{4\pi} \sigma_\pi , \tag{XII.4}$$

where $T(0) = T_{\alpha\alpha}$, and k is the particle momentum in the center-of-mass system. Equation (XII.4) is called the optical theorem. It is a logical and necessary consequence of $\hat{S}$-matrix theory in that it expresses the conservation of probability: the departure of particles from the initial state α is compensated for by their appearance in other possible states γ. The unitarity condition (XII.2) is not as necessary as (XII.3) which leads to the optical theorem. However if we give up the unitarity condition, we loose the mathematical basis for (XII.3). Thus giving up or violating the unitariness of the $\hat{S}$-matrix leads to a difficult situation.

APPENDIX XIII (TO SECTION 37)

In order to compute the elastically scattered wave Ψ_{out}, we start from Equation (35.10) where the function $\mathfrak{G}$ is defined in terms of the function $\mathfrak{G}_0$, (37.13). We begin by considering the first case; the second is simpler to compute. We write the four-dimensional δ-function in the form

$$\delta^4 (p_4 + p_3 - p_2 - p_1) = \int \mathrm{d}^4\alpha \exp \mathrm{i}\alpha (p_4 + p_3 - p_2 - p_1) \tag{XIII.1}$$

and the function $\tilde{D}_c(q)$ in the form

$$\tilde{D}_c(q) = \int d^4\beta D_c(\beta)\, e^{i\beta q}, \tag{XIII.2}$$

where $q = p_3 - p_1$ or $p_2 + p_1$. Inserting these expressions in (37.13), differentiating (37.13) with respect to $\partial/\partial x_2^0$ and $\partial/\partial x_1^0$, and using (35.10) we obtain

$$\begin{aligned}\Psi_{\text{out}}(x_4, x_3) = u_2(x_4)\, u_1(x_3) + i \int & d^4\alpha\, d^4\beta \\ & \times \exp i\,[\alpha(p_4 + p_3 - p_2 - p_1) + p_4 x_4 \\ & + p_3 x_3 - p_2 x_2 - p_1 x_1] \\ & \times \{D_c(\beta) \exp[i\beta(p_3 - p_1)] \\ & + D_c(\beta) \exp[i\beta(p_2 + p_1)]\} \\ & \times \frac{d^3 p_4}{2p_4^0} \frac{d^3 p_3}{2p_3^0} d^3 p_1\, d^3 p_2 u_2(x_2)\, u_1(x_1)\, d^3 x_2\, d^3 x_1,\end{aligned} \tag{XIII.3}$$

where we assume that the original wave $\Psi_{\text{in}}(x_2, x_1)$ represents two wave packets $u_2(x_2)$ and $u_1(x_1)$. We first compute the integral over x_1 and obtain

$$\int u_1(x_1) \exp(-ip_1 x_1)\, d^3 x_1 = \frac{\tilde{u}_1(p_1)}{2p_1^0}. \tag{XIII.4}$$

Using (35.4), we now integrate over p_1 and obtain

$$\int \frac{\tilde{u}_1(p_1)}{2p_1^0} \exp[-ip_1(\alpha + \beta)]\, d^3 p_1 = u_1(-\alpha - \beta) \tag{XIII.5}$$

for the first term in the curly brackets, and

$$\int \frac{\tilde{u}_1(p_1)}{2p_0} \exp[-ip_1(\alpha - \beta)]\, d^3 p_1 = u_1(\beta - \alpha) \tag{XIII.6}$$

for the second term. We similarly integrate over x_2 and p_2, and finally, by integrating over p_3, we obtain

$$\int \exp[ip_3(\alpha + \beta + x_3)] \frac{d^3 p_3}{2p_3^0} = D_+(\alpha + \beta + x_3) \tag{XIII.7}$$

and

$$\int \exp\left[\mathrm{i}p_3(\alpha + x_3)\right]\frac{\mathrm{d}^3 p_3}{2p_3^0} = D_+(\alpha + x_3). \tag{XIII.7'}$$

The integration over p_4 is carried out similarly. By assuming that, after integrating the first term in the curly brackets, $\alpha = -x$, $\alpha + \beta = -y$, and after integrating the second term, $\alpha - \beta = -x$, $(\alpha + \beta) = -y$, we obtain the final form

$$\begin{aligned}\Psi_{\text{out}}(x_4, x_3) &= u_2(x_4)\, u_1(x_3) \\ &+ \mathrm{i}\int \mathrm{d}^4 x\, \mathrm{d}^4 y D_+(x_4 - y)\, D_+(x_3 - x)\, D_c(x - y)\, u_1(x)\, u_2(y) \\ &+ \mathrm{i}\int \mathrm{d}^4 x\, \mathrm{d}^4 y D_+\left(x_4 + \frac{x+y}{2}\right) D_+\left(x_3 - \frac{x+y}{2}\right) \\ &\times D_c\left(\frac{x-y}{2}\right) u_1(x)\, u_2(x).\end{aligned} \tag{XIII.8}$$

APPENDIX XIV (TO SECTION 37)

The computations of the function $\Delta^+(t-r, A, M)$ for $A \to \infty$ will be reduced to integrals found in tables. When we set $2m^2 = M^2$, $t - r = \varepsilon$, $x = m\varepsilon$, the integral

$$\Delta^+(\varepsilon, \infty, M) = M^2 \int_0^\infty \frac{e^{\mathrm{i}v\varepsilon}}{2v^2 + M^2}\,\mathrm{d}v \tag{XIV.1}$$

may be rewritten in the form

$$\Delta^+(\varepsilon, \infty, M) = \delta^+(\varepsilon) + m\frac{\mathrm{d}^2}{\mathrm{d}x^2} I(x), \tag{XIV.2}$$

where

$$\begin{aligned} I(x) &= \int_0^\infty \frac{e^{\mathrm{i}zx}}{(z^2+1)}\,\mathrm{d}z \\ &= \frac{1}{2\mathrm{i}}\left[\int_0^\infty \frac{z+\mathrm{i}}{z^2+1} e^{\mathrm{i}zx}\,\mathrm{d}z - \int_0^\infty \frac{z-\mathrm{i}}{z^2+1} e^{\mathrm{i}zx}\,\mathrm{d}z\right] \\ &= \frac{1}{2\mathrm{i}}\left[-e^{-x}E^+\mathrm{i}(x) + e^x Ei(-x)\right].\end{aligned} \tag{XIV.3}$$

where

$$Ei(-x) = -\int_x^\infty \frac{e^{-t}}{t}\, dt, \tag{XIV.4}$$

$$E^+i(x) = \bar{E}i - i\pi \tag{XIV.4'}$$

These formulas may be found in tables.†

The asymptotic expansion of these functions has the form

$$Ei(-x) = -\frac{e^{-x}}{x}\left[1 - \frac{1}{x} + \frac{2!}{x^2} - \cdots\right], \tag{XIV.5}$$

$$E^+i(x) = -i\pi + \frac{e^x}{x}\left[1 + \frac{1}{x} + \frac{2!}{x^2} + \cdots\right]. \tag{XIV.5'}$$

Therefore, the integral $I(x)$, for $x \to \infty$, is equal to

$$I(x) = \frac{\pi}{2} e^{-x} - \frac{2}{x}\left[1 - \frac{2!}{x} + \frac{1!}{x^4} - \cdots\right], \tag{XIV.6}$$

and, therefore,

$$\frac{d^2 I(x)}{dx^2} = \frac{\pi}{2} e^{-x} + O\left(\frac{1}{x^3}\right). \tag{XIV.7}$$

Using (XIV.2) we obtain

$$\Delta^+(\varepsilon, \infty, M) = \delta^+(\varepsilon) + O\left(\frac{1}{m^2\varepsilon^3}\right) + \frac{\pi m}{2} e^{-m\varepsilon}. \tag{XIV.8}$$

Remembering that $\delta^+(\varepsilon)$ is defined as

$$\delta^+(\varepsilon) = \delta(\varepsilon) + \frac{1}{\pi\varepsilon} \tag{XIV.9}$$

[see (I.9)], when $m = M/(2)^{1/2} \to \infty$, the last two terms in (X.9) become small corrections. Recalling that $m = M/(2)^{1/2} = 1/a(2)^{1/2}$ and assuming that $\varepsilon = t - r \gg a$ we are led to the result

$$\Delta^+\left(t - r, \infty, \frac{1}{a}\right) = \delta^+(t - r) + O\left[\frac{2a^2}{(t-r)^3}\right] + \frac{\pi}{a2\sqrt{2}} \exp\left(-\frac{t-r}{a\sqrt{2}}\right). \tag{XIV.10}$$

† E. Jahnke and F. Emde, *Table of Functions*, Section V, Dover, New York, 1945.

For $t-r \ll 2-a$, the same equation is obtained except that $(t-r)$ is replaced by $-(t-r)$. Then from (XIV.1) it follows that for $t-r \to 0$, the function $\Delta^+(t-r, \infty, 1/a)$ is regular and tends to the constant value

$$\Delta^+\left(0, \infty, \frac{1}{a}\right) = \frac{\pi}{4} \frac{M}{\sqrt{2}}. \tag{XIV.11}$$

APPENDIX XV (TO SECTION 43)

Equation (43.7) may be derived in the following manner.† Let the $\hat{S}$-matrix, in the coordinate representation, be given by the usual exponentials, ordered in time

$$\hat{S} = T \exp\left(\mathrm{i} \int \hat{L}(x)\, \mathrm{d}x\right), \tag{XV.1}$$

where $\hat{L}(x)$ is the Lagrangian and T denotes the ordering in time (cf. Appendix VII).

By assuming

$$\hat{S} = \hat{I} + \mathrm{i}\hat{T}, \tag{XV.2}$$

we obtain from (XV.1)

$$\hat{T} = \sum_{s=1}^{\infty} \frac{\mathrm{i}^{s-1}}{s!} \int T(\hat{L}(x_1) \cdots \hat{L}(x_s))\, \mathrm{d}x_1 \cdots \mathrm{d}x_s$$
$$= \sum_{s=1}^{\infty} \mathrm{i}^{n-1} \int \theta(t_1 - t_2) \cdots \theta(t_{s-1} - t_s)\, \hat{L}(x_1) \cdots \hat{L}(x_s)$$
$$\times\, \mathrm{d}x_1 \cdots \mathrm{d}x_s = \sum_{s=1}^{\infty} \hat{T}_s. \tag{XV.3}$$

We write (XV.3) in its full four-dimensional form by replacing $\theta(t)$ by the invariant functions $\theta(nx)$ where

$$nx = n_4 t - \mathbf{nx}, \quad \text{with} \quad n^2 = 1, \quad n_4 > 0. \tag{XV.4}$$

† In this appendix we follow references [100, 101]

The result is

$$\sum_{s=1}^{\infty} \hat{T}_s = \sum \mathrm{i}^{s-1} \int \theta(n, x_1 - x_2) \cdots \theta(n, x_{s-1} - x_s) \times \hat{L}(x_1) \cdots \hat{L}(x_s)\, \mathrm{d}x_1 \cdots \mathrm{d}x_s . \tag{XV.5}$$

As is well known, the dependence of $\hat{T}_s$ on n in (XV.5) is purely fictitious because, for $(x_i - x_{i+1})^2 > 0$, $\theta(t_i - t_{i+1})$ is always equal to $\theta(n, x_i - x_{i+1})$ and for $(x_i - x_{i+1})^2 < 0$ the function $\theta(n, x_i - x_{i+1})$ does not contribute because the "locality" condition

$$[\hat{L}(x_i), \hat{L}(x_{i-1})]_- = 0 , \tag{XV.6}$$

is satisfied in this region.

For definiteness we set

$$\hat{L}(x) = g : \hat{\varphi}^3 : , \tag{XV.7}$$

where $\hat{\phi}$ is a scalar field, and apply Fourier transform (XV.5) assuming

$$\theta(n, x) = \frac{1}{2\pi \mathrm{i}} \int_{-\infty}^{\infty} \frac{\mathrm{e}^{\mathrm{i}\tau(nx)}}{\tau - \mathrm{i}\varepsilon} \mathrm{d}\tau , \tag{XV.8}$$

$$\begin{aligned} \hat{\tilde{L}}(p) &= \int \mathrm{e}^{-\mathrm{i}px} \hat{L}(x)\, \mathrm{d}x \\ &= \frac{g}{\sqrt{2\pi}} \int \delta(p - k_1 - k_2 - k_3) : \hat{\varphi}(k_1)\, \hat{\varphi}(k_2) \\ &\quad \times \hat{\varphi}(k_3) : \mathrm{d}k_1\, \mathrm{d}k_2\, \mathrm{d}k_3 , \end{aligned} \tag{XV.9}$$

where

$$\hat{\varphi}(k) = \frac{1}{(2\pi)^{5/2}} \int \mathrm{e}^{-\mathrm{i}kx} \hat{\varphi}(x)\, \mathrm{d}x . \tag{XV.10}$$

The result is

$$\hat{T}_1 = \hat{\tilde{L}}(0) , \tag{XV.11}$$

$$\hat{T}_2 = \frac{1}{2\pi} \int_{-\infty}^{\infty} \hat{\tilde{L}}(-n\tau) \frac{\mathrm{d}\tau}{\tau - \mathrm{i}\varepsilon} \hat{\tilde{L}}(n\tau), \tag{XV.11'}$$

$$\hat{T}_s = \frac{2}{(2\pi)^{s-1}} \int_{-\infty}^{\infty} \hat{\hat{L}}(-n\tau) \frac{d\tau_1}{\tau_1 - i\varepsilon} \hat{\hat{L}}(n\tau_1 - \lambda\tau_2) \frac{d\tau_2}{\tau_2 - i\varepsilon} \cdots$$

$$\times \frac{d\tau_{s-1}}{\tau_{s-1} - i\varepsilon} \hat{\hat{L}}(n\tau_{s-1}). \qquad \text{(XV.11'')}$$

Equation (XV.11) can be considered as a successive iteration on some linear integral equation

$$\hat{R}(n\tau) = \hat{\hat{L}}(n\tau) = \frac{1}{2\pi} \int_{-\infty}^{\infty} \hat{\hat{L}}(n\delta - n\tau') \frac{d\tau'}{\tau' - i\varepsilon} \hat{R}(n\tau') \qquad \text{(XV.12)}$$

with the condition that

$$\hat{T} = \sum_{s=1}^{\infty} \hat{T}_s = \hat{R}(0). \qquad \text{(XV.13)}$$

Equation (XV.12) is the invariant equation of motion for the scattering matrix in the p-representation. Using (XV.12), we can for example show that the $\hat{S}$-matrix is unitary. In order to do this, we rewrite (XV.12) in the form

$$\hat{R}(n\tau) = \hat{\hat{L}}(n\tau) + \int \hat{\hat{L}}(n\tau') \hat{F}(\tau - \tau') \, d\tau', \qquad \text{(XV.14)}$$

where

$$\hat{F}(\tau - \tau') = \frac{1}{2\pi} \frac{\hat{R}(n\tau - n\tau')}{\tau - \tau' - i\varepsilon}. \qquad \text{(XV.15)}$$

Remembering that the Hermite conjugate is

$$\hat{\hat{L}}^+(n\tau) = \hat{\hat{L}}(-n\tau), \qquad \text{(XV.16)}$$

the conjugate of Equation (XV.14) may be written as

$$\hat{R}^+(-n\tau) = \hat{\hat{L}}(n\tau) + \int \hat{F}^+(\tau' - \tau) \hat{\hat{L}}(n\tau') \, d\tau', \qquad \text{(XV.17)}$$

where

$$\hat{F}^+(\tau' - \tau) = \frac{1}{2\pi} \frac{\hat{R}^+(\tau' - \tau)}{\tau' - \tau + i\varepsilon}. \qquad \text{(XV.18)}$$

From (XV.14) and (XV.17), it follows that

$$\hat{R}(n\tau) - \hat{R}^{+}(-n\tau) = \int \hat{\hat{L}}(n\tau')\,\hat{F}(\tau - \tau')\,\mathrm{d}\tau' - \int \hat{F}^{+}(\tau' - \tau)\,\hat{\hat{L}}(n\tau')\,\mathrm{d}\tau'. \tag{XV.19}$$

On the other hand, by using these equations after simple transformation, we can show that

$$\int \hat{R}^{+}(-n\tau')\,\hat{F}(\tau - \tau')\,\mathrm{d}\tau' - \int \hat{F}^{+}(\tau' - \tau)\,\hat{R}(n\tau')\,\mathrm{d}\tau' = \int \hat{\hat{L}}(n\tau')\,\hat{F}(\tau - \tau')\,\mathrm{d}\tau - \int \hat{F}^{+}(\tau' - \tau)\,\hat{\hat{L}}(n\tau)\,\mathrm{d}\tau' + \hat{\Phi}_1(\tau) - \hat{\Phi}_2(\tau), \tag{XV.20}$$

where

$$\hat{\Phi}_1(\tau) = \int \hat{F}^{+}(\tau' - \tau'')\,\hat{\hat{L}}(n\tau')\,\hat{F}(\tau - \tau'')\,\mathrm{d}\tau'\,\mathrm{d}\tau'', \tag{XV.21}$$

$$\hat{\Phi}_2(\tau) = \int \hat{F}^{+}(\tau'' - \tau)\,\hat{\hat{L}}(n\tau')\,\hat{F}(\tau'' - \tau')\,\mathrm{d}\tau'\,\mathrm{d}\tau''. \tag{XV.21'}$$

We now successively replace the integration variables in (XV.21) by

$$\tau' - \tau'' \to \xi, \quad \tau' - \tau'' \to \eta \quad \text{in} \quad \hat{\Phi}_1(\tau), \tag{XV.22}$$

$$\tau'' - \tau \to \eta, \quad \tau'' - \tau \to \xi \quad \text{in} \quad \hat{\Phi}_2(\tau). \tag{XV.22'}$$

and as a result obtain

$$\hat{\Phi}_1(\tau) = \hat{\Phi}_2(\tau) = \int \hat{F}^{+}(\xi)\,\hat{\hat{L}}(n, \tau + \xi - \eta)\,\hat{F}(\eta)\,\mathrm{d}\xi\,\mathrm{d}\eta. \tag{XV.23}$$

From (XV.19) and (XV.20), using (XV.23), it follows that

$$\hat{R}(n\tau) - \hat{R}^{+}(-n\tau) = \int [\hat{R}^{+}(-n\tau)\,\hat{F}(\tau - \tau') - \hat{F}(\tau' - \tau)\,\hat{R}(n\tau')]\,\mathrm{d}\tau'$$

or

$$\hat{R}(n\tau) - \hat{R}^{+}(-n\tau)$$
$$= \frac{1}{2\pi} \int \frac{\mathrm{d}\tau'}{\tau' - \mathrm{i}\varepsilon} [\hat{R}^{+}(-n\tau')\hat{R}(n\tau - n\tau')$$
$$+ \hat{R}^{+}(n\tau' - n\tau)\hat{R}(n\tau')]. \tag{XV.24}$$

Assuming that $\tau = 0$ in (XV.24) and using the equation

$$\frac{1}{\tau - \mathrm{i}\varepsilon} = \mathrm{P}\frac{1}{\tau} + \mathrm{i}\pi\delta(\tau), \tag{XV.25}$$

we obtain

$$\hat{R}(0) - \hat{R}^{+}(0) = \mathrm{i}\hat{R}^{+}(0)\hat{R}(0), \tag{XV.26}$$

i.e., the condition that the matrix $\hat{S} = 1 + \mathrm{i}\hat{R}(0)$ is unitary. Equation (XV.24) may be considered as the condition that the $\hat{S}$-matrix is unitary for $\tau \neq 0$.

APPENDIX XVI (TO SECTION 47)

According to (38.14′) and (38.14″), the additional terms Ψ_1 and Ψ_2 have the form

$$\Psi_1 = \frac{\omega^2 - \omega_0^2}{M^2 + \omega_0^2}\tilde{\rho}(\omega)\,\mathrm{d}_1(\mathrm{i}M), \tag{XVI.1}$$

$$\Psi_2 = \frac{\omega^2 - \omega_0^2}{M^2 + \omega_0^2}\tilde{\rho}(\omega)\,\mathrm{d}_2(\mathrm{i}M). \tag{XVI.1′}$$

In this form the equations cannot be easily compared to observations because the real part, d and the imaginary part, a of the function N are taken at the point $\omega = \mathrm{i}M$.

In order to evaluate the functions Ψ_1 and Ψ_2 we can use the analytical amplitudes in an angle from 0° to 90° and with a large magnitude M. In order to do this, we consider the function

$$f(z) = \frac{T(z)}{z} = \frac{\tilde{\rho}(z)N(z)}{z}, \tag{XVI.2}$$

where

$$\tilde{\rho}(z) = \frac{M^2}{z^2 + M^2}. \tag{XVI.3}$$

On the real axis, for $z=\omega\to\infty$, the function $f(z)$ tends to a constant value which we shall denote by $i\alpha$. This assumption is equivalent to assuming that the total asymptotic cross section $\sigma_\pm$ is constant. In particular, according to the optical theorem [see (XII.4)]

$$\operatorname{Re}\alpha_\pm = \frac{\sigma_\pm}{4\pi}. \tag{XVI.4}$$

From (XVI.2) and (XVI.3), for $|z|=V$, $V\to+\infty$, using Lindelöf's theorem [63] we have

$$f(V) = \frac{\tilde{\rho}(V)\,N(V)}{V} \approx \frac{N(V)}{V^3} = i\alpha \tag{XVI.5}$$

and for $z=+iV$

$$f(iV) = \frac{N(iV)}{(iV)^3} = i\alpha. \tag{XVI.6}$$

Thus

$$N(iV) = (i)^3\,N(V) = -\,iN(V). \tag{XVI.7}$$

and, therefore,

$$d(iV) = \operatorname{Re} N(iV) = +\,a(V), \tag{XVI.8}$$

$$a(iV) = \operatorname{Im} N(iV) = -\,d(V) \tag{XVI.8'}$$

for large V. If the characteristic frequency M is large enough so that the asymptotic value of $f(z)$ is reached for $V=M$, then we may set

$$d(iM) = a(M), \tag{XVI.9}$$

$$a(iM) = d(M), \tag{XVI.9'}$$

or, remembering that $\tilde{\rho}(M)=\frac{1}{2}$, we obtain

$$\tfrac{1}{2}\,d(iM) = A(M), \tag{XVI.10}$$

$$\tfrac{1}{2}\,a(iM) = -\,D(M). \tag{XVI.10'}$$

Therefore, we can rewrite (38.13) and (38.13') in the form

$$\Psi_1(\omega) = \frac{\omega^2 - m^2}{M^2 + m^2}\,\tilde{\rho}(\omega)\,[A_+^0(M) + A_-^0(M)], \tag{XVI.11}$$

$$\Psi_2(\omega) = -\frac{\omega^2 - m^2}{M^2 + m^2}\,\tilde{\rho}(\omega)\,\frac{\omega}{M}\,\left[D_+^0(M) + D_-^0(M)\right], \quad \text{(XVI.11')}$$

or, by expressing the amplitudes $A_\pm^0$ in terms of the total cross sections $A_\pm^0(M) = [\mathrm{k}(\mathrm{M})/4\pi]\,\sigma_\pm(M)$, we obtain

$$\Psi_1(\omega) = \frac{\omega^2 - m^2}{4\pi\sqrt{M + m^2}}\,\tilde{\rho}(\omega)\left[\sigma_+(M) + \sigma_-(M)\right]. \quad \text{(XVI.12)}$$

BIBLIOGRAPHY

[1] Einstein, A.: 1921, 'Geometrie und Erfahrung', *Sitzber. Preuss. Akad. Wiss.* **1**, 123.

[2] Einstein, A.: 1923, 'Electrodynamics of Moving Bodies' in: *Principles of Relativity*, Methuen, London.

[3] Minkowski, H.: 1923, 'Space and Time' in: *Principles of Relativity*, Methuen, London.

[4] Klein, F.: 1956, 'Erlaugen Program' in: *On the Foundations of Geometry*, State Technical and Theoretical Press, Moscow.

[5] *Gravitation and Relativity*, W. A. Benjamin Inc., New York, Amsterdam.

[6] Synge, I. H.: 1960, The General Theory of Relativity, North-Holland Publ. Co., Amsterdam, Wiley (Interscience), New York.

[7] Newton, I.: 1947, *Mathematical Principles of Natural Philosophy*, Univ. of California, Berkeley, California, p. 6.

[8] Helmholtz, H.: 1956, 'Concerning the Facts; Laws in the Foundations of Geometry', in: *On the Foundations of Geometry*, State Technical and Theoretical Press, Moscow.

[9] Smirnov, V. I.: 1958, *A Course in Higher Mathematics*, Vol. IV, Fizmatgiz, Moscow, p. 91

[10] Courant, R. and Friedrichs, K.: 1948, *Supersonic Flow and Shock Waves*, Wiley (Interscience), New York.

[11] Blokhintsev, D. I.: 1965, 'Metrics of Space-Time', preprint of Joint Institute for Nuclear Research (R-2152), U.S.S.R.

[12] Petrov, A. Z.: 1961, *Einstein Spaces*, Fizmatgiz, Moscow.

[13] Fok, V. A.: 1959, *The Theory of space, Time and Gravitation*, Pergamon Press, New York.

[14] Landau, L. D. and Lifshitz, E. M.: 1962, *The Classical Theory of Fields*, Addison-Wesley, Reading, Mass.

[15] Riemann, B.: 1956, 'On the Hypotheses, Laws of Basic Geometry', in *On the Foundations of Geometry*, State Technical and Theoretical Press, Moscow.

[16] Weyl, H.: 1956, 'Commentary to Riemann's paper', in: *On the Foundations of Geometry*, State Technical and Theoretical Press, Moscow.

[17] Weyl, H.: 1950, Space, Time and Matter, Dover, New York.

[18] Martske, R. and Euler, D.: 1965, in: Gravitation and Relativity (ed. by Hong-Yee Chin and William F. Hoffmann), W. Benjamin Inc., New York, Amsterdam.

[19] Born, M.: 1934, *Proc. Roy. Soc.* **A 143**, 410.

[20] Born, M. and Infeld, L.: 1934, *Proc. Roy Soc.* **A 144**, 425.

[21] Blokhintsev, D. I.: 1952, *Doklady* **82**, 553.

[22] Smirnov, V. I : 1958, *A Course in Higher Mathematics*, Vol. IV, Fizmatgiz, Moscow, p. 311.

[23] Von Karmann, Th.: 1947, 'Supersonic Aerodynamics', *J. Aeronaut. Sci* **14**, No. 7.

[24] Blokhintsev, D. I.: 1963, *Atomic Energy* **14**, 105

[25] Blokhintsev, D. I. and Orlov, V. I.: 1953, *ZhETF* **25**, 513.

[26] Pauli, W.: 1958, *Theory of Relativity*, Pergamon Press, New York.

[27] Du Mond, I. W. M.: 1959, *Ann Phys.* **7**, 365.
[28] Blokhintsev, D. I.: 1964, *Quantum Mechanics*, Reidel, Dordrecht, The Netherlands.
[29] Blokhintsev, D. I.: 1968, *The Philosophy of Quantum Mechanics*, Reidel, Dordrecht, The Netherlands.
[30] Blokhintsev, D. I.: 1967, On the Localization of Particles, in: *High Energy Physics and the Theory of Elementary Particles*, 'Naukova Dumka', Kiev, 1967.
[31] Pauli, W.: 1946, *General Principles of Wave Mechanics*, Univ. of Michigan Press, Ann Arbor, Michigan.
[32] Shirokov, M. I.: 1965, preprint of Joint Institute for Nuclear Research (E-2478), U.S.S.R.
[33] Mandelshtam, L. I.: 1950, *Collected Works*, Vol. III, Academy of Sciences of the U.S.S.R., Moscow, p. 397.
[34] Mandelshtam, L. I.: 1950, *Collected Works*, Vol. V, Academy of Sciences of the U.S.S.R., Moscow, p. 347.
[35] Mott, N. and Massey, H. 1949, *Theory of Atomic Collisions*, Clarendon Press, Oxford.
[36] Feynmann, R. and Hibbs, A.: 1965, *Quantum Mechanics and Path Integrals*, McGraw Hill Co., New York.
[37] Akhiezer, A. I. and Berestetski, B.: 1959, *Quantum Electrodynamics*, Fizmatgiz, Moscow.
[38] Bogolyubov, N. N. and Shirkov, D. V.: 1957, *Introduction to the Theory of Quantum Fields*, State Technical and Theoretical Press, Moscow.
[39] Newton, T. and Wigner, E.: 1949, *Rev. Mod. Phys.* **21**, 400.
[40] Hofstadter, R.: 1957, *Ann. Rev. Nucl. Sci.* **7**, 231; Hofstadter, R., Bumiller, F., and Yearian, M.: 1958, *Rev. Mod. Phys.* 482.
[41] Drell, S. D. and Zachariasen, F.: 1961, *Electromagnetic Structure of Nucleons*, Oxford University Press.
[42] Blokhintsev, D. I., Barashenkov, V. S. and Barbashov, B. M.: 1959, *Usp. Fiz. Nauk* **68** (1959) 417; *Sov. Phys. Uspekhi* **2**, 505.
[43] Chew, K. W. *et al.*: 1964, *XII International Conference on High Energy Physics in Dubna*, Vol. I, Atomizdat, Moscow, 1964, p. 861.
[44] Ohun, L. B.: 1963, *Weak Interactions of Elementary Particles*, Fizmatgiz, Moscow.
[45] Blokhintsev, D. I., Barashenkov, V. and Grishin, V.: 1958, *Nuovo Cimento* **9**, 249.
[46] Blokhintsev, D. I.: 1946, *Doklady* **53**, 205.
[47] Blokhintsev, D. I.: 1964, *Nuovo Cimento* **33**, 1094.
[48] Logunov, A. A. *et al.*: 1963, *Nuovo Cimento* **30**, 1.
[49] Puzikov, L. D., Ryndin, R. M. and Smorodinski, J. A.: 1957, *ZhETF* **32** 592; 1957, *Sov. Phys. JETP* **5**, 489.
[50] Blokhintsev, D. I.: 1962, *Nucl. Phys.* **31**, 628.
[51] Blokhintsev, D. I : 1959, *Usp. Fiz. Nauk* **69**, 3; 1959, Sov. Phys. *Uspekhi* **2**, 505.
[52] Fierz, M.: 1950, *Helv. Phys. Acta* **23**, 731.
[53] Stuekelberg, E and Rivier, D.: 1949, *Helv. Phys. Acta* **22**, 215.
[54] Blokhintsev, D. I.: 1964, in: *Space, Time and Causality* preprint of the Joint Institute for Nuclear Research (D-1735), U.S.S.R
[55] Bloch F.: 1943, *Sov. Phys.* **5**, 301.
[56] Mayer, A. A.: 1945, Doctoral Thesis, Moscow State University.
[57] Kronig, R.: 1926, *J Amer. Opt. Soc.* **12**, 547.
[58] Kramers, H. A.: 1927, *Atti Congr Inst. Fisici Como* **2**, 545.
[59] Goldberger, M. L.: 1955, *Phys. Rev.* **97**, 508.

[60] Symanzik, K. and Jost, D.: 1956, Reports on the Conference in Seattle, Sept.
[61] Bogolyubov, N. N., Medvjedev, B. V. and Polivanov, M. K.: 1958, *Questions on the Theory of Dispersion Relations*, Fizmatgiz, Moscow.
[62] Pomeranchuk, I.: 1958, *ZhETF* **34**, 725; 1958, *Sov. Phys. JETP*, **7**, 449.
[63] Meiman, N. N.: 1962, in: *Questions on Elementary Particle Physics*, Erevan.
[64] Logunov, A. A., Todorov, I. and van Hieu, N.: 1966, *Usp. Fiz. Nauk* **88**, No. 1; 1966, *Sov. Phys. Uspekhi* **9**, 31.
[65] Chernikov, N. A.: 1961, preprint of the Joint Institute for Nuclear Research (R-723), U.S.S.R.
[66] Smorodinski, Ya. A.: 1965, *Fortschr. Phys.* **13**, 4; 1962, preprint of the Joint Institute for Nuclear Research (R-1109), U.S.S.R.
[67] Rosenfeld, A. H. *et al.*: 1967, preprint of the Joint Institute for Nuclear Research (R1-3129), U.S.S.R.
[68] Shirkov, Yu. M.: 1957, *ZhETF* **33**, 861; 1957, *Sov. Phys. JETP* **6**, 664.
[69] Blokhintsev, D. I. and Kolerov, H. I.: 1966, *Nuovo Cimento* **44**, 974.
[70] Blokhintsev, D. I.: 1967, Lectures in Trieste, Macroscopic Causality, Intern. Center Theor. Phys., Trieste, IC/67/36.
[71] Blokhintsev, D. I.: 1967, *Acta Phys. Acad. Sci. Hung.* **22**, 307.
[72] Blokhintsev, D. I. and Kolerov, G. I.: 1964, *Nuovo Cimento* **34**, 163.
[73] Blokhintsev, D. I. and Vinogradov, V. M.: 1967, preprint of the Joint Institute for Nuclear Research (R2-3506), U.S.S.R.
[74] Blokhintsev, D I.: 1945, Lecture notes from Moscow State University, Physics, Vol. III, p. 101.
[75] Blokhintsev, D. I.: 1946, *ZhETF* **16**, 480.
[76] Blokhintsev, D. I.: 1946, *J. Phys. USSR* **10**, 167.
[77] Blokhintsev, D. I.: 1947, *ZhETF* **17**, 266.
[78] Meiman, N. N.: 1964, *ZhETF* **47**, 1966; 1965, *Sov. Phys. JETP* **20**, 1320.
[79] Efimov, G. V.: 1967, preprint of the Joint Institute for Nuclear Research (R2-3390), U.S.S.R.
[80] Blokhintsev, D. I.: 1947, *ZhETF* **17**, 116.
[81] Blokhintsev, D. I.: 1947, *J. Phys. USSR* **11**, 72.
[82] Watagin, G.: 1934, *Zs. Phys.* **88**, 92.
[83] Markov, M. A.: 1940, *J. Phys. USSR* **2**, 453.
[84] Blokhintsev, D. I.: 1948, *ZhETF* **18**, 566.
[85] Blokhintsev, D. I.: 1952, *ZhETF* **22**, 254.
[86] Menger, K.: 1951, *Proc. Nat. Acad. Soc. USA* **37**, 226.
[87] Schweizer, B. and Sklar, A.: 1960, *J. Math.* **10**, 313.
[88] March, A.: 1934, *Zs. Phys.* **104**, 93, 161; 1937, *Zs. Phys.* **105**, 620.
[89] Markov, M. A.: 1958, *Hyperons and K-mesons*, Fizmatgiz, Moscow.
[90] Ingraham, R. I.: 1964, *Nuovo Cimento* **34**, 182.
[91] Ingraham, R. I.: 1967, *Renormalization Theory of Quantum Field with a Cut-Off*, Gordon and Beach, New York.
[92] Yukawa, H.: 1966, *Res. Inst. Fund. Phys. Kyoto Univ.* **RIFP-55**.
[93] Shirokov, Yu. M.: 1951, *ZhETF* **21**, 748.
[94] Ambartzumian, V. and Ivanenko, D.: 1930, *Zs. Phys.* **64**, 563.
[95] Darling, B.: 1953, *Phys. Rev.* **80**, 460; 1953, *Phys. Rev.* **91**, 1252.
[96] Das, A.: 1960, *Nuovo Cimento* **18**, 482.
[97] Vyaltsev, A. N.: 1965, *Discrete Space-Time*, Nauka, Moscow.
[98] Snyder, H.: 1947, *Phys. Rev.* **71**, 38.

[99] Golfand, Yu. A.: 1959, *ZhETF* **37**, 504; 1959, *Sov. Phys. JETP* **10**, 356.
[100] Kadyshevski, V. G.: 1961, *ZhETF* **41**, 1885; 1962, *Sov. Phys. JETP* **14**, 1340; 1962, *Doklady* **147**, 588; 1963, *Sov. Phys. Doklady* **7**, 1031.
[101] Golfand, Yu. A.: 1962, *ZhETF* **43**, 256; 1963, *Sov. Phys. JETP* **16**, 184; 1963, *ZhETF* **44**, 1248; 1963, *Sov. Phys JETP* **17**, 842.
[102] Kadyshevski, V. G.: 1962, *Doklady* **147**, 1336; 1963, *Sov. Phys. Doklady* **7**, 1138.
[103] Tamm, I. E.: 1964, in: *XII International Conference on High Energy Physics in Dubna*, Vol. II, Atomizdat, Moscow, p. 229; 1965, *Proc. of Int. Conf. on Element*, p. 314, Kyoto.
[104] Kadyshevski, V. G.: 1964, in: *Space, Time and Causality*, preprint of the Joint Institute of Nuclear Research (D-1735), U.S.S.R., 1964.
[105] Blokhintsev, D. I.: 1964, *Phys. Letts.* **12**, 272.
[106] Blokhintsev, D. I.: 1960, *Nuovo Cimento* **16**, 382.
[107] Markov, M. A.: 1966, preprint of the Joint Institute for Nuclear Research (E2-2973), U.S.S.R.
[108] Barbashov, B. M. and Chernikov, N. A.: 1965, preprint of the Joint Institute for Nuclear Research (R-2151), U.S.S.R.
[109] Barbashov, B. M. and Chernikov, N. A.: 1966, *ZhETF* **50**, No. 5; 1966, *Sov. Phys. JETP* **23**, 861.
[110] Von Neuman, J.: 1967, *Mathematical Foundations of Quantum Mechanics*, Nauka, Moscow.
[111] Barashenkov, V. S.: 1962, *Fortschr. Phys.* **10**, 205.
[112] Barashenkov, V. S. and Dedyu, V. I.: 1966, *Nucl. Phys.* **64**, 636.
[113] Hohler, G., Ebel, G. and Giesiche, I.: 1964, *Zs. Phys* **108**, 430.
[114] Foley, K. J. *et al.*: 1963, *Phys. Rev. Letters* **11**, 425.
[115] Kirillova, L. *et al.*: 1964, Phys. Letters **13**, 93.
[116] Nikitin, V. A. *et al.*: 1965, School for Physicists in Erevan.
[117] Gabraith, W. *et al.*: 1964, in: *XII International Conference on High Energy Physics in Dubna*, Vol. I, Atomizdat, Moscow, p. 109.
[118] Lindenbaum, S. I.: 1964, in: *XII International Conference on High Energy Physics in Dubna*, Vol. I, Atomizdat, Moscow, p. 188
[119] de Pagter, I. *et al.*: 1964, *Phys. Rev. Letters* **12**, 739.
[120] de Pagter, I. *et al.*: 1965, *Proc. Int. Symp. on Electron and Photon Interactions* **11**, 360.
[121] Drell, S. D.: 1967, report on *Proc. XII Int. Conf. on High Energy Physics*, p. 85, Berkeley.
[122] Drell, S. D.: 1958, *Ann. Phys.* **4**, 75.
[123] Christensen, I. H., Cronin, I. W., Fitch, V. L. and Tarlay, R.: 1964, in: *XIII International Conference on High Energy Physics in Dubna*, Vol. II, Atomizdat, Moscow, p. 105.
[124] Terentev, M. V.: 1965, *Usp. Fiz Nauk* **86**, 231; 1965, *Sov. Phys. Uspekhi* **8**, 445.
[125] Dorfman, D. *et al.*: 1967, *Phys. Rev. Letters* **19**, 987.
[126] Bennett, S. *et al.*: 1967, *Phys. Rev. Letters* **19**, 993.
[127] Feynman, R.: 1949, *Phys. Rev.* **76**, 749.
[128] Pauli, W. (ed.): 1955, *Niels Bohr and the Development of Physics*, Pergamon Press, New York.
[129] Jost, R.: 1957, *Helv. Phys. Acta* **30**, 409.
[130] Blokhintsev, D. I.: 1947, *ZhETF* **17**, 924.
[131] Shapiro, I. S.: 1957, *Usp. Fiz. Nauk* **61**, 313.
[132] Arbusov, B. A and Filipov, A T.: 1966, preprint of the Joint Institute for Nuclear

Research. (R2-2578), U.S.S.R.
[133] Arbusov, B. A. and Filipov, A. T.: 1966, preprint of the Joint Institute for Nuclear Research (R2-3067), U.S.S.R.
[134] Christensen, I. H. *et al.*: 1964, *Phys. Rev. Letters* **13**, 138.
[135] Bonard, X. D. *et al.*: 1965, *Phys. Rev. Letters* **15**, 58.
[136] Nikitin, V. A. *et al.*: 1963, *Instr. Experim. Techn.* **6**, 18.
[137] Blokhintsev, D. I.: 1946, *C. R. U.S.S.R. (Doklady)* **53**, No. 3.
[138] Landau, L. D. and Lifshitz, E. M.: 1958, *Quantum Mechanics*, Pergamon Press, London.
[139] Kadyshevski, V. G.: 1963, preprint of the Joint Institute for Nuclear Research (R-1327), U.S.S.R.
[140] Kadyshevski, V. G.: 1967, preprint of the Joint Institute for Nuclear Research (R-1328), U.S.S.R.
[141] Smorodinski, Ya. A.: 1949, *Usp. Fiz. Nauk* **39**, 325.
[142] Ivanenko, D. D.: 1950, in: *Changes in the Levels of Atomic Electrons*, Foreign Lit. Press, Moscow.
[143] Panofsky, W.: 1968, 'Electromagnetic interactions', *Proc. XIV Int. Conf. on High Energy Physics*, CERN, p. 23.
[144] Blokhintsev, D. I.: 1958, *Nuovo Cimento* **9**, 925.
[145] Blokhintsev, D. I.: 1960, *Proc. Berkeley Conf. on Instrumentation*, Berkeley, California, p. 198.
[146] Blokhintsev, D. I.: 1969, *Fortschr. Phys.* (in press).
[147] Cronin, I. W.: 1968, *Proc. XIV Int. conf. on High Energy Physics*, CERN, p. 300.

9 789027 702401

GPSR Compliance

The European Union's (EU) General Product Safety Regulation (GPSR) is a set of rules that requires consumer products to be safe and our obligations to ensure this.

If you have any concerns about our products, you can contact us on ProductSafety@springernature.com

In case Publisher is established outside the EU, the EU authorized representative is:

Springer Nature Customer Service Center GmbH
Europaplatz 3
69115 Heidelberg, Germany

Batch number: 10823846

Printed by Printforce, the Netherlands